NONLINEAR
CIRCUITS

The Artech House Microwave Library

Introduction to Microwaves by Fred E. Gardiol

Microwaves Made Simple: Principles and Applications by W. Stephen Cheung and Frederic H. Levien

Microwave Tubes by A. S. Gilmour, Jr.

Electric Filters by Martin Hasler and Jacques Neirynck

Nonlinear Circuits by Martin Hasler and Jacques Neirynck

Microwave Technology by Erich Pehl

Receiving Systems Design by Stephen J. Erst

Microwave Mixers by Stephen A. Maas

Feedback Maximization by B.J. Lurie

Applications of GaAs MESFETs by Robert Soares, et al.

GaAs Processing Techniques by Ralph E. Williams

GaAs FET Principles and Technology by James V. DiLorenzo and Deen D. Khandelwal

Dielectric Resonators, Darko Kajfez and Pierre Guillon, eds.

Modern Spectrum Analyzer Theory and Applications by Morris Engelson

Design Tables for Discrete Time Normalized Lowpass Filters by Arild Lacroix and Karl-Heinz Witte

Microwave Materials and Fabrication Techniques by Thomas S. Laverghetta

Handbook of Microwave Testing by Thomas S. Laverghetta

Microwave Measurements and Techniques by Thomas S. Laverghetta

Principles of Electromagnetic Compatibility by Bernhard E. Keiser

Linear Active Circuits: Design and Analysis by William Rynone, Jr.

The Design of Impedance-Matching Networks for Radio-Frequency and Microwave Amplifiers by Pieter L.D. Abrie

Microwave Filters, Impedance Matching Networks, and Coupling Structures by G.L. Matthaei, Leo Young, and E.M.T. Jones

Analysis, Design, and Applications of Fin Lines by Bharathi Bhat and Shiban Koul

Microwave Engineer's Handbook, 2 vol., Theodore Saad, ed.

Microwave Integrated Circuits, Jeffrey Frey and Kul Bhasin, eds.

Computer-Aided Design of Microwave Circuits by K.C. Gupta, Ramesh Garg, and Rakesh Chadha

Microstrip Lines and Slotlines by K.C. Gupta, R. Garg, and I.J. Bahl

Advanced Mathematics for Practicing Engineers by Kurt Arbenz and Alfred Wohlhauser

Microstrip Antennas by I.J. Bahl and P. Bhartia

Antenna Design Using Personal Computers by David M. Pozar

Microwave Circuit Design Using Programmable Calculators by J. Lamar Allen and Max Medley, Jr.

Stripline Circuit Design by Harlan Howe, Jr.

Microwave Transmission Line Filters by J.A.G. Malherbe

Electrical Characteristics of Transmission Lines by W. Hilberg

Microwave Diode Control Devices by Robert V. Garver

Tables for Active Filter Design by Mario Biey and Amedeo Premoli

Ferrite Control Components, 2 vol., Lawrence Whicker, ed.

Microwave Remote Sensing, 3 vol., by F.T. Ulaby, R.K. Moore, and A.K. Fung

NONLINEAR CIRCUITS

Martin Hasler
Jacques Neirynck

International Standard Book Number: 0-89006-208-0
Library of Congress Catalog Card Number: 86-71747

Translation of *Circuits non linéaires*, originally published in French as a supplement to the *Traité d'Électricité* by the Presses Polytechniques Romandes, Lausanne, Switzerland. © 1985.

10 9 8 7 6 5 4 3 2 1

Introduction

A nonlinear circuit is a circuit complying with Kirchhoff's laws, in which at least one of the components has a nonlinear characteristic. All circuits comprised of physical components are nonlinear, because any component to which high enough voltages or currents are applied will eventually fail to comply with linear relations, if only by being destroyed by excessive voltage or current.

Thus, we can say that any linear circuit is only the simplified model of a nonlinear circuit which we choose to ignore. Through suitable circuit realization and signal selection, it is possible to remain within the range of application of the linear model, which is very attractive due to the resulting simplicity in calculations. In such circuits, e.g., filters, amplifiers, and phase correcting or power distribution networks, nonlinear effects must be considered as parasitic. Nevertheless, they are worth studying so as to define the areas in which the circuit behavior remains close to linearity.

However, these "linear" circuits cannot fulfill all the functions desired by the engineer. Any linear circuit complies with the superposition theorem and the isomorphic response theorem. Concretely, this means that the signals of a linear circuit can be decomposed into sums or integrals of sinusoids whose responses can be studied separately in order to construct the global response by superposing the partial responses. Such is the theoretical basis of the analysis in the frequency domain. This calculation convenience has an obvious drawback: in a linear circuit, it is impossible to generate any frequency other than those already existing in the signal.

Therefore, nonlinear effects are absolutely necessary to carry out functions such as frequency rectification, multiplication or division, oscillation, and modulation or limitation, not to mention logic gates whose functioning is highly nonlinear. The methods developed for linear circuits are not applicable for any of these circuits. Such a fundamental concept as impedance is meaningless.

Surely, the analysis of a nonlinear circuit is a difficult problem, for which approximate analytical methods have been used in the past. With the advent of powerful computers, in the early 1970s, analytical methods have been increasingly replaced by numerical methods. In principle, numerical methods make it possible to calculate the response of any given circuit as accurately as desired. However, the disadvantage of the numerical calculation method is that it is far from being transparent for the user. To know one computed response is not enough. We should also be certain that no other responses of radically different nature exist, resulting from slightly different initial conditions.

Obviously, the problem of nonlinear circuits has nothing in common with that of linear ones, where we need only find one response and we can be sure that there will not be any other responses. This book is not meant to provide numerical circuit analysis methods that can be found in specialized texts. On the contrary, our purpose is to study the qualitative properties of the solutions.

With this in mind, we consider successively: the existence and uniqueness of the solution, starting from initial conditions, in chapter 2; the existence and uniqueness of the solution for a circuit without any capacitors or inductors in chapter 3; the asymptotic behavior for solutions with t tending toward infinity in chapter 5; the harmonic components of periodic and quasiperiodic solutions in chapter 6; the study of only these components in the case of filtered nonlinear circuits in chapter 7.

Also, this book definitely departs from many other texts that deal with ordinary differential equations or the theory of systems. By dealing with a nonlinear circuit through the general methods of these fields, we would lose very valuable information; namely, the fact that half the equations describing the circuit are linear because they result from Kirchhoff's laws. The true interest and specificity of the nonlinear circuit theory lies precisely in this peculiarity of the system of equations, which allows us to obtain results that are of special interest to the engineer. Indeed, the engineer is much less interested in tests of a mathematical nature related to the circuit equations, such as the Liapunov method, than by the theorems which allow him to say *a priori*, based on the nature of the circuit components and connections, something about the type of response, for example, its uniqueness or its asymptotic behavior.

This particular approach to nonlinear circuits has first been used by some American researchers, among whom Leon Chua, Erwin Sandberg, Allan Willson, and Charles Desoer are surely the most well known. This book is meant to present in a coherent manner the results of research that began some twenty years ago. This domain is still very active; indeed, in recent years, some very powerful theorems about the existence and uniqueness of the solution of resistive circuits have been proved and the existence of chaotic solutions has been discovered for very simple circuits. Thus, this text may also be considered as a starting point for broader and deeper research in this field.

Contents

Chapter 1

Fundamentals

1.1 PHYSICAL CIRCUIT AND MODEL CIRCUIT

1.1.1 Kirchhoff's Laws

This work is concerned with the electrical behavior of physical devices apt to be based on the Kirchhoff model. This is the case if they comply with the ***Kirchhoff's laws*** with fair approximation:

- the sum of voltages in a loop is zero any time;
- the sum of incident currents in a node is zero any time.

These two laws are based on certain concepts, the application of which must be explained. The concepts of voltage and current do not pose any problem. They are quantities that can be measured, respectively, by a voltmeter or an ammeter. On the contrary, the loops and nodes must be identified in the physical device. In other words, the locations where voltmeters or ammeters must be connected to verify Kirchhoff's laws must be specified.

For that purpose, the circuit is decomposed into *elements* and *connections*. Measurements are made at the connections, whereas it is supposedly impossible to obtain access inside the elements with the measuring instruments. The contact points between an element and the connections are called the *terminals* of the element.

In conventional discrete circuits, the decomposition into elements and connections is imposed by their physical structure. In integrated circuits, the decomposition is less obvious, as will be seen in subsection 1.1.2.

The notions of node and loop refer to the connections. They will be introduced in subsection 1.1.3.

1.1.2 Example

We illustrate the decomposition into elements and connections by means of an inverting amplifier implemented as a circuit integrated in silicon-grid CMOS technology [1]. The mask layout of figure 1.1 gives a fairly good idea of the circuit as physical object, which we call the *physical circuit*. In figure 1.2, we show the connections and

2

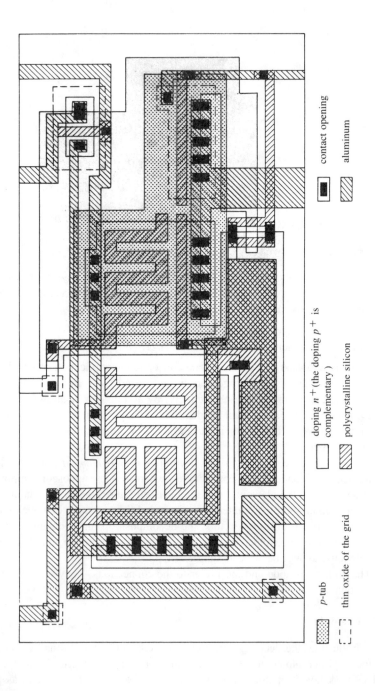

doping n^+ (the doping p^+ is complementary)

polycrystalline silicon

contact opening

aluminum

p-tub

thin oxide of the grid

Fig. 1.1

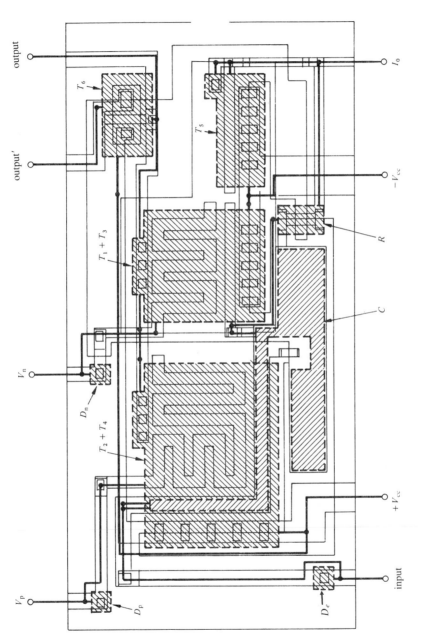

Fig. 1.2

4

11 physical devices acting as elements. The terminals of the elements are indicated by dots, and the connections are shown by lines linking the dots. It is not always possible to follow the circuit's physical connections along these lines, since the layout is a two-dimensional representation of a three-dimensional circuit designed to occupy a minimum space. Passing from figure 1.1 to figure 1.2 adds another difficulty due to the fact that the polycrystalline silicon is part of both the elements and the connections.

In figure 1.3 the same circuit is redrawn in more schematic form. Also, sources and charges outside the integrated circuit as well as the connections with the substrate, omitted in figure 1.2, have been added.

1.1.3 Verification of Kirchhoff's Laws

As shown in example 1.1.2, the connections can be represented as a set of links connecting the terminals two by two. Also, from each terminal, a link enters the element to which the terminal belongs.

The current is measured in a link by inserting an ammeter, and the voltage is measured between two links by connecting a voltmeter. For example, figure 1.3 shows

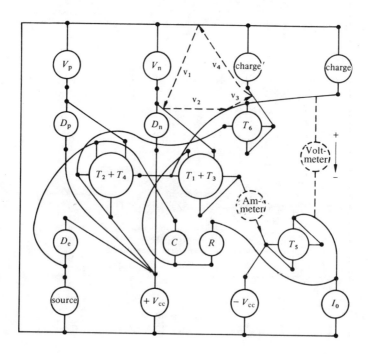

Fig. 1.3

the locations of an ammeter and a voltmeter by dotted lines. Naturally, such measurements inside an integrated circuit are only possible if the circuit has been designed so as to make the internal points accessible from the outside. Here, we are only interested in the principles of measurement.

We allow currents and voltages to depend on time, but we exclude their dependence on the particular location where the ammeter is inserted into the link and where the voltmeter is connected to the links. Therefore, the contact points could also be the terminals.

Otherwise, the circuit—at least as decomposed into elements and connections —is not apt to be described by the Kirchhoff model. Such a situation occurs in particular when a wave propagation phenomenon is observed along the links. In that case, the Maxwell model (vol. III) is more appropriate than the Kirchhoff model.

In the representation of the connections in figures 1.2 and 1.3, the nodes can be identical to the element terminals. Therefore, for each terminal of each element, there is a Kirchhoff current law to be verified.

A voltage measurement can be symbolized by an arrow connecting the links between which the voltmeter is connected. Several arrows can form a loop. An example of a voltage loop is shown by the dotted line in figure 1.3. For each loop of this type, there is a Kirchhoff voltage law to be verified.

1.1.4 Potentials and Modeling of Connections

If Kirchhoff's voltage law is verified, we can associate with each link a quantity, the *potential*, such that the voltage between two links will be equal to the difference of the potentials. We associate the same potential with the connected terminals. It follows that all the terminals which are connected to one another by links have the same potential. Then, logically, they can be gathered into a single node. By doing so, Kirchhoff's current law applied to any individual terminal becomes the Kirchhoff law for the common node.

This connection modification is introduced in figure 1.4 for the example of subsection 1.1.2. It is interesting to follow the evolution of this circuit from the representation of the physical device in figure 1.1 to the diagram of figure 1.4. Together with Kirchhoff's law, this process constitutes the modeling of the connections.

It must be noted that the potential is not a magnitude that can be measured as such, but a purely mathematical construction. In fact, its value may only be determined up to a constant. Similar situations occur in electromagnetism and mechanics.

In figure 1.5 we show the usual diagram of the circuit of figure 1.1, complemented by external sources and charges. Compared with figure 1.4, it contains additional information on the nature of the elements and their terminals.

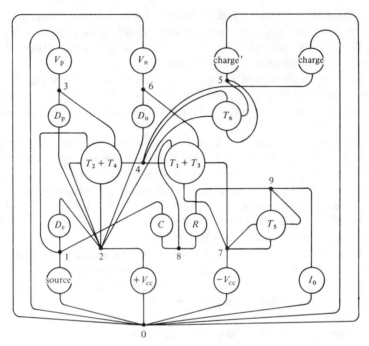

Fig. 1.4

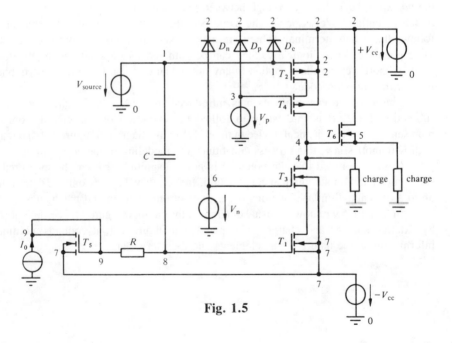

Fig. 1.5

1.1.5 Measuring and Modeling of an Element

The currents entering an element through its terminals, as well as the voltages between the terminals of an element, depend on the element itself and the circuit surrounding it. In order to determine the constraints on currents and voltages arising from the element, we set the element in various circuits and record the currents and voltages at its terminals as functions of time.

Because the number of measurements that can be actually carried out is not without limit, the excitation circuits for the element must be carefully chosen. The results should permit us to construct a model for the element. Choosing the measurements and the model is a critical operation requiring a good deal of intuition and experience. This art, simply called *modeling*, constitutes an important discipline in the theory of nonlinear circuits. It is, nevertheless, not treated in this text.

From this point on, the theory of nonlinear circuit departs from the theory of linear circuits (vol. IV). Linear circuit theory is *a priori* only concerned with elements imposing linear relations, either differential or not, between currents and voltages at their terminals. In other words, the elements must be constructed to fit into the framework of this theory.

The viewpoint of nonlinear circuit theory is completely different. The physical devices are accepted as they are, and modeled according to a list of constraints that they impose on currents and voltages or on potentials at the terminals.

If an element is modeled by constraints involving potentials, we must bear in mind that these potentials are only determined up to a constant. Therefore, these constraints must not be dependent on such a constant.

Nonetheless, a general hypothesis is made concerning the elements. We assume that an element cannot store charge. It follows that the sum of the currents entering an element from its terminals vanishes. Thus, this relation is necessarily part of the constraints modeling an element.

1.1.6 Study of a Physical Circuit by a Model Circuit

So far, we have described how a physical circuit evolves toward its model. It is important to distinguish clearly between these two objects. The former is a *reality*, whereas the latter is a *concept*.

The physical circuit is *assembled* in a laboratory or *produced* in a factory. It can only be examined by measurements. In a physical circuit, voltages and currents are a reality, just like the circuit itself.

A model of the physical circuit introduced according to the principles of subsections 1.1.1 to 1.1.5 is called a *model circuit*, or simply a circuit. It is *defined* by an axiomatic formalism, which will be introduced in section 1.2.

The study of a physical circuit based on a model circuit follows three steps (fig. 1.6). In the first step, modeling, a model circuit is chosen, for which the engineer's experience and intuition are more useful than the mathematician's exactitude.

8

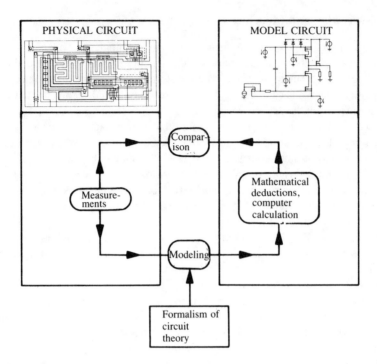

Fig. 1.6

Conversely, the second step, the study of the model circuit, is carried out by mathematical deduction. It is the main theme of this work, wherein mathematical deduction is taken in its broadest sense. It makes use of algebraic derivations and theorems as well as computerized calculations. Currents and voltages that are explicitly or implicitly obtained, or which are the object of theorems, are mathematical quantities related to the model circuit.

Finally, in the third step, the currents and voltages obtained for the model circuit are compared with those measured in the physical circuit. The results, whether acceptable or not, are a piece of information on the functioning of the physical circuit. If they do not compare satisfactorily, a more adequate model must be found.

It is imperative to avoid confusion between the two parts of figure 1.6. In particular, it is out of the question to prove any properties of the physical circuit's currents and voltages. On the other hand, any current or voltage satisfying the equations of the model circuit cannot be rejected because it contradicts the principles of physics, even fundamental ones. What is in question is not the solution, but the model circuit.

1.1.7 Example

A very simple example can illustrate the methodology of figure 1.6. Figure 1.7 shows a physical circuit composed of a 4.5 V battery and a switch. Let us assume that this switch is open for $t < 0$. At $t = 0$, we short-circuit the battery by closing the switch.

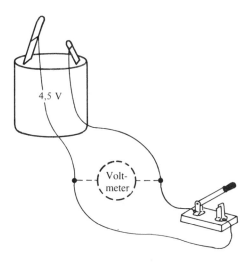

4,5 V

Volt-meter

Fig. 1.7

A model that is widely used for a battery is an ideal voltage source. The resulting model circuit is represented in figure 1.8. For $t < 0$, the voltage u satisfies $u(t) = 4.5$ V. On the other hand, for $t > 0$, $u(t)$ should be equal to 4.5 V because of the battery, and 0 V because of the short circuit. In conclusion, this model does not have any solution for $t > 0$.

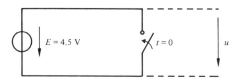

$E = 4.5$ V $\quad t = 0 \quad u$

Fig. 1.8

Obviously, by connecting a voltmeter to the links as shown in figure 1.7 by the dotted lines, it is possible to measure $u(t)$ in the physical circuit, even after $t = 0$. Therefore, there is a fundamental disagreement between the measured voltage in the physical circuit and the corresponding voltage in the model circuit, which must be a solution of the model circuit's equations.

We conclude that the circuit of figure 1.8 is a bad model of the physical circuit of figure 1.7. By adding a resistor to the model of the battery, a better agreement will be obtained between the measured and calculated voltages.

Without going that far, two conclusions can already be made regarding the negative outcome of our modeling. First, the secondary effects, such as the internal resistance of the battery, which we deemed negligible, in fact, determine the solution. Second, the physical circuit can be expected to show a very special behavior. Indeed, a relatively high peak current will occur after the switch is closed.

1.2 AXIOMATICS OF MODEL CIRCUITS

1.2.1 Introduction

Section 1.1 described the transition from the physical circuit to the model circuit. The purpose of this section is to establish the formal framework of model circuits. As its name implies, the model circuit must reflect the structure of the physical circuit. The same terms will be found for both the physical and model circuits, but they refer to objects of a different nature. Let us mention some examples:

- An element of the physical circuit is simply one part of it. Its terminals are the contact points with its surroundings. An element of the model circuit is a black box having terminals with which currents and potentials are associated. It is defined by a number of relations between its currents and potentials as functions of time.
- The connections of the physical circuit are its other parts. It consists of all the links between the elements, whether they are metal wires of discrete circuits or strips of conducting material in integrated circuits. The connections of model circuits are sets of terminals linked to a node.
- Kirchhoff's laws, within the framework of the physical circuit, are natural laws verified by measurements. By contrast, as in the case of model circuits, they are either axioms or consequences of axioms. We shall formulate the axioms in terms of potentials rather than voltages. Thus, Kirchhoff's laws hence will not be axioms, but are derived from the axioms.

Most of the concepts in this section have already been introduced by Boite and Neirynck in volume IV.

1.2.2 Definitions

A (*Kirchhoff network*) *circuit* is a connection of a finite number of elements.

An *element* has a certain number, M, of *terminals*. Each terminal is associated with two real-valued functions of a real argument, time: the *potential* $v(t)$ of the terminal and the *current i(t)* entering the terminal. The element is characterized by M independent relations between the potentials and the currents of its terminals. An element with two terminals is called a *two-pole or one-port*; with three terminals, a *three-pole*, with four terminals, a *four-pole*; with N terminals, an *N-terminal element, et cetera.*

We say that M *relations* are *dependent* if there are, among the relations, M-1 relations implying the remaining one. This is the case if any set of currents and potentials satisfying the M-1 relations also satisfies the remaining relation. If such a set of M-1 relations cannot be found, the M relations are *independent.*

A *node* is a set of element terminals. If a terminal is part of this set, we say that it is *connected to the node.* The *circuit connections* are defined by the decomposition of the set of all the terminals of all elements into nodes.

1.2.3 Axioms

The basic hypotheses for the connections are:

- the terminals connected to the same node have the same potential;
- the sum of the currents entering the elements through the terminals connected to the same node is zero.

The basic hypotheses for the elements are as follows:

- The relations defining the element imply that the sum of the currents entering through its terminals is zero.
- The relations are independent, at any time, from the common level of the potentials. Therefore, they involve only the differences of potentials.

1.2.4 Definition

If we eliminate from among the M relations the one expressing that the sum of the currents entering through the terminals becomes zero, the remaining M-1 relations are called the *constitutive relations of the M-terminal element.*

1.2.5 Definitions

The *voltage* between two terminals (or nodes) is the difference of the two potentials. Because this is not symmetrical between the two terminals, a positive direction for the voltage must be selected.

Two terminals of an element, which are such that the sum of the two entering currents becomes zero, form a *port.* We associate a voltage and a current with a port in an obvious way. A positive direction for these two quantities must be selected. While one of the orientations is arbitrary, the other is chosen according to the convention of figure 1.9.

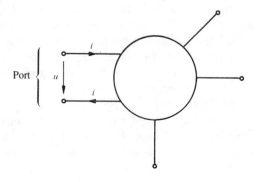

Port u i i

Fig. 1.9

An *N-port* is an element with $2N$ terminals having the following properties:

- The $2N$ terminals can be grouped into N ports;
- The relations defining the element are independent from the level of potential for each port. Thus, they involve only the voltages of the ports.

Consequently, among the $2N$ relations defining an N-port, there are N of them which express that the sum of the currents entering through the terminals of each port becomes zero, and N of them which involve only the currents and voltages of the ports. The latter are the *constitutive relations* of the N-port.

In the case where $N = 1$, both conditions are identical to the axioms of subsection 1.2.3, which means that a two-terminal element is always a one-port. This is no longer true from $N = 2$, the *two-port*, onward.

An N-port is *linear* if all its constitutive relations are linear.

1.2.6 Comment

The definition of the N-port is very restrictive and, except for the two-terminal elements that are always one-ports, only a few devices satisfy both conditions. Nevertheless, without restriction of generality, we could confine ourselves to circuits composed of N-ports, for two reasons.

First, a $2N$-terminal element often functions as an N-port inside a circuit, even though it is not an intrinsic N-port. It is by way of the circuit surrounding the $2N$-terminal element that the terminals are grouped into ports; for instance, when N one-ports link the $2N$ terminals by pairs. The constitutive relations of this N-port are obtained from the relations of the $2N$-terminal element in the following way: given the conditions for the ports, the number of different currents in the $2N$ relations of the $2N$-terminal element is simply N. When one of the potentials of each port is replaced as a variable by the voltage of the port, there remain only N potentials. After these remaining potentials are eliminated, there normally remain N equations, the unknowns of which are the currents and voltages of the ports. Note that the voltages between different port terminals are well defined, but they are irrelevant outside the element.

If an N-terminal element does not function as an $N/2$-port, either because of its internal structure or because N is odd, it can still be associated with an $N-1$-port which will not modify the behavior of the circuit when it replaces the N-terminal element. Such a construction is shown in figure 1.10 for $N = 3$. A terminal of each port is connected to the same node of the circuit. The choice of the common terminal is arbitrary. In this case, it is by the connection to the circuit that the voltages between terminals of different ports are imposed, so that the $N-1$-port will be equivalent to the N-terminal element.

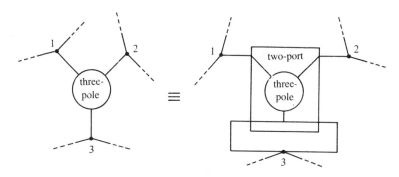

Fig. 1.10

Hereafter, we shall primarily use the concept of the N-port. According to the preceding comments, all developments and results can be transcribed without difficulty for the M-terminal elements.

1.2.7 Comment

By requiring N constitutive relations for an N-port, we deliberately exclude singular elements such as the *nullator* and *norator*.

The *nullator* (fig. 1.11) is a one-port defined by the constitutive relations:

$$u = i = 0 \tag{1.1}$$

Thus, it has two constitutive relations instead of a single one. On the other hand, the *norator* (fig. 1.12) does not have any. This means that its current and voltage are not subject to any constraint.

14

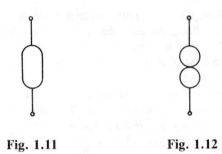

Fig. 1.11   **Fig. 1.12**

A nullator and norator put together in a one-port, according to figure 1.13, constitute an *ideal operational amplifier*. Because it has the two constitutive relations:

$$u_1 = i_1 = 0 \tag{1.2}$$

it is thus a two-port, according to definition 1.2.5.

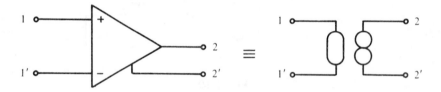

Fig. 1.13

If nullators and norators are used in that which follows, they will always appear in pairs, and hence must be considered as parts of an ideal operational amplifier.

1.2.8   Definition

We associate an *oriented graph* with a circuit composed of multiports in the following way: the nodes (branches) of the graph are the nodes (ports) of the circuit. A branch is incident with a node in the graph if a port terminal is connected to the node in the circuit. The voltage and current of a branch are concomitantly the voltage and current of the port. The orientation of a branch is the positive direction chosen for the port current, according to figure 1.9.

If the circuit also comprises multipoles, we convert them into multiports before associating a graph with it. Because the reference node is arbitrary, the associated graph is not unique.

1.2.9   Example

In figure 1.14, we represent a circuit and its associated graph. The node 7 has been chosen to transform the three-pole into a two-port.

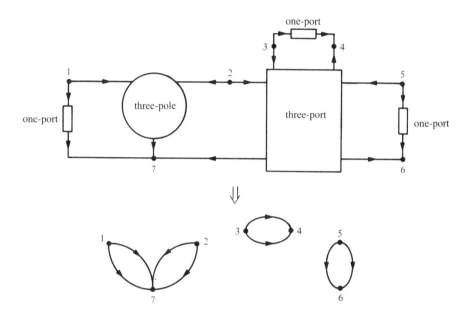

Fig. 1.14

1.2.10   Kirchhoff Lemmas

The *Kirchhoff lemmas* are contained in the definitions of subsection 1.2.2 and the axioms of subsection 1.2.3. We use the language of the graphs to state them:

- the sum of the voltages of all the branches constituting a loop is zero;
- the sum of the currents of all the branches incident with a node is zero.

From the first Kirchhoff lemma, we derive the *Kirchhoff voltage equations*, and from the second one we get the *Kirchhoff current equations*.

1.2.11   Definitions

The set of all the Kirchhoff equations combined with the constitutive relations of all the elements forms the *system of circuit equations*. The unknowns are the voltages and currents u_k, i_k, $k = 1, \ldots, b$ of all the ports of all the elements. A set of functions $\{u_k(t), i_k(t); k = 1, \ldots, b\}$, defined in the interval $t_0 \leq t < \infty$, is called a solution of the circuit in the interval $[t_0, \infty]$, if it satisfies the system of equations of the circuit for $t_0 \leq t < \infty$.

Instead of gathering all the Kirchhoff equations, we can confine ourselves to a maximum subset of independent equations, without changing the set of solutions. Thereby, we reduce the number of Kirchhoff equations to b (vol. IV, sec. 4.1). Thus, the system of circuit equations is composed of b independent Kirchhoff equations and b constitutive relations.

In section 2.3, we shall give an explicit form to the system of circuit equations by using the capacitor charges along with the fluxes in the inductors as auxiliary variables.

A circuit is *linear* if its system of equation is linear, otherwise it is a *nonlinear* circuit.

1.2.12   Comment

The constitutive relations of the elements studied in this book involve currents, voltages, and, possibly, their derivatives and integrals at the same time. Thus, the system of circuit equations will be composed of differential or integrodifferential equations and nondifferential equations. At first glance, the theory of ordinary nonlinear differential equations and the study of nonlinear circuits seem identical. In fact, however, the theory of nonlinear circuits is much more specific; at least half the equations are linear. We shall exploit this property to the maximum. It will form a basis for concepts, methods, and results that are specific to nonlinear circuits.

1.3   BASIC ELEMENTS

1.3.1   Introduction

The concept of elements that we introduced in subsection 1.1.1 is too general to serve as a basis for a powerful theory. In the case of linear circuits, we have confined ourselves to a very small set of elements. The elementary one-ports are the resistor, the capacitor, the inductor, and the independent current and voltage sources. As a basic two-port, we use the ideal transformer, the gyrator, the controlled sources, and the ideal operational amplifier. The more complicated N-ports are constructed by connecting the basic elements.

For nonlinear circuits, we operate in a similar way. Because there are many more nonlinear N-ports than linear ones, and because we would like to compose them from the basic elements, we should also expect a wider range of basic elements.

To start with, we introduce nonlinear versions of the resistor, the capacitor, and the inductor. Later, we shall see that the independent sources are already included in this list. On the other hand, we shall discover a fourth elementary one-port. Of course, the linear elementary one-ports remain special cases, which are included in the set of basic elements.

This list will be extended by some elementary two-ports, all of which are linear.

It must be noted that the set of basic elements that we introduce in this section is but a selection. Depending on the intended applications, other selections could be chosen. Also, computer programs for circuit analysis are usually limited to a well defined set of basic elements. Generally, it does not coincide with the set to be presented in this section. The list of basic elements of the SPICE program will be given in section 1.4. We shall also show to what extent the basic elements of section 1.3, on which this entire text is built, can be obtained by connecting SPICE basic elements.

1.3.2   Definition: Resistor

A *resistor* is a one-port defined by a constitutive relation of the form:

$$f(u, i) = 0 \qquad\qquad (1.3)$$

The symbol for the resistor is shown in figure 1.15. We refer to the *characteristic* of the resistor as the set of points in the plane (u,i) that satisfies (1.3). Therefore, it is the curve of level 0 of the function f. Consequently, different functions f can define the same characteristic. Their curves of level 0 simply must be identical.

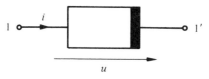

Fig. 1.15

Note that a thick bar distinguishes one terminal from the other in figure 1.15. Such a distinction is necessary because there are nonsymmetric resistors. A resistor is *symmetric* if, for any point (u,i) of the characteristic, the point $(-u, -i)$ belongs to the characteristic as well. Thus, the characteristic shows symmetry with respect to the origin.

If time is explicitly involved in the constitutive relation:

$$f(u, i, t) = 0 \qquad\qquad (1.4)$$

the resistor is *time dependent,* whereas equation (1.3) defines an *autonomous* resistor.

A voltage is called *admissible voltage for the resistor* if there is at least one current i such that (1.3) is satisfied. The concept of *admissible current for the resistor* is similarly defined.

If, for any value u of the voltage, there is not more than one value i of the current such that (1.3) is satisfied, we can rewrite (1.3) in the equivalent form:

$$i = g(u) \tag{1.5}$$

where the function g is defined for the set of admissible voltages for the resistor. In this case, the resistor is said to be *voltage controlled*.

Similarly, a resistor is *current controlled* if it has a constitutive relation in the form:

$$u = h(i) \tag{1.6}$$

where the function h is defined for any admissible current for the resistor. In the case of a time-dependent resistor, concepts such as admissible currents and voltages are separately defined for each moment.

1.3.3 Comments

The concepts introduced in subsection 1.3.2 have a very concrete significance.

If we connect a current (voltage) source of value $i(u)$ in series (parallel) with a resistor, and if $i(u)$ is not an admissible current (voltage) of the resistor, then there is no solution for the system of circuit equations. Thus, such a circuit does not constitute a suitable model of a physical circuit.

Additionally, if the current (voltage) of the source is admissible, but the resistor is controlled by the current (voltage), we might find several solutions of the circuit. In this case, choosing the suitable one among them remains an open question.

This kind of problem was previously mentioned in subsection 1.1.7 and will be more thoroughly discussed in chapter 2.

In order to permit more rigorous development later, some hypotheses must be made regarding the functions f, g, and h.

1.3.4 Definition

A function $f: \mathbb{R}^n \to \mathbb{R}$ is *smooth* if it is continuous and has continuous partial derivatives of all orders.

1.3.5 Definitions

A function $f: \mathbb{R}^n \to \mathbb{R}$ is *piecewise linear* if the entire space $\mathbb{R}^n$ can be divided into areas called *linear domains* of f, such that f is an affine function in each domain. A function is *affine* if it is of the form:

$$f(x_1, ..., x_n) = \sum_i a_i x_i + b \tag{1.7}$$

This function is linear if $b = 0$. The coefficients a_i and b, which define the affine function, vary depending on the domains. The linear domains of a piecewise linear function are, by definition, divided by a finite number of hyperplanes. A *hyperplane* in $\mathbb{R}^n$ is defined by an equation in the form:

$$\sum_i c_i x_i + d = 0 \tag{1.8}$$

It is a set of dimension $n - 1$. Thus, a linear domain is defined by a finite number of relations (1.8), where the equality is replaced by the inequality ">" or "<". The various domains differ according to the inequalities chosen.

1.3.6 Comment

We use the term "piecewise linear" because it is well established. However, "piecewise affine" would be a better choice.

A piecewise linear function that has a single domain extending over the entire space is an affine function. This clearly shows the inconsistency of the terminology. If there is no more than one piece, we would expect a piecewise linear function to become in fact a linear function. However, this is only true if the constant term b of (1.7) vanishes.

Linear circuit theory faces the same terminology problem. One-ports in which the constitutive relation is given by an affine function are common. Nevertheless, the affine-linear conflict is neatly skipped. The one-port defined by an affine relation has no name. It is an element composed of a linear resistor whose voltage-current relation is actually linear and of an independent source.

1.3.7 Hypothesis

We adopt the following hypothesis for the functions f, g, and h involved in the various forms of the constitutive relation of a resistor: the functions are either smooth or piecewise linear. Similarly, we introduce the terms *smooth resistor* and *piecewise linear resistor*. This hypothesis concerns the dependence of u and i. The dependence of t, if any, is assumed to be smooth.

The same hypothesis will be valid for nonlinear capacitors and inductors.

1.3.8 Comments

This choice of function types should not be considered rigid. Modeling of nonlinear resistors, either by elementary functions, such as exponentials or sines, or by piecewise linear functions, is simply common practice. Sometimes, the piecewise linear functions constitute an interpolation between measured points.

Additionally, both types of functions represent opposite extremes. The smooth functions can be derived without any problem. It is their nonlinearity that can cause trouble. On the other hand, the piecewise linear functions cause problems at the boundaries of the linear domains, where their derivatives are discontinuous.

Affine functions are the only functions that are simultaneously smooth and piecewise linear. Therefore, only linear circuits belong to both classes of nonlinear circuits with which we are dealing in this work.

Hereafter, the only time-dependent elements taken into consideration will be the independent sources. For the sake of simplicity, we assume that the source-injected signals are smooth. Conversely, Boite and Neirynck (vol. IV) allowed distributions to be treated as signals for the linear circuits, but such would exceed the scope of this book.

1.3.9 Examples

The *linear resistor* of value R is defined by the constitutive relation:

$$u = Ri \tag{1.9}$$

It can also be written according to form (1.3):

$$u - Ri = 0 \tag{1.10}$$

or form (1.5):

$$i = u/R \tag{1.11}$$

Thus, it is a symmetric resistor controlled both by voltage and current, and all the current and voltage values are admissible if R is different from 0 and ∞. Its characteristic is represented in figure 1.16.

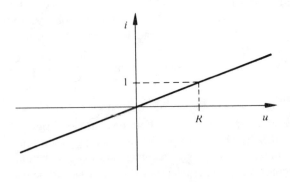

Fig. 1.16

The *dc-voltage source* of value E is defined by the constitutive relation:

$$u = E \tag{1.12}$$

Its graphic symbol and characteristic are respectively represented in figures 1.17 and 1.18. Any current is admissible, whereas the only admissible voltage is of value E, as is to be expected. On the other hand, it is surprising to find that an independent source is a resistor. What may seem even more strange is the fact that a voltage source is controlled by current and not by voltage. Indeed, we need only set $h(i) \equiv E$ for (1.12) to take the form (1.6).

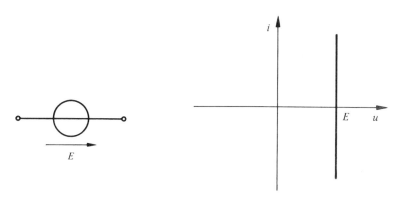

Fig. 1.17   **Fig. 1.18**

A nonautonomous or *time-dependent voltage source* is defined by the constitutive relation:

$$u = e(t) \tag{1.13}$$

Rewriting relation (1.13) as

$$u - e(t) = 0 \tag{1.14}$$

leads us back to form (1.4). Consequently, it is a current-controlled time-dependent resistor. Any current is admissible, but only the voltage $e(t)$ is admissible at the instant t. According to hypothesis 1.3.7, the function $e(t)$ of (1.13) is always assumed to be smooth.

Similarly, a *dc-current source* and a *time-dependent current source* are voltage-controlled resistors. The graphic symbol and the characteristic of the continuous current source of value I are respectively represented in figures 1.19 and 1.20.

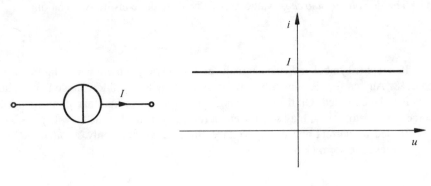

Fig. 1.19   Fig. 1.20

The physical device *junction diode* is normally modeled by a nonlinear resistor defined through the constitutive relation:

$$i = I_s(\exp(u/nU_T) - 1) \tag{1.15}$$

which is called an *exponential diode*, where I is the *inverse saturation current*, U_T is the *volt-equivalent of temperature*, and n is the *emission coefficient* (vol. VII, chap. 3, vol. VIII, chap. 1). The symbol of the exponential diode is shown in figure 1.21, and its characteristic is given in figure 1.22. It is a nonsymmetric resistor, controlled by voltage and current. Any voltage is admissible, while the admissible currents must satisfy $i > -I_s$.

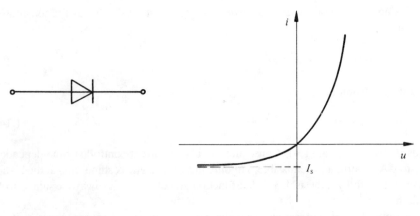

Fig. 1.21   Fig. 1.22

A model that accounts for the junction diode at higher inverse voltages is the *Zener diode*. Its symbol is that of figure 1.23 and its characteristic is represented in figure 1.24. Contrary to the exponential diode, all the currents, but only the voltages higher than $-U_z$, are admissible.

The *tunnel diode* is a nonlinear resistor, the characteristic of which is given in figure 1.25. The symbol is that of figure 1.26. This resistor is voltage controlled. It is not current controlled because, for currents $i_1 < i < i_2$, there are three different voltages u such that (u,i) is part of the characteristic.

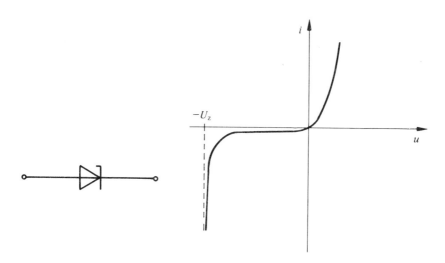

Fig. 1.23   **Fig. 1.24**

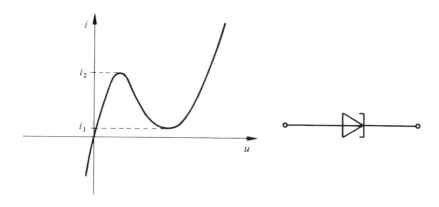

Fig. 1.25   **Fig. 1.26**

The *thyristor with disconnected gate*, or, in other words, the *four-layer diode*, is modeled by a resistor having a characteristic as represented in figure 1.27 (vol. VIII, chap. 6, and vol. XV, chap. 2). Its symbol is given in figure 1.28. It is an example of nonlinear resistor controlled by current instead of voltage.

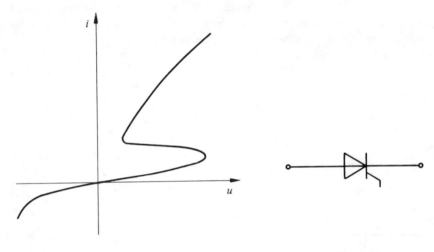

Fig. 1.27   **Fig. 1.28**

Until now, we have only introduced examples of smooth resistors. Let us also introduce one piecewise linear resistor.

The junction diode is often modeled, more roughly than with (1.15), by using the constitutive relation (fig. 1.29):

$$i = g(u) = \begin{cases} g_s u & \text{for} \quad u > 0 \\ g_o u & \text{for} \quad u < 0 \end{cases} \tag{1.16}$$

We call this model the *piecewise linear diode*.

The *passing* (*blocking*) *conductance*, $g_s(g_o)$, has a high (low) value. The function g is of form (1.7) in the two linear domains $u > 0$ and $u < 0$, which are separated by the point $u = 0$. In this example, $n = 1$ and the hyperplane dimension is $n - 1 = 0$, which indeed corresponds to a point. Thus, it is actually a piecewise linear resistor. Note as well that if $g_s \neq 0$ and $g_o \neq 0$, this resistor is controlled by current and voltage, as was the case for (1.15). On the other hand, not all currents were admissible for model (1.15).

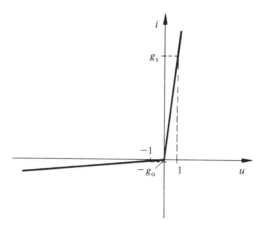

Fig. 1.29

The limit case of (1.16), with $g_s = \infty$ and $g_o = 0$, is called an *ideal diode*. Its characteristic, represented in figure 1.30, shows that this resistor is controlled by neither current nor voltage. Consequently, neither form (1.5) nor form (1.6) can be used as a constitutive relation, and we must return to (1.3). Normally, an ideal diode is described by the two relations $u < 0 \rightarrow i = 0$ and $i > 0 \rightarrow u = 0$. However, in order to obtain a constitutive relation of form (1.3), we must divide the plane (u,i) into two pieces, for instance, along the intersection of the second and fourth quadrants. The corresponding piecewise linear function:

$$f(u,i) = \begin{cases} u & \text{for} & u + i > 0 \\ -i & \text{for} & u + i < 0 \end{cases} \tag{1.17}$$

is continuous and its curve of level 0 is the characteristic shown in figure 1.30. Note that our choice of the two plane pieces and the function f could have been different without altering the characteristic. For example,

$$f(u,i) = \begin{cases} 2u & \text{for} & u + i/3 > 0 \\ -2i/3 & \text{for} & u + i/3 < 0 \end{cases} \tag{1.18}$$

In order to distinguish ideal diodes from other diodes, we use the symbol of figure 1.31, represented as the solid triangle.

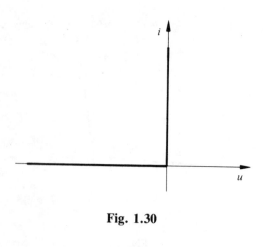

Fig. 1.30

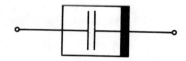

Fig. 1.31

1.3.10 Definition: Capacitor

A *capacitor* is a one-port, which is defined by a constitutive relation of the form:

$$f(u, q) = 0 \tag{1.19}$$

where the *charge q* and the current *i* are related by

$$i = dq/dt \tag{1.20}$$

Its symbol is represented in figure 1.32. The characteristic of the capacitor is the set of points in the plane (u, q) which satisfies (1.19). The hypotheses concerning f are given in subsection 1.3.7. The *time-dependent capacitor* is similarly defined, as is the concept of *symmetric capacitor*.

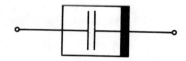

Fig. 1.32

A voltage u is called an *admissible voltage for the capacitor* if there is at least one charge q such that (1.19) is satisfied. Similarly, we introduce the concept of *admissible charge for the capacitor*.

If (1.19) can be put into one of the forms:

$$q = g(u) \tag{1.21}$$
$$u = h(q) \tag{1.22}$$

then it is a *voltage-controlled capacitor* or a *current-controlled capacitor*, respectively.

1.3.11   Examples

The *linear capacitor* of value C, which is defined by the constitutive relation:

$$q = Cu \tag{1.23}$$

is symmetric and controlled by both voltage and charge if $C \neq 0, \infty$. Any value of voltage or charge is admissible.

A *varactor diode* is modeled by a nonlinear capacitor with the constitutive relation:

$$q(u) = -\tfrac{3}{2} C_0 U_0 (1 - u/U_0)^{2/3}, \, u < U_0 \tag{1.24}$$

where the constants C_0 and U_0 are peculiar to each varactor diode. The symbol and characteristic of this element are shown in figures 1.33 and 1.34. It is a voltage-controlled capacitor, with admissible voltages $u < U_0$. It is controlled by the charge as well, the negative values of which are admissible.

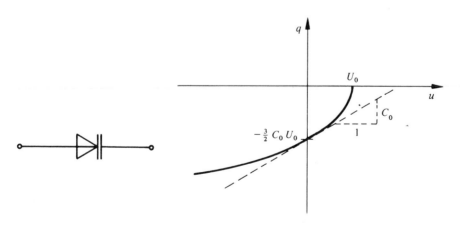

Fig. 1.33   Fig. 1.34

1.3.12 Definition: Inductor

An *inductor* is a one-port defined by a constitutive relation of the form:

$$f(\varphi, i) = 0 \tag{1.25}$$

where the *flux* φ and the voltage u are related by

$$u = d\varphi/dt \tag{1.26}$$

Its symbol is shown in figure 1.35.

Fig. 1.35

The *characteristic* of the inductor is the set of points in the plane (φ, i) which satisfies (1.25). The hypotheses for f are those of subsection 1.3.7. Similarly, we define the following concepts: *symmetric inductor, time dependent inductor, admissible current and flux, and inductor controlled by flux and by current*.

1.3.13 Examples

The *linear inductor* of value L, which is defined by the constitutive relation:

$$\varphi = L\,i \tag{1.27}$$

is symmetric and simultaneously controlled by flux and current, provided that $L \neq 0, \infty$. All the currents and fluxes are admissible.

An *inductor with a ferromagnetic core* can be modeled, if an abstraction of the hysteresis phenomenon is made, by a nonlinear inductor having the characteristic shown in figure 1.36. It is sometimes approximated by a constitutive relation of the form:

$$i = a\varphi + b\varphi^3 \tag{1.28}$$

where a and b are constants peculiar to the inductor. A rougher approximation is provided by the piecewise linear function:

$$i = \begin{cases} \varphi/L_0 & \text{for} \quad |\varphi| \leqslant \varphi_0 \\ \dfrac{\varphi}{L_1} - \varphi_0 \left(\dfrac{1}{L_1} - \dfrac{1}{L_0} \right) & \text{for} \quad \varphi > \varphi_0 \\ \dfrac{\varphi}{L_1} + \varphi_0 \left(\dfrac{1}{L_1} - \dfrac{1}{L_0} \right) & \text{for} \quad \varphi < -\varphi_0 \end{cases} \qquad (1.29)$$

This characteristic is shown in figure 1.37.

In both cases, (1.28) and (1.29), the inductor is flux and current controlled, and all the currents and fluxes are admissible.

A *superconducting junction* or *Josephson junction* is modeled by a smooth symmetric inductor defined according to the constitutive relation:

$$i = I_0 \sin k_0 \varphi \qquad (1.30)$$

Its characteristic is shown in figure 1.38. It is a flux-controlled inductor, all the values of which are admissible. However, it is not current controlled, and the admissible currents are in the interval $[-I_0, +I_0]$.

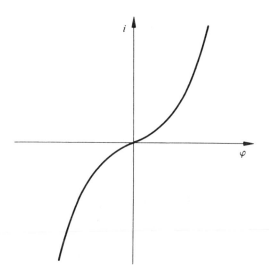

Fig. 1.36

30

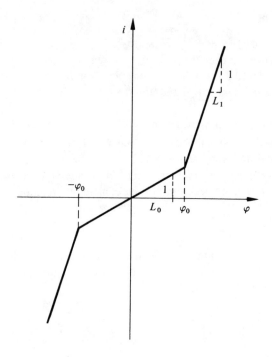

Fig. 1.37

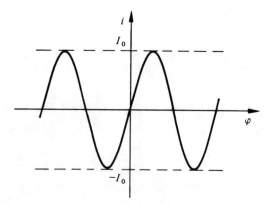

Fig. 1.38

1.3.14 Comments

The resistor is defined by a simultaneous relation between current and voltage, i.e., *at the same time*. It cannot connect the present to the past. Thus, it is an element without memory. On the other hand, capacitors link the present voltage to the past current, although very summarily, through the current's integral, i.e., the charge. The same is true for inductors.

It is instructive to represent graphically the links between the different variables used to define the basic elements (fig. 1.39). The four variables u, i, φ, and q form six pairs. Two pairs of variables (u,φ) and (i,q) are trivially linked by (1.20) and (1.26). The resistor, the capacitor, and the inductor respectively correspond to the pairs (u,i), (u,q), and (φ,i). As we can see, a basic element corresponding to the pair (φ,q) is still missing. It looks like the resistor, but it is capable of memorizing, and thus is called "memristor."

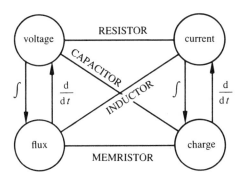

Fig. 1.39

1.3.15 Definition: Memristor

A *memristor* [2] is a one-port defined by a constitutive relation:

$$f(\varphi, q) = 0 \tag{1.31}$$

Its symbol is shown in figure 1.40. The concepts of *characteristic, admissible flux,* and *charge, et cetera* are defined as for the other basic elements.

Fig. 1.40

1.3.16 Example

A *linear memristor* is defined by the constitutive relation:

$$\varphi = Rq \tag{1.32}$$

Deriving (1.32) with respect to time, we obtain

$$u = Ri \tag{1.33}$$

Consequently, a linear memristor is equivalent to a linear resistor, which explains why the memristor does not appear in the theory of linear circuits.

On the other hand, if the memristor is given by a nonlinear relation:

$$\varphi = f(q) \tag{1.34}$$

then deriving with respect to time leads to

$$u = (df/dq)i \tag{1.35}$$

and df/dq depends on q. Roughly speaking, we could call df/dq a charge-dependent resistor.

1.3.17 Property

The memristor is so called because it usually has a memory, while resistors do not.

Let us check this property by considering the circuit of figure 1.41, which contains a memristor with the characteristic of figure 1.42. Let us assume that at time $t = 0$, the circuit is in the state $u = i = \varphi = q = 0$. When a rectangular impulse of amplitude I and duration $\Delta t = q_0/I_0$ is supplied by the current source, the voltage u remains zero because we remain in the area of zero "resistance" (fig. 1.43). On the other

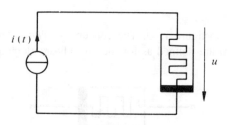

Fig. 1.41

hand, the memristor charge takes the value q_0 at the end of the current impulse. If we later apply a second impulse, the voltage on the memristor reacts as well with a rectangular impulse (fig. 1.44). Thus, the memristor remembers the first impulse. This example will be verified by numeric simulation in subsection 1.4.12.

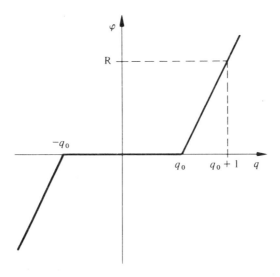

Fig. 1.42

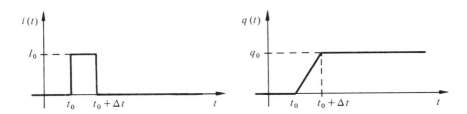

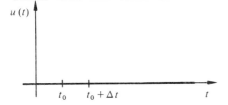

Fig. 1.43

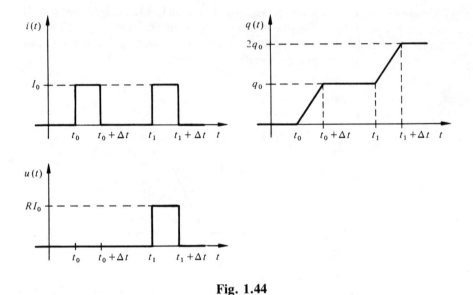

Fig. 1.44

1.3.18 Comment

Only a few examples of actual devices suitably modeled by a memristor have been proposed. Most of these are in the field of electrochemistry. Thus, the memristor is, above all, a theoretical concept.

1.3.19 Definitions

Let us repeat here the definitions of the following linear two-ports:

The *voltage-controlled voltage source* (fig. 1.45) is defined by

$$i_1 = 0 \qquad (1.36)$$

$$u_2 = \alpha u_1 \qquad (1.37)$$

where α is a constant without dimension.

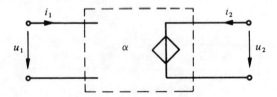

Fig. 1.45

The *current-controlled voltage source* (fig. 1.46) is defined by

$$u_1 = 0 \tag{1.38}$$
$$u_2 = r i_1 \tag{1.39}$$

where r is a constant with a resistance dimension.

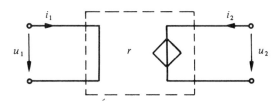

Fig. 1.46

The *voltage- or current-controlled current sources* are defined analogously. The *ideal operational amplifier* (fig. 1.47) is a two-port defined by (vol. XIX, subsec. 9.1.2)

$$u_1 = 0 \tag{1.40}$$
$$i_1 = 0 \tag{1.41}$$

We have already encountered this two-port in subsection 1.2.6.

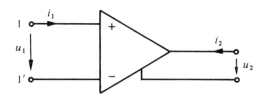

Fig. 1.47

1.3.20 Comments

The representation of the controlled sources as two-ports may come as a surprise. In figure 1.48, a circuit including controlled sources is first represented in the usual way, then in a way that identifies the two-ports.

The two-ports introduced in subsection 1.3.18 relate the voltages and currents at the **same time**. Thus, they do not have any memory. From this point of view, they behave as resistors.

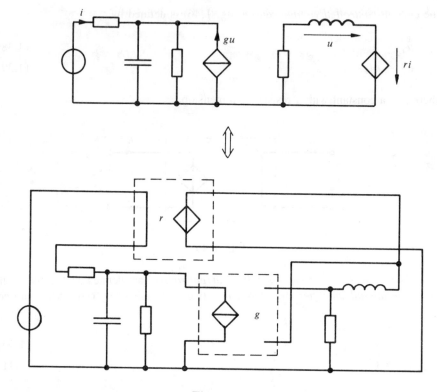

Fig. 1.48

1.3.21 Independence of the Basic Elements

Now, we may wonder whether the basic elements have been properly chosen; not too few, not too many.

Are there enough basic elements? By connecting them, can we model any physical circuit which we intend to study? This question is both vague and difficult. It belongs to the domains of synthesis and modeling of nonlinear circuits, which are not treated in this book. Let us simply mention that all the usual models of nonlinear devices, included in circuits for which the Kirchhoff model is applicable, fit into the formalism of this section.

Are there too many basic elements? Could any basic element be replaced by a connection of other basic elements and thus become dispensable? We shall show in subsection 1.3.22 that actually only one of the three one-ports with memory—the capacitor, the inductor, and the memristor—is indispensable. Also, we shall see in subsection 1.3.23 that we can simply use the ideal operational amplifier as a single basic linear two-port.

1.3.22 Dependence among the Capacitor, Inductor, and Memristor

Let us consider the nonlinear two-port of figure 1.49. By closing its ports on different elements, we obtain different one-ports. The operational amplifier imposes the constraint on the potentials:

$$v_1 = v_3 \tag{1.42}$$

so that

$$u_2 = u \tag{1.43}$$

and

$$u_1 = v_3 - v_4 \tag{1.44}$$

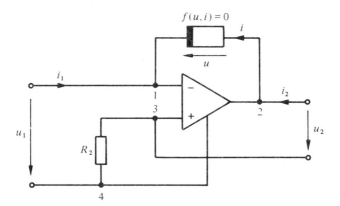

Fig. 1.49

Because the input current of the operational amplifier is zero, the currents in the resistors are i_1 and i_2, respectively, from which we deduce the constitutive relations of the two-port:

$$f(u_2, -i_1) = 0 \tag{1.45}$$
$$u_1 = -R_2 i_2 \tag{1.46}$$

In figure 1.50, we have closed the two-port onto a linear capacitor, for which the constitutive relation is

$$q = Cu_2 \tag{1.47}$$

The charge on the capacitor is expressed as

$$q(t) = q(t_0) - \int_{t_0}^{t} i_2(t')\,dt' \tag{1.48}$$

and, based on (1.46),

$$q(t) = q(t_0) + \frac{1}{R_2} \int_{t_0}^{t} u_1(t')\,dt' \tag{1.49}$$

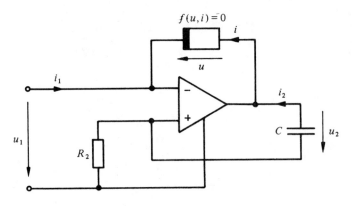

Fig. 1.50

The integral in (1.49) is nothing but the flux associated with port 1. By specifying the undetermined constant in the flux definition by

$$\varphi_1(t_0) = R_2 q(t_0) \tag{1.50}$$

we find

$$q(t) = \varphi_1(t)/R_2 \tag{1.51}$$

for any t. By combining (1.45), (1.47), and (1.51), we obtain the constitutive relation of the one-port of figure 1.50:

$$f(\varphi_1/CR_2, -i_1) = 0 \tag{1.52}$$

Because it is of the form (1.25), the circuit of figure 1.50 implements a nonlinear inductor. Conversely, any nonlinear inductor can be synthesized in the same way. A numeric example is given in subsection 1.4.11.

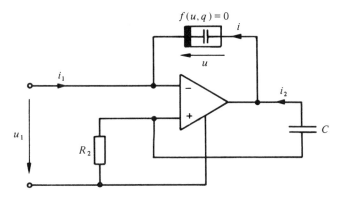

Fig. 1.51

If we had closed the other port on the two-port of figure 1.49 onto a linear inductor, we would have obtained a nonlinear capacitor.

By replacing the nonlinear resistor of figure 1.50 by a nonlinear capacitor (fig. 1.51) having the constitutive relation:

$$f(u,q) = 0 \tag{1.53}$$

then, instead of (1.52), we obtain the constitutive relation:

$$f(\varphi_1/CR_2, -q_1) = 0 \tag{1.54}$$

It is a nonlinear memristor. Another memristor implementation is given in subsection 1.4.12.

1.3.23 Dependence of Linear Two-Ports

The linear two-ports that we introduced in subsection 1.3.19 are independent as well.

By cascading a voltage-controlled current source and a current-controlled voltage source, we obtain either a voltage-controlled voltage source or a current-controlled current source (fig. 1.52.).

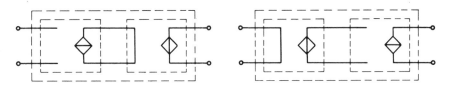

Fig. 1.52

The ideal operational amplifier is the limiting case of a voltage-controlled voltage source when the voltage gain tends toward infinity.

On the other hand, the controlled sources are obtained from the ideal operational amplifier. Indeed, if we replace the nonlinear resistor of the two-port in figure 1.49 by a linear resistor of value R_1, the constitutive relations become

$$u_2 = -R_1 i_1 \tag{1.55}$$

$$u_1 = -R_2 i_2 \tag{1.56}$$

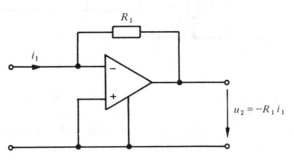

Fig. 1.53

If $R_2 = 0$, then (1.55) and (1.56) are the constitutive relations of a current-controlled voltage source (fig. 1.53). Similarly, by setting $R_1 = \infty$, we obtain a voltage-controlled current source:

1.3.24 Comments

In order to be a good model of the operational amplifier, which is a mass-produced physical device, the ideal operational amplifier must have an output terminal connected to ground. Consequently, a nonlinear inductor implemented by the circuit of figure 1.50 through an operational amplifier necessarily has a ground terminal. This is a limitation often encountered with RC-active filters in the case of direct simulation of LC filters (vol. XIX, chap. 9).

The constitutive relations (1.45) and (1.46) of the two-port of figure 1.49 are valid if and only if the circuit actually functions as a two-port. This condition must be ensured by the surrounding circuit because the circuit itself is not an intrinsic two-port.

1.3.25 Circuit Hypotheses

We have shown in subsections 1.3.22 and 1.3.23 that inductors, memristors, and controlled sources could be taken off the list of basic elements. It would even be possible to restrict the list to linear capacitors only; this would leave the resistor as the sole nonlinear element.

Such a radical choice does not seem to be advisable. The inductor is a good model for a physical device comprised of a coil, and it would be a pity to have to model a coil by using a circuit as complicated as that of figure 1.50. It is also convenient to retain all the basic two-ports of subsection 1.3.19, partly by tradition, partly because of the reservations made in subsection 1.3.24. However, we can easily do without the memristor.

This leads us to the following hypotheses concerning the circuits studied in this book. Unless otherwise mentioned, a circuit is composed of

- autonomous linear and nonlinear capacitors;
- autonomous linear and nonlinear inductors;
- time-dependent voltage and current sources;
- autonomous linear and nonlinear resistors;
- autonomous linear controlled sources and ideal operational amplifiers.

The circuit elements are either all smooth or all piecewise linear. Accordingly, we speak of a *smooth circuit* or a *piecewise linear circuit*.

To simplify, let us assume that all elements are autonomous, except for the independent sources which play a special role as clarified in section 1.5. If a circuit does not include any time-dependent elements, it is *autonomous*; otherwise, it is *time dependent*.

The differential equations among the circuit equations come from capacitors and inductors. If such elements are absent, the circuit is *resistive*.

1.4 BASIC ELEMENTS AND CONNECTIONS OF THE SPICE PROGRAM

1.4.1 Introduction

A circuit, as defined by sections 1.2 and 1.3, is a connection of basic elements. In the computer analysis of circuits we encounter the same definition, be it explicit or not. The computer program is capable of analyzing a certain class of circuits. This class is defined by a list of basic elements and a specification of the connections that are admissible. Each program has its own list of basic elements.

Such a conceptual construction comes with a descriptive language. While the theory of circuits primarily uses a graphic language (i.e, the circuit diagrams), the computer needs a special language, often heavily coded. Each program uses a special syntax for the data, which is communicated to the user in the user's manual.

The purpose of this section is to define the circuits that are capable of being analyzed by the SPICE program, and to compare them with the class of circuits defined in section 1.3. For the documentation of SPICE, we shall refer to [3]. We have chosen SPICE—version 2 of which was designed in 1975 by a group of researchers at the

University of California, Berkeley—because it is at present the most widely used general circuit analysis program. It is basically meant for electronic circuits with semiconductors, but, with some precautions, its range of application can be substantially extended.

1.4.2 Basic One-Ports

The basic one-ports are

- the autonomous linear capacitor, of positive value;
- the autonomous linear inductor, of positive value;
- the time-dependent and autonomous independent voltage and current source. The functions of time that can be specified are a pulse, a damped or undamped sinusoid, an exponential, a continuous piecewise linear signal, and a frequency modulated sinusoid;
- the autonomous linear resistor, of positive or negative value;
- the junction diode in the form of a sophisticated model.

1.4.3 Basic Two-Ports

The basic two-ports are

- the four types of controlled sources, with nonlinear control functions;
- the mutual inductor;
- the transmission line.

Nonlinear controlled sources are defined in the following way:
If there is one control variable x only, the value y of the source is given by

$$y = a_0 + a_1 x + a_2 x^2 + ... + a_n x^n \tag{1.57}$$

where the α_i are positive, negative, or zero, and n is an arbitrary positive integer. The variables x and y are voltages or currents, resulting in four types of controlled sources.
If there are two control variables, x_1 and x_2, the value of the source is given by

$$y = a_0 + a_1 x_1 + a_2 x_2 + a_3 x_1^2 + a_4 x_1 x_2 + a_5 x_2^2 + a_6 x_1^3 + ... \tag{1.58}$$

Here, the variables x_1 and y_2 must be either both voltages or currents. The sources controlled by more than two variables are similarly defined.

1.4.4 Basic Multipoles

The basic three-poles are

- a sophisticated model of the bipolar transistor. By specifying certain parameters, this model becomes a four-pole element;

- a sophisticated model of the JFET transistor;

and the basic four-pole is

- a sophisticated model of the MOS transistor.

1.4.5 Comment

It follows that the nonlinear elements considered by SPICE are the diode, the polynomial controlled sources, and the three types of transistors. As the examples of nonlinear circuits for this text, we shall primarily use diodes and controlled sources.

1.4.6 Connections

According to definition 1.2.1, a connection is defined by gathering the terminals of the elements into nodes. This definition finds a direct application in SPICE. The elements are introduced by an identifier, a value, possibly other parameters, and one integer for each terminal. This integer specifies to which node the terminal is connected.

A connection must satisfy the following conditions:

- The circuit does not include any loops (vol. IV, subsec. 1.2.4) of voltage sources or inductors, nor any cut sets (vol. IV, subsec. 1.2.4) of current sources or capacitors.
- If we short-circuit the inductors and eliminate the capacitors, the circuit must remain connected.
- Each node must have at least two elements connected to it. An exception is made for the nodes of transmission lines and substrates of MOS transistors. In the MOS case, however, it is the model of the transistor which provides the necessary elements connected to the substrate.

1.4.7 Comment

Any circuit analysis made by SPICE begins with the calculation of dc operating point (definition 2.3.8). To that effect, we short-circuit the inductors and the time-dependent voltage sources, and we open the capacitors and the time-dependent current sources. If the first condition were not satisfied, we would have voltage source loops or current source cut sets. In nearly all cases, this results in an equation system without solution, implying that the circuit does not have any dc operating point. On the other hand, if the second condition is not satisfied, it produces an infinity of dc operating points because the potential level of the various connected parts are not determined by the equations for the dc operating point.

We shall see in chapter 3 that these conditions are necessary, but not sufficient, for the existence and uniqueness of the dc operating points.

1.4.8 Analyses

SPICE primarily carries out three types of analysis:

- Calculation of the dc operating point, whereby to obtain element and transfer characteristics, we can calculate a sequence of dc operating points by varying a constant source. The sensitivities of the dc operating point with respect to the circuit parameters can be determined as well.
- Frequency analysis of the circuit linearized around the dc operating point, in the presence of dc sources and sinusoidal sources, wherein, accessorily, the noise and distortion analysis can be carried out.
- Calculation of the time-dependent solution starting from initial conditions, in the presence of constant sources and time dependent sources; a Fourier analysis of this solution is also possible.

The concepts of dc operating point and a circuit linearized around a dc operating point will be introduced in chapter 2.

1.4.9 Simulation of Basic Elements

The basic elements defined in section 1.3 and the basic elements of SPICE are only partially matching. To the question of whether any basic element of section 1.3 can be obtained by connecting basic elements of SPICE, there are two possible answers.

Very strictly speaking, the answer is no. Consideration of the nonlinear resistors makes this clear. With linear resistors and nonlinear controlled sources, it is possible to simulate polynomial, rational, algebraic characteristics. By using, in addition, the exponential diode, we can combine powers and exponentials. Still, it is clearly impossible to implement any characteristic exactly through a finite connection of SPICE elements. For instance, the resistor defined by

$$i = I_0 \sin(u/u_0) \qquad (1.59)$$

cannot be implemented. Likewise, the piecewise linear characteristics cannot be exactly described by SPICE because the elements have characteristics with continuous derivatives, and this fact remains true for their connection.

On the other hand, if we only want to approximate the basic elements within a finite range of the variables, the SPICE elements are amply sufficient. We should not forget that the basic elements themselves are only used to model physical devices and, consequently, their characteristics do not represent absolute data.

1.4.10 Example

This first example shows how a nonlinear resistor can be simulated by means of a nonlinear source. A sequence of operating points in the circuit of figure 1.54 is calculated by SPICE. This is equivalent to recording the nonlinear resistor characteristic, the area surrounded by a dashed line in figure 1.54, the constitutive relation of which is

$$i = a_0 + a_1 u + a_2 u^2 + a_3 u^3 \qquad (1.60)$$

The input data and the resulting output are represented in figure 1.55 for

$$a_0 = 0 \qquad (1.61)$$
$$a_1 = 6\,\Omega^{-1} \qquad (1.62)$$
$$a_2 = -4{,}5\,\text{V}^{-1}\Omega^{-1} \qquad (1.63)$$
$$a_3 = 1\,\text{V}^{-2}\Omega^{-2} \qquad (1.64)$$

For the input syntax, refer to [3]. The only task of the VDUMMY source, having a value of zero, is to identify the current for the controlled source G.

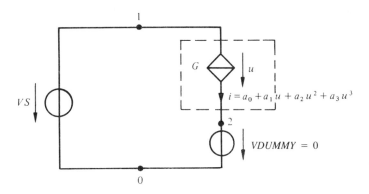

Fig. 1.54

The downward vertical axis of the diagram identifies the voltage, and the rightward horizontal axis identifies the resistor current. Note that the characteristic of figure 1.55 looks similar to that of the tunnel diode (fig. 1.25).

```
Resistor with a polynomial characteristic of degree 3.

VS      1 0        1
G       1 2   1 2    0  6  -4.5  1
VDUMMY  2 0        0

.DC    VS    0  2.5  0.05
.PLOT  DC   I(VDUMMY)
.END
```

VS	I(VDUMMY)					
		0.000D+00	1.000D+00	2.000D+00	3.000D+00	4.000D+00
0.000D+00	0.000D+00	*	.	.	.	.
5.000D-02	2.889D-01	. *	.	.	.	.
1.000D-01	5.560D-01	. *	.	.	.	.
1.500D-01	8.021D-01	. *	.	.	.	.
2.000D-01	1.028D+00	.	.*	.	.	.
2.500D-01	1.234D+00	.	. *	.	.	.
3.000D-01	1.422D+00	.	. *	.	.	.
3.500D-01	1.592D+00	.	. *	.	.	.
4.000D-01	1.744D+00	.	. *	.	.	.
4.500D-01	1.880D+00	.	. *	.	.	.
5.000D-01	2.000D+00	.	.	*	.	.
5.500D-01	2.105D+00	.	.	.*	.	.
6.000D-01	2.196D+00	.	.	. *	.	.
6.500D-01	2.273D+00	.	.	. *	.	.
7.000D-01	2.338D+00	.	.	. *	.	.
7.500D-01	2.391D+00	.	.	. *	.	.
8.000D-01	2.432D+00	.	.	. *	.	.
8.500D-01	2.463D+00	.	.	. *	.	.
9.000D-01	2.484D+00	.	.	. *	.	.
9.500D-01	2.496D+00	.	.	. *	.	.
1.000D+00	2.500D+00	.	.	. *	.	.
1.050D+00	2.496D+00	.	.	. *	.	.
1.100D+00	2.486D+00	.	.	. *	.	.
1.150D+00	2.470D+00	.	.	. *	.	.
1.200D+00	2.448D+00	.	.	. *	.	.
1.250D+00	2.422D+00	.	.	. *	.	.
1.300D+00	2.392D+00	.	.	. *	.	.
1.350D+00	2.359D+00	.	.	. *	.	.
1.400D+00	2.324D+00	.	.	. *	.	.
1.450D+00	2.287D+00	.	.	.*	.	.
1.500D+00	2.250D+00	.	.	*	.	.
1.550D+00	2.213D+00	.	.	*	.	.
1.600D+00	2.176D+00	.	.	*	.	.
1.650D+00	2.141D+00	.	.	*	.	.
1.700D+00	2.108D+00	.	.	*	.	.
1.750D+00	2.078D+00	.	.	*	.	.
1.800D+00	2.052D+00	.	.	*	.	.
1.850D+00	2.030D+00	.	.	*	.	.
1.900D+00	2.014D+00	.	.	*	.	.
1.950D+00	2.004D+00	.	.	*	.	.
2.000D+00	2.000D+00	.	.	*	.	.
2.050D+00	2.004D+00	.	.	*	.	.
2.100D+00	2.016D+00	.	.	*	.	.
2.150D+00	2.037D+00	.	.	.*	.	.
2.200D+00	2.068D+00	.	.	. *	.	.
2.250D+00	2.109D+00	.	.	. *	.	.
2.300D+00	2.162D+00	.	.	. *	.	.
2.350D+00	2.227D+00	.	.	. *	.	.
2.400D+00	2.304D+00	.	.	. *	.	.
2.450D+00	2.395D+00	.	.	. *	.	.
2.500D+00	2.500D+00	.	.	.	*	.

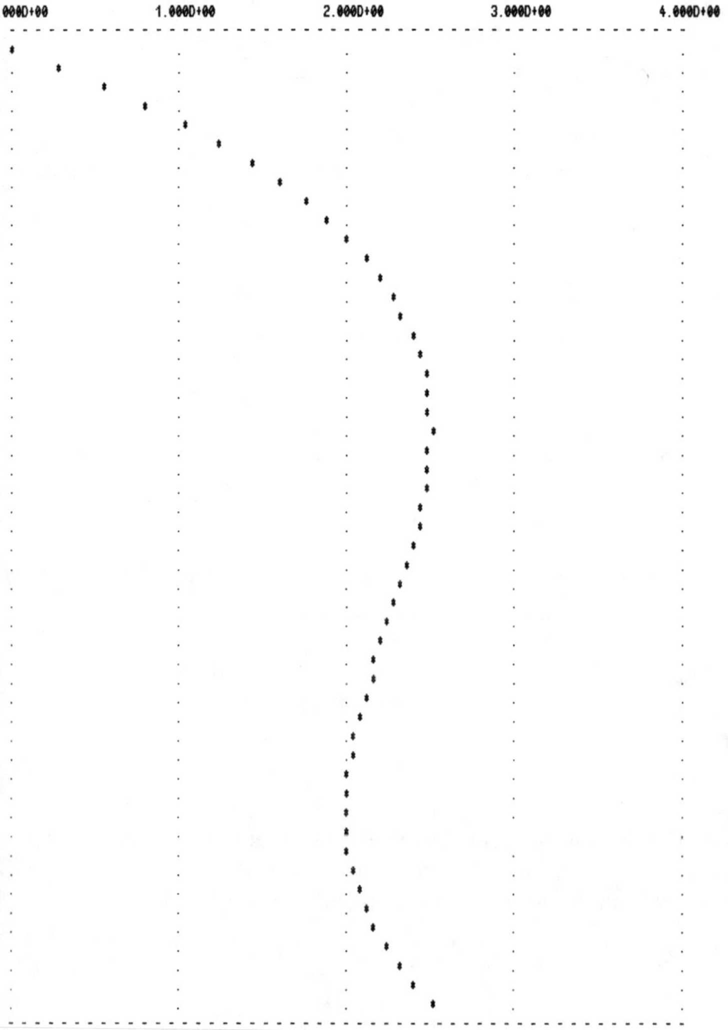

Fig. 1.55

1.4.11 Example

We intend to implement an inductor having a constitutive relation:

$$i = a\varphi + b\varphi^7 \tag{1.65}$$

with

$$a = 0{,}03 \text{ A/Vs} \tag{1.66}$$
$$b = 0{,}061 \text{ A/(Vs)}^7 \tag{1.67}$$

by means of the circuit of figure 1.50. In figure 1.58, the input data for SPICE describe the circuit of figure 1.56. The dashed area implements the inductor. The operational amplifier is simulated by the voltage-controlled voltage source E1, with a gain $A = 10$. The RDUMMY resistor had to be inserted to prevent a current source cut set, which is forbidden, according to subsection 1.14.6. Its value is sufficiently large so as not to interfere with the inductor characteristic.

An inductor is equivalent to a short circuit at dc. Therefore, its characteristic cannot be obtained by dc operating points as was the case for a resistor. We have connected a time-dependent current source to the inductor with a current proportional to time. The flux is transformed into voltage by an integrator, according to figure 1.57. In the circuit of figure 1.56, the integrator is located to the right of the inductor.

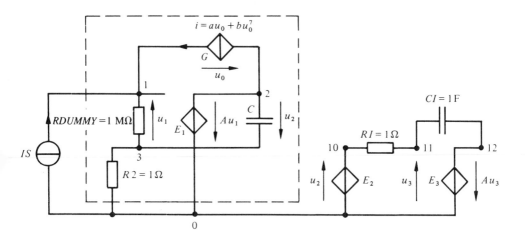

Fig. 1.56

The diagram of figure 1.58 represents the inductor characteristic in the first quadrant: on the vertical axis the time is proportional to the current, and on the horizontal axis the voltage v_{12} is proportional to the inductor flux.

48

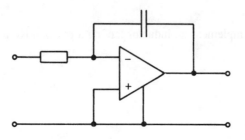

Fig. 1.57

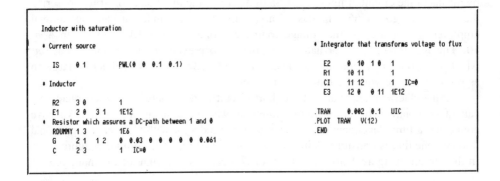

```
Inductor with saturation

* Current source

  IS    0 1      PWL(0 0 0.1 0.1)

* Inductor

  R2    3 0      1
  E1    2 0  3 1   1E12
* Resistor which assures a DC-path between 1 and 0
  RDUMMY 1 3     1E6
  G     2 1  1 2   0 0.03 0 0 0 0 0 0.061
  C     2 3      1  IC=0
```

```
* Integrator that transforms voltage to flux

  E2    0 10 1 0   1
  R1    10 11      1
  CI    11 12      1  IC=0
  E3    12 0  0 11  1E12

.TRAN   0.002 0.1  UIC
.PLOT   TRAN  V(12)
.END
```

```
TIME      V(12)

                    -5.000D-01       0.000D+00        5.000D-01        1.000D+00        1.500D+00
                     - - - - - - - - - - - - - - - - - - - - - - - - - - - - - - - - - - - - - -
0.000D+00  -3.333D-11   .                .   *            .                .                .
2.000D-03   6.667D-02   .                .    *           .                .                .
4.000D-03   1.333D-01   .                .       *        .                .                .
6.000D-03   2.000D-01   .                .         *      .                .                .
8.000D-03   2.664D-01   .                .            *   .                .                .
1.000D-02   3.323D-01   .                .              * .                .                .
1.200D-02   3.966D-01   .                .                *                .                .
1.400D-02   4.576D-01   .                .                .  *             .                .
1.600D-02   5.134D-01   .                .                .     *          .                .
1.800D-02   5.627D-01   .                .                .       *        .                .
2.000D-02   6.051D-01   .                .                .         *      .                .
2.200D-02   6.414D-01   .                .                .           *    .                .
2.400D-02   6.725D-01   .                .                .            *   .                .
2.600D-02   6.994D-01   .                .                .             *  .                .
2.800D-02   7.228D-01   .                .                .              * .                .
3.000D-02   7.436D-01   .                .                .               *.                .
3.200D-02   7.622D-01   .                .                .                *                .
3.400D-02   7.789D-01   .                .                .                .*                .
3.600D-02   7.942D-01   .                .                .                .*                .
3.800D-02   8.082D-01   .                .                .                . *               .
4.000D-02   8.211D-01   .                .                .                . *               .
4.200D-02   8.331D-01   .                .                .                .  *              .
4.400D-02   8.443D-01   .                .                .                .  *              .
4.600D-02   8.548D-01   .                .                .                .   *             .
```

Fig. 1.58

```
TIME      V(12)

              -5.000D-01        0.000D+00         5.000D-01         1.000D+00         1.500D+00
            - - - - - - - - - - - - - - - - - - - - - - - - - - - - - - - - - - - - - - - - - - - -
4.800D-02   8.647D-01   .                 .                 .                 *   .
5.000D-02   8.740D-01   .                 .                 .                 *   .
5.200D-02   8.829D-01   .                 .                 .                 *   .
5.400D-02   8.913D-01   .                 .                 .                 *   .
5.600D-02   8.993D-01   .                 .                 .                 *   .
5.800D-02   9.069D-01   .                 .                 .                 *   .
6.000D-02   9.142D-01   .                 .                 .                 *   .
6.200D-02   9.213D-01   .                 .                 .                 *   .
6.400D-02   9.280D-01   .                 .                 .                 *   .
6.600D-02   9.345D-01   .                 .                 .                 *   .
6.800D-02   9.407D-01   .                 .                 .                 *   .
7.000D-02   9.468D-01   .                 .                 .                 *   .
7.200D-02   9.526D-01   .                 .                 .                 * . .
7.400D-02   9.582D-01   .                 .                 .                 * . .
7.600D-02   9.637D-01   .                 .                 .                 * . .
7.800D-02   9.690D-01   .                 .                 .                 * . .
8.000D-02   9.741D-01   .                 .                 .                 *.  .
8.200D-02   9.791D-01   .                 .                 .                 *.  .
8.400D-02   9.840D-01   .                 .                 .                 *.  .
8.600D-02   9.887D-01   .                 .                 .                 *.  .
8.800D-02   9.933D-01   .                 .                 .                 *   .
9.000D-02   9.978D-01   .                 .                 .                 *   .
9.200D-02   1.002D+00   .                 .                 .                 *   .
9.400D-02   1.006D+00   .                 .                 .                 *   .
9.600D-02   1.011D+00   .                 .                 .                  .* .
9.800D-02   1.015D+00   .                 .                 .                  .* .
1.000D-01   1.019D+00   .                 .                 .                  .* .
            - - - - - - - - - - - - - - - - - - - - - - - - - - - - - - - - - - - - - - - - - - - -
```

Fig. 1.58 (cont'd)

1.4.12   Example

The circuit of figure 1.59 constitutes a nonlinear memristor to which a time dependent source has been connected in order to check the memorization phenomenon explained in subsection 1.3.17.

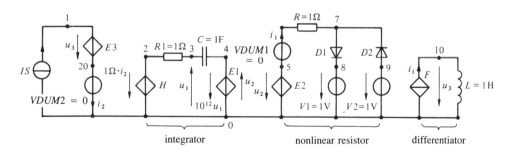

Fig. 1.59

50

The memristor is implemented by an integrator, according to figure 1.57, having an output voltage equal to the input flux; by a nonlinear resistor; and by a differentiator, according to figure 1.60, having an output charge equal to the input current. The nonlinear resistor and the resulting memristor have the same characteristic, after replacement of φ by u, and of q by i. It is a smooth characteristic, which approximates the piecewise linear characteristic of figure 1.42, with $q_0 = 1C$. In order to ensure a quick change of the derivative at $q = q_0$, we have chosen a very small emission coefficient for the diode.

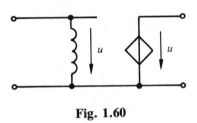

Fig. 1.60

In order to avoid continuity problems, we apply two signals of triangular current, rather than rectangular current, to the memristor. Because the charge supplied by a triangular signal is equal to q_0, nothing is changed regarding the reasoning of subsection 1.3.17.

```
Nonlinear memristor                              D2    9 7          DIODE
                                                 V1    8 0          1
* Current source                                 V2    0 9          1

   IS    0 1      PWL(0 0  1 0  2 1  3 0  4 0  5 1  6 0)   * Differentiator

* Integrator                                     F     0 10  VDUM1  1
                                                 L     10 0         1
   H     2 0  VDUM2   1
   R1    2 3          1                           * Transfer of the signal to the input
   C     3 4          1    IC=0
   E1    4 0  0 3     1E12                            E3    1 20  10 0   1
                                                     VDUM2 20 0        0
* Nonlinear resistor
                                                 .MODEL DIODE D (N=0.001)
   E2    5 0  0 4     1
   VDUM1 5 6          0                           .TRAN   0.1  7      UIC
   R     6 7          1                           .PLOT   TRAN   V(1)
   D1    7 8          DIODE                        .END
```

Fig. 1.61(a)

```
  TIME      V(1)

                 -5.000D-01        0.000D+00        5.000D-01        1.000D+00        1.500D+00
                - - - - - - - - - - - - - - - - - - - - - - - - - - - - - - - - - - - - - - - -
0.000D+00    0.000D+00    .                *                .                .                .
1.000D-01    0.000D+00    .                *                .                .                .
2.000D-01    0.000D+00    .                *                .                .                .
3.000D-01    0.000D+00    .                *                .                .                .
4.000D-01    0.000D+00    .                *                .                .                .
5.000D-01    0.000D+00    .                *                .                .                .
6.000D-01    0.000D+00    .                *                .                .                .
7.000D-01    0.000D+00    .                *                .                .                .
8.000D-01    0.000D+00    .                *                .                .                .
9.000D-01    0.000D+00    .                *                .                .                .
1.000D+00    0.000D+00    .                *                .                .                .
1.100D+00    1.992D-13    .                *                .                .                .
1.200D+00    4.006D-13    .                *                .                .                .
1.300D+00    5.984D-13    .                *                .                .                .
1.400D+00    8.009D-13    .                *                .                .                .
1.500D+00    1.000D-12    .                *                .                .                .
1.600D+00    1.199D-12    .                *                .                .                .
1.700D+00    1.402D-12    .                *                .                .                .
1.800D+00    1.600D-12    .                *                .                .                .
1.900D+00    1.799D-12    .                *                .                .                .
2.000D+00    2.002D-12    .                *                .                .                .
2.100D+00    1.800D-12    .                *                .                .                .
2.200D+00    1.601D-12    .                *                .                .                .
2.300D+00    1.399D-12    .                *                .                .                .
2.400D+00    1.200D-12    .                *                .                .                .
2.500D+00    1.000D-12    .                *                .                .                .
2.600D+00    7.986D-13    .                *                .                .                .
2.700D+00    6.008D-13    .                *                .                .                .
2.800D+00    4.003D-13    .                *                .                .                .
2.900D+00    2.001D-13    .                *                .                .                .
3.000D+00    1.215D-14    .                *                .                .                .
3.100D+00   -9.470D-16    .                *                .                .                .
3.200D+00    1.266D-15    .                *                .                .                .
3.300D+00   -6.797D-16    .                *                .                .                .
3.400D+00   -5.381D-16    .                *                .                .                .
3.500D+00    1.629D-15    .                *                .                .                .
3.600D+00   -9.205D-16    .                *                .                .                .
3.700D+00   -9.913D-17    .                *                .                .                .
3.800D+00    8.072D-16    .                *                .                .                .
3.900D+00   -1.459D-15    .                *                .                .                .
4.000D+00    1.994D-15    .                *                .                .                .
4.100D+00    8.576D-02    .            *                    .                .                .
4.200D+00    2.097D-01    .                    *            .                .                .
4.300D+00    2.812D-01    .                        *        .                .                .
4.400D+00    4.087D-01    .                            *    .                .                .
4.500D+00    5.020D-01    .                .                *                .                .
4.600D+00    5.868D-01    .                .                    *            .                .
4.700D+00    7.143D-01    .                .                        *        .                .
4.800D+00    7.966D-01    .                .                            *    .                .
4.900D+00    8.923D-01    .                .                .                *                .
5.000D+00    9.807D-01    .                .                .                    *            .
5.100D+00    9.000D-01    .                .                .                *                .
5.200D+00    8.000D-01    .                .                .            *        .           .
5.300D+00    7.000D-01    .                .                .        *            .           .
5.400D+00    6.000D-01    .                .                .    *                .           .
5.500D+00    5.000D-01    .                .                *                    .           .
5.600D+00    4.000D-01    .                .            *                        .           .
5.700D+00    3.000D-01    .                .        *                            .           .
5.800D+00    2.000D-01    .                .    *                                .           .
5.900D+00    1.000D-01    .                *                                     .           .
6.000D+00    4.520D-07    .                *                .                .                .
6.100D+00   -2.725D-12    .                *                .                .                .
6.200D+00   -3.563D-12    .                *                .                .                .
```

Fig. 1.61(b)

TIME	V(1)

		-5.000D-01	0.000D+00	5.000D-01	1.000D+00	1.500D+00
6.300D+00	-4.254D-12	.	*	.	.	.
6.400D+00	-5.001D-12	.	*	.	.	.
6.500D+00	-5.806D-12	.	*	.	.	.
6.600D+00	-6.482D-12	.	*	.	.	.
6.700D+00	-7.258D-12	.	*	.	.	.
6.800D+00	-8.035D-12	.	*	.	.	.
6.900D+00	-8.709D-12	.	*	.	.	.
7.000D+00	-9.550D-12	.	*	.	.	.

Fig. 1.61(b) (cont'd)

The diagram of figure 1.61 represents the memristor voltage on the horizontal axis with respect to time on the vertical axis, calculated by SPICE. We observe that the first triangular signal, which is injected between $t = 1$s and $t = 3$s, is not perceived, whereas the second signal injected between $t = 4$s and $t = 6$s is exactly reproduced, as expected.

1.5 SYSTEMS

1.5.1 Systems and Circuits

A circuit is never an end in itself, but it fulfills a certain function. More precisely, a *circuit* is the implementation of a *system*.

A system (fig. 1.62) transforms a signal $s(t)$ into a response $r(t)$ (vol. IV, sect. 2.1). If the system is implemented by a circuit, the signal is injected by a time-dependent independent source, and the response is a voltage or a current; otherwise, it is a quantity deduced from voltages and currents. It is not always simple to detect which one of the quantities on the diagram of a circuit is the output quantity. Thus, the correspondence between circuits and systems is not unique in either event. Nevertheless, it is common to use the word circuit for both the circuit itself and the system it implements.

In this book, we primarily concentrate on circuits, often without referring to systems.

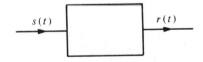

Fig. 1.62

1.5.2 Examples

The circuit of figure 1.63 is a *filter*. The signal is the voltage $e(t)$ of the source and the response is the power dissipation in the resistor R.

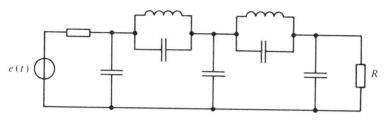

Fig. 1.63

The circuit of figure 1.64 is an *amplifier*. The signal is the source voltage $u_s(t)$, but the voltage of the other source is not a signal, it is the dc-supply voltage. The response is the potential of the node r.

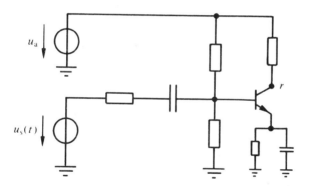

Fig. 1.64

The circuit of figure 1.65 is a *modulator*. The signal is the source voltage $u_s(t)$. The other source does not supply any information to be transformed, it is the *carrier source*. The response is the power dissipation in the load resistor R.

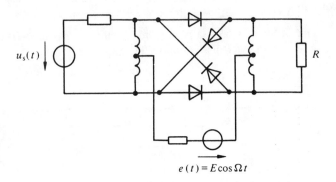

$$e(t) = E \cos \Omega t$$

Fig. 1.65

The circuit of figure 1.66 is a *multivibrator*. As in the case of the amplifier, the signal is supplied by the source voltage $u_s(t)$, u_a is the constant supply voltage, and the response is the potential of the node r.

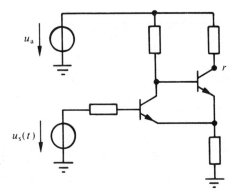

Fig. 1.66

1.5.3 Definition: Superposition Principle

A system satisfies the *superposition principle* if the response to a linear combination of signals is the same linear combination of the responses.

1.5.4 Fourier and Laplace Analysis

In an autonomous system which satisfies the superposition principle, the response to a complex exponential is an exponential with the same complex frequency (vol.

IV, subsect. 2.1.11). Thusly stated is the *isomorphic response theorem*. The proportionality factor between this signal and its response is the *response function* or *transfer function* of the system (vol. IV, subsect. 2.1.12).

The idea of the Fourier and Laplace analysis is to decompose an arbitrary signal into a sum or an integral of exponentials. When the transfer function is known, we can immediately find the response in the same form.

It is worth noting that the Fourier and Laplace analysis is **based on the superposition principle and the theorem of isomorphic response.** Neither is valid for nonlinear circuits in general.

1.5.5 Superposition Principle for Circuits

Let us consider a circuit including an independent source, e.g., a voltage source. Based on subsection 1.5.1, we associate a system with it by defining the source voltage as signal and by specifying, for example, a voltage u_r as response (fig. 1.67).

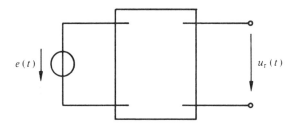

Fig. 1.67

If the circuit is linear, if there are **no other independent sources,** and if the circuit is in the **zero state** before the signal is injected, then the associated system satisfies the superposition principle.

In the presence of other sources and initial conditions different than zero, we can preserve the superposition principle by redefining the response of the associated system. We specify the new response as the difference between the old response in the presence of the signal and the response in its absence. We call this the *incremental response*. If a system satisfies the superposition principle for the incremental response, we call it a *linear system*.

A system associated with a linear circuit is always linear. Such is the case, in particular, for the filter of figure 1.63. Even a nonlinear circuit can implement a linear system, at least approximately. This is the case for the amplifier of figure 1.64. The supply source fixes a dc operating point. As long as the amplitude of the signal $u_r(t)$ is not too large, the voltage u remains close to the dc operating point. Then, the incremental response is, with fair approximation, a linear function of the signal. This example will be treated in chapter 2.

For the same reason, the system associated with the modulator of figure 1.65 satisfies the superposition principle if the signal amplitude is small with respect to the carrier amplitude. However, unlike the case for the amplifier, the theorem of isomorphic response is no longer satisfied. In fact, the response to an exponential is the sum of two exponentials, both of which are different than the signal. Is this a contradiction? It is not at all, since the system which contains the source of the carrier in its interior is no longer autonomous. Nevertheless, under certain conditions, a frequency analysis is possible. This topic will be treated in chapter 6.

Contrary to the above examples, the system associated with the multivibrator of figure 1.66 is far from being linear and a frequency analysis thus becomes impossible.

Chapter 2

Existence and Uniqueness of the Solutions

2.1 MOTIVATION THROUGH EXAMPLES

2.1.1 Introduction

According to the definition of subsection 1.2.11, a solution for a circuit is a solution for its system of equations, which is a combination of the Kirchhoff equations and the constitutive relations of the elements.

Two questions arise as a matter of course. With voltages and currents existing at the instant $t = t_0$, does the circuit have a solution for $t \geqslant t_0$ that coincides with the initial voltages and currents given in t_0? If so, is such a solution unique?

It is tempting to skip these questions, which seem to be without practical significance because, in practice, voltages and currents evolve over time in a precise way. This misleading impression arises from confusing the physical circuit with the model circuit. Their difference, which is shown in figure 1.6, must always be borne in mind. It is the solution of the model circuit that may not exist or may not be unique. If such is the case, the model circuit is an unsatisfactory model for the physical circuit. Most of the time, this fact provides valuable information about the physical circuit. It tells us that some effects deemed insignificant during modeling are actually crucial to the behavior of the physical circuit. It tells us as well that the solution of the physical circuit might have unexpected properties.

In Boite and Neirynck's study of linear circuits (vol. IV), the question of the solutions' existence and uniqueness was hardly mentioned—and rightly so, since it is guaranteed by the theorems of linear algebra, except for a few cases. Said exceptions normally have no practical interest. Curiously, these exceptions become important in the context of nonlinear circuits, as we shall see in section 2.5 and chapter 3.

The purpose of this section is to show, with some simple examples, what happens concerning the solutions' existence and uniqueness when dealing with nonlinear circuits. It is thus not surprising that nonlinear circuits, the operation of which does not differ greatly from that of linear circuits, have unique solutions as well. This is the case, in particular, for the amplifier of subsections 2.1.2 to 2.1.5. On the other hand, the behavior of circuits with a tunnel diode becomes increasingly nonlinear.

The amplifier example will illustrate the usual method of analysis of nonlinear circuits used in electronics. The limitations of this method will be set forth in the study of circuits with a tunnel diode.

The classical theorems of existence and uniqueness, to which we will refer in section 2.2, are inadequate to resolve the problem posed in this chapter. Additional conditions will be required, and these shall be described in sections 2.4 and 2.5. In our opinion, their most intuitive expression is found in a geometric language, which shall be introduced in section 2.3.

2.1.2 Amplifier

The circuit of figure 2.1 is a *single-stage amplifier in common-emitter config-uration* (vol. VIII, chap. 2). It amplifies the signal u_s, which leads to the response u_r. The dc supply voltage is u_a.

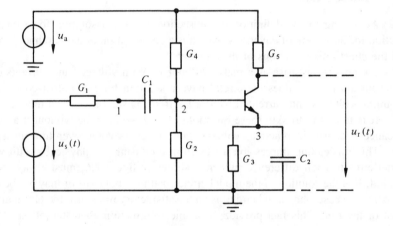

Fig. 2.1

The only nonlinear element is an *npn bipolar transistor* (fig. 2.2), which we model using the *Ebers-Moll equations* (vol. VII, sec. 5.3; vol. VIII, subsec. 1.3.5):

$$\begin{pmatrix} i_1 \\ i_2 \end{pmatrix} = \begin{pmatrix} 1 + 1/\beta_F & -1 \\ -1 & 1 + 1/\beta_R \end{pmatrix} \begin{pmatrix} g(u_1) \\ g(u_2) \end{pmatrix} \tag{2.1}$$

where

$$g(u) = -I_s(\exp(-u/U_T) - 1) \tag{2.2}$$

is the expression defining the exponential diode in (1.15), with $n = 1$. However, the current and voltage have changed sign, which amounts to inverting the two terminals of the diode. The *pnp bipolar transistor* would be modeled by the same expressions, but with the original signs of (1.15) for (2.2). The positive constant β_F (β_R) is called *current gain in region F(R)*.

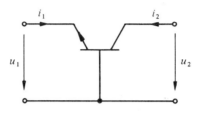

Fig. 2.2

By writing the Kirchhoff equations of the currents for the output of nodes 1, 2, and 3, and by replacing the currents with the potentials, using constitutive relations of the elements, we obtain

$$G_1(v_1 - u_s) + C_1 \frac{d}{dt}(v_1 - v_2) = 0 \tag{2.3}$$

$$C_1 \frac{d}{dt}(v_2 - v_1) + G_2 v_2 + G_4(v_2 - u_a) - (1/\beta_F)g(v_3 - v_2) -$$
$$- (1/\beta_R)g(u_r - v_2) = 0 \tag{2.4}$$

$$G_3 v_3 + C_2 \frac{d}{dt}v_3 + (1 + 1/\beta_F)g(v_3 - v_2) - g(u_r - v_2) = 0 \tag{2.5}$$

$$G_5(u_r - u_a) - g(v_3 - v_2) + (1 + 1/\beta_R)g(u_r - v_2) = 0 \tag{2.6}$$

First, we consider the solution of the circuit in the absence of any signal, given the assumption that a constant operation has been reached. It is the *dc operating point* of the circuit. It amounts to cancelling u_s and the time derivatives in (2.3) and (2.6). From (2.3), we obtain $v_1 = 0$, and from (2.5) and (2.6):

$$G_3 v_3 + (1 + 1/\beta_F)g(v_3 - v_2) - g(u_r - v_2) = 0 \tag{2.7}$$
$$G_5(u_r - u_a) - g(v_3 - v_2) + (1 + 1/\beta_R)g(u_r - v_2) = 0 \tag{2.8}$$

and the sum of equations (2.4) to (2.6) leads to

$$v_2 = -\frac{G_3}{G_2 + G_4}v_3 + \frac{G_4 + G_5}{G_2 + G_4}u_a - \frac{G_5}{G_2 + G_4}u_r \tag{2.9}$$

Thus, taking (2.9) into account, we get a system of two transcendental equations, (2.7) and (2.8), in the unknowns v_3 and u_r. As we will show in chapter 3, this system has exactly one solution. Unfortunately, there is no analytical expression for this solution.

Such is the case for almost all nonlinear circuits. Consequently, we are forced to make use of numerical calculations. Occasionally, graphical methods can be applied as well. Although less accurate, they allow us to study how the dc operating point depends on the value of the elements and the supply voltage, which is very useful for synthesis.

2.1.3 Approximate Calculation

As an alternative, an approximate calculation often provides results that are sufficiently accurate. The approximations to be chosen depend on the order of magnitude of the currents and voltages, which is known or assumed *a priori*. When the solution is found, we must still check to see that the starting hypotheses are satisfied.

In our example, we want the transistor to operate in the *forward-active region* (vol. VII, sec. 5.2; vol. VIII, subsec. 1.3.3). In this case, we use the approximation:

$$g(u_r - v_2) \approx 0 \tag{2.10}$$

in (2.7) and (2.8), which makes it possible to eliminate v_3 and to get a single transcendental equation. By using (2.2) and approximating $\exp(x) - 1$ by $\exp(x)$ for $x > 0$, we obtain

$$I \exp(-\alpha(u_a - u_r)/U_T) = G_5(u_a - u_r) \tag{2.11}$$

with

$$I = I_s \exp \frac{G_4}{G_2 + G_4} \cdot \frac{u_a}{U_T} \tag{2.12}$$

$$\alpha = \frac{G_5}{G_3} \left(1 + \frac{G_2 + G_3 + G_4}{G_2 + G_4} \cdot \frac{1}{\beta_F}\right) \tag{2.13}$$

The two members of (2.11) are shown in figure 2.3 as functions of $u_a - u_r$. It is easy to see that both graphs have only one intersection point which corresponds to a solution $\bar{u}_r$ of (2.11). By substituting this value in the other circuit equations, we find all the currents and voltages of the dc operating point. *A posteriori*, we check to see that the starting hypothesis is verified, i.e., that the transistor operates in forward-active region, or if the conductances $G_2, \ldots, G_5$ must be changed.

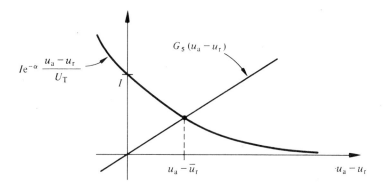

Fig. 2.3

2.1.4 Calculation of the Increments

The solution in the presence of the signal is calculated by linearizing the equation system (2.3) to (2.6) around the dc operating point. The underlying hypothesis for this procedure is that the signal is small.

Let us consider the transistor of figure 2.2. We shall assume that at the dc operating point, the voltages are $\bar{u}_1$ and $\bar{u}_2$ and the currents are $\bar{i}_1$ and $\bar{i}_2$. The voltage and current increments are defined by

$$\left. \begin{array}{l} \Delta u_k = u_k - \bar{u}_k \\ \Delta i_k = i_k - \bar{i}_k \end{array} \right\} \; k = 1,2 \qquad \begin{array}{l}(2.14)\\[1em](2.15)\end{array}$$

The expansion of (2.1) and (2.2) around the dc operating point up to the first order gives

$$\begin{pmatrix} \Delta i_1 \\ \Delta i_2 \end{pmatrix} = \begin{pmatrix} \beta_F + 1 & -\beta_R \\ -\beta_F & \beta_R + 1 \end{pmatrix} \begin{pmatrix} G_F \, \Delta u_1 \\ G_R \, \Delta u_2 \end{pmatrix} \qquad (2.16)$$

where

$$G_F = \frac{I_s}{\beta_F \, U_T} \, e^{-\bar{u}_1 / U_T} \qquad (2.17)$$

$$G_R = \frac{I_s}{\beta_R \, U_T} \, e^{-\bar{u}_2 / U_T} \qquad (2.18)$$

Equations (2.16) to (2.18) are the constitutive relations of the linear two-port in figure 2.4. Note that the values of the two conductances, (2.17) and (2.18), depend on the dc operating point.

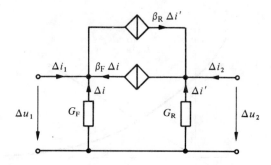

Fig. 2.4

Therefore, the signal response is obtained, as a first approximation, by a linear circuit calculation. Assuming, as in subsection 2.1.3, a transistor operating in forward-active region, we set $G_R \approx 0$ in (2.16), and hence the linear circuit for the increment calculation becomes that of figure 2.5.

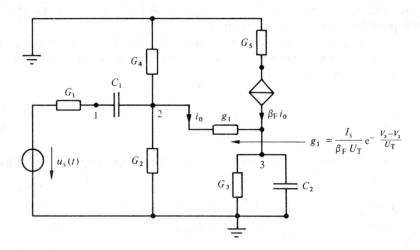

Fig. 2.5

Assuming, additionally, that C_1 and C_2 are sufficiently large and the frequency components of Δu_s are high enough so that both capacitors can be approximated by a short circuit in figure 2.5, we obtain

$$\Delta u_r = -\beta_F \frac{G_1 G_F}{G_5 (G_1 + G_2 + G_4 + G_F)} u_s \qquad (2.19)$$

where

$$G_F = \frac{I_s}{\beta_F U_T} \exp\left(\frac{\bar{v}_2 - \bar{v}_3}{U_T}\right) \tag{2.20}$$

2.1.5 Numerical Example

Consider an amplifier with the following element values: $R_2 = R_4 = 10 \text{ k}\Omega$, $R_3 = 4.4 \text{ k}\Omega$, $R_5 = 5 \text{ k}\Omega$, $C_1 = 4.2 \text{ μF}$, $C_2 = 10 \text{ μF}$, and $u_a = 10 \text{ V}$. The pertinent parameters of the transistor are $I_s = 10^{-14}$ A, $\beta_F = 100$. The solution of (2.11) is $\bar{u}_r = 5.1645$ V, from which we deduce the potentials of the dc operating point $v_2 = 4.9516$ V and $v_3 = 4.2978$ V, and the gain for the small signals: $\Delta u_r = 118.6 \, u_s$.

Figure 2.6 shows the dc operating point and the transfer function, absolute value and phase, calculated by SPICE. Note the excellent agreement with the approximate calculation of subsection 2.1.3. As might be expected, the gain only reaches its value (2.19) where frequencies are sufficiently high, because of the capacitors C_1 and C_2.

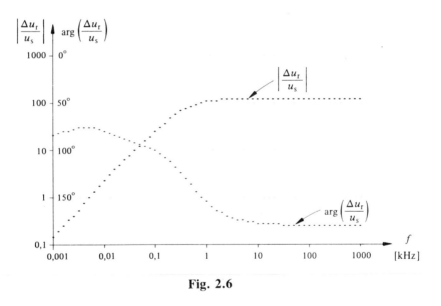

Fig. 2.6

In order to show that the circuit of figure 2.1 implements a linear system for small signals only, we have calculated the time response to a sinusoid of 10 kHz, with 40 mV peak voltage. The result is represented in figure 2.7. The response of the linearized circuit around the dc operating point would be a sinusoid oscillating between 9.91 V and 0.42 V. Figure 2.7 shows that, in fact, half the sinusoid is virtually cut and the other half does not reach the expected maximum.

64

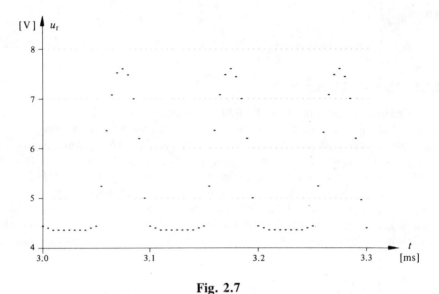

Fig. 2.7

2.1.6 Circuit with Tunnel Diode

The second example of this section is the circuit of figure 2.8. When form (1.5) is used for the constitutive relation of the tunnel diode, the system of circuit equations, after simplification, becomes

$$i = g(u) \tag{2.21}$$
$$i = G(E - u) \tag{2.22}$$

Because the circuit includes neither a capacitor nor an inductor, there is no differential equation.

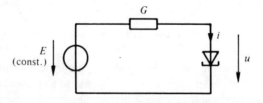

Fig. 2.8

In the plane (u,i), (2.21) and (2.22) represent two curves, the characteristic of the tunnel diode and a straight line (fig. 2.9). A solution of the circuit $u(t)$, $i(t)$,

$t \geqslant t_o$ corresponds at each instant t to an intersection point of the two curves. In the case of figure 2.9, there are three intersections, P_1, P_2, and P_3.

Starting from an initial point $u(t_0), i(t_0)$, compatible with (2.21) and (2.22) (i.e., $u(t_0) = U_k$, $i(t_0) = I_k$, for $k = 1$, 2, or 3), what is the solution $u(t)$, $i(t)$, for $t \geqslant t_0$? Does it remain at P_k, or does it jump to another point of intersection after a certain time? Both cases are compatible with (2.21) and (2.22).

Therefore, the solution of (2.21) and (2.22), based on initial voltages and currents, is not unique.

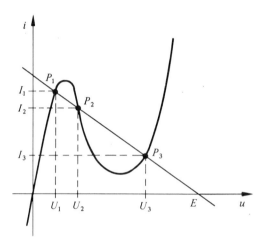

Fig. 2.9

It is tempting to resolve this indeterminacy with the following argument: because there is no reason for the solution to jump from one point of intersection to another, it will remain close to the initial point of intersection. However, this rule cannot be applied to the circuit obtained by adding a time-dependent source of voltage $u_s(t)$ in series with the constant source. The solutions of this circuit are the intersections of the characteristic of the tunnel diode with the straight line defined by (2.22), after replacing E with $e(t) = E + u_s(t)$. Thus, we have to imagine the straight line in figure 2.9 to be moving parallel as a function of time. If we start from the position of the straight line of figure 2.9 while increasing e, the two solutions, P_1 and P_2, come closer, merge, and disappear. Therefore, it is not possible to follow the solution P_1 continuously. At a given time, we are forced to jump from P_1 to P_3.

We can still modify the rule as follows: the solution is a continuous function of time as long as this is possible. However, this interpretation goes beyond the context of the model circuit. It is not advisable to add arbitrary rules whenever a model leads to indeterminacy. It is preferable to admit that the circuit is an inadequate model of a physical circuit, and investigate which effect, neglected during modeling, is, in fact, critical to the circuit's behavior.

2.1.7 Circuit with an Additional Capacitor

When modeling the tunnel diode, we have not taken into consideration a parasitic capacitive effect. This effect is modeled by a parallel capacitor (fig. 2.10). Furthermore, we maintain the time-dependent source in series with the constant source, which was introduced at the end of subsection 2.1.6. The equation system of this circuit is

$$i_1 = C\,du/dt \tag{2.23}$$

$$i_2 = g(u) \tag{2.24}$$

$$i_1 + i_2 = G(e - u) \tag{2.25}$$

where

$$e(t) = E + u_s(t) \tag{2.26}$$

Let us try to follow the same procedure as is used for the amplifier. The first step is to determine the dc operating point. It is the constant solution when $e(t) = E$. By hypothesis, the voltage u is constant, which implies that the current i_1 is zero. Consequently, the system (2.23) to (2.25) boils down to (2.21) and (2.22). Therefore, depending on the value of the circuit elements, there are one, two, or three dc operating points. In the case represented in figure 2.9, the dc operating points are P_1, P_2, and P_3.

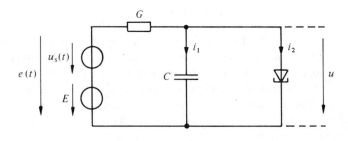

Fig. 2.10

2.1.8 Calculation of Increments

Let us assume that at time $t = 0$ we have $u_s(t) = 0$ and the circuit is at a dc operating point. Suppose that the signal $u_s(t)$ is small. As was done in subsection 2.1.4, we introduce the increments Δi_k and Δu_k with respect to the dc operating point.

Let u_0 be the voltage at the terminals of the tunnel diode at the dc operating point. Up to the first approximation, its constitutive relation becomes

$$\Delta i = \frac{\mathrm{d}g}{\mathrm{d}u}(u_0) \cdot \Delta u \tag{2.27}$$

which stands for a linear resistor whose conductance, called *differential conductance*, is

$$g_0 = \frac{\mathrm{d}g}{\mathrm{d}u}(u_0) \tag{2.28}$$

The linear resistor of conductance G and the capacitor maintain their constitutive relation in terms of increments, and the constant source must be replaced by a short circuit. By a simple linear circuit calculation, we obtain the solution Δu in the frequency domain:

$$\Delta U(p) = T(p) \cdot U_s(p) \tag{2.29}$$

with the transfer function:

$$T(p) = \frac{G}{C} \cdot \frac{1}{p + (G+g_0)/C} \tag{2.30}$$

Therefore, for small signals, the circuit of figure 2.10 is equivalent to a simple RC circuit having a resistor dependent on the dc operating point (fig. 2.11).

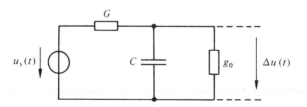

Fig. 2.11

2.1.9 Stability of the Small-Signal Circuit

The preceding comments notwithstanding, we must express a reservation about the equivalence of the nonlinear circuit of figure 2.10 with a linear RC circuit in the neighborhood of a dc operating point. The differential resistance of the tunnel diode is sometimes negative, which could render the circuit unstable.

Indeed, if

$$g_0 + G < 0 \qquad\qquad\qquad (2.31)$$

then the pole of the transfer function T is located in the right half-plane, and the small-signal circuit is thus unstable. A signal, however small it is, will produce a response that increases exponentially. Therefore, it quickly leaves the neighborhood of the dc operating point where the small-signal circuit model is valid.

The inequality (2.31) may be easily discussed by means of figure 2.9 because $g_0 = dg/du$ is the slope of the tunnel diode characteristic and $-G$ is the slope of the straight line, which represents the source in series with the linear resistor. We conclude that the small-signal circuit is unstable if the slope of the diode is smaller than that of the straight line. Such is the case at point P_2 of figure 2.9. On the other hand, if the slope of the diode is larger than that of the straight line, the small-signal circuit is stable. The points P_1 and P_3 of figure 2.9 satisfy this condition.

We shall show, in subsection 2.1.11 for this circuit and in chapter 5 in general, that this stability criterion is not only valid for the small-signal circuit, but is also capable of determining the local stability of the operating point of the nonlinear circuit. In other words, in the neighborhood of a dc operating point, the linear approximation of the nonlinear circuit is sufficient to detect the stability or instability of the dc operating point. It is only when the linear circuit is at the stability limit, meaning when the pole of the transfer function T is located on the imaginary axis, that nonlinearities will control the stability.

2.1.10 Numerical Example

A numerical example illustrates the properties of the circuit of figure 2.10, which have been identified in subsections 2.1.7 to 2.1.9. Instead of the tunnel diode, we use the nonlinear resistor of example 1.4.10, having characteristic of similar form. The other elements have the values $E = 15.9$ V, $G = \frac16$ Ω, $C = 0.1$ F. Using the graphic representation of figure 2.12, we find three dc operating points, with $u = 0.94$ V, 1.2 V, and 2.37 V. Let us superpose to the dc voltage E a signal $u_s(t)$ of 0.1 V amplitude and 2 s duration (fig. 2.13). The solutions $u(t)$, calculated by the SPICE program and starting at the three dc operating points, are represented in figures 2.14 to 2.16. Observe that $u(t)$ does not differ greatly from the dc operating points P_1 and P_3, and it returns to the starting point once the signal becomes zero. On the contrary, starting from P_2 (fig. 2.15), $u(t)$ escapes and converges toward P_3. This agrees with the fact that P_1 and P_3 are stable, and P_2 is unstable.

Moreover, $\Delta u(t)$ actually resembles the response of a first-order RC circuit in the case of the dc operating points P_1 and P_3. In P_1, the time constant is manifestly larger than at P_3. Indeed, the slope of the nonlinear resistor characteristic is smaller at P_1 that at P_3, and consequently the differential resistance is larger.

On the other hand, in the case of the dc operating point P_2, the response of the nonlinear circuit is qualitatively different from that of its linear approximation. Indeed, while the solution of an unstable linear circuit tends toward infinity, the solution represented in figure 2.15 tends toward the finite point P_3.

69

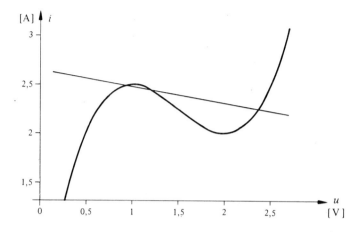

Fig. 2.12

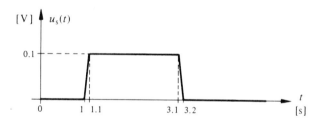

Fig. 2.13

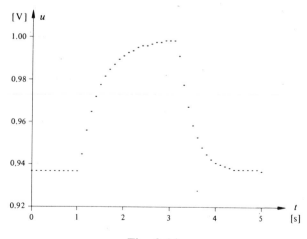

Fig. 2.14

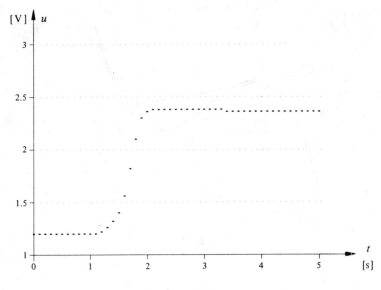

Fig. 2.15

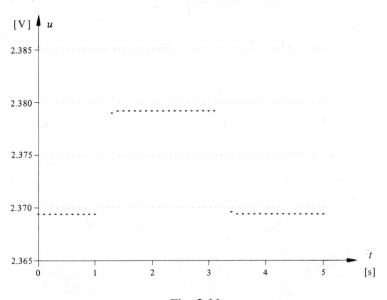

Fig. 2.16

2.1.11 Global Behavior of the Solutions

The circuit of figure 2.10 is convenient for a qualitative analysis of the solutions, without need of any special hypothesis about the voltage source. From (2.23) to (2.25), we find the *state equation:*

$$C(du/dt) = G(e-u) - g(u) \qquad (2.32)$$

Consequently,

$$du/dt > 0 \iff G(e-u) > g(u) \qquad (2.33)$$

Relation (2.33) can be interpreted in the plane (u, i_2). The orbit $(u(t), i_2(t))$ of an arbitrary solution from (2.23) to (2.25) follows the characteristic of the diode at any moment. When this characteristic is located below (or above) the straight line $i_2 = G(e - u)$, then $du/dt > 0$. This property is shown in figures 2.17 to 2.19, by means of arrows, for three different values of e.

When the source is constant, the figures 2.17 to 2.19 give a complete picture of a solution with an arbitrary initial condition. In the case of figure 2.17, all the solutions converge toward P_1, which we call a *globally asymptotically stable dc operating point*. Point P_3 of figure 2.19 has the same property. The case of figure 2.18 is more complicated. Any solution that starts to the left (or right) of P_2 converges toward $P_1(P_3)$. This confirms the results of subsection 2.1.9. However, the stability of P_1 and P_3, described in figure 2.17 and 2.19, is stronger; it is a **global** property, not merely a **local** one.

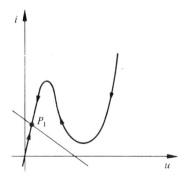

Fig. 2.17

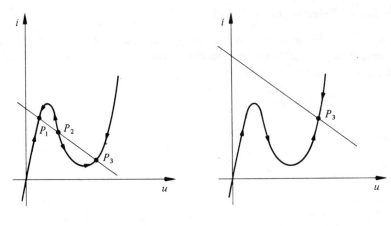

Fig. 2.18   Fig. 2.19

When the voltage source is a function of time, the straight line $i_2 = G(e - u)$ of the plane (u,i) moves in parallel, together with the points of intersection of the straight line with the characteristic of the diode. If $e(t)$ does not vary too quickly, the solution follows, with some delay, the movement of the point of intersection, P_1 or P_3, toward the attraction zone to which it belongs.

Let us assume that the solution is in the neighborhood of P_1. If we increase e, it will follow the movement of P_1. At a given moment, P_1 disappears; P_3 remains as the only intersection between the straight line and the characteristic of the diode. From that time on, the solution tends toward P_3. If e is reduced again, the solution will follow P_3, provided that P_3 still exists. Thus, such a variation of e has caused the solution to change from P_1 to P_3.

According to the terminology of Chatelain and Dessoulavy (vol. VIII, chap. 6), the circuit of figure 2.10, with a constant source corresponding to figure 2.18 ((2.17), (1.19)), is a *bistable* or *(monostable) multivibrator*.

2.1.12   Numerical Example

The switching effect from the dc operating point P_1 to point P_3 is illustrated by the same example 2.1.10. Indeed, if we increase the signal u_s of figure 2.13 from 0.1 V to 0.2 V, we obtain the response of figure 2.20, which is to be compared with that of figure 2.14.

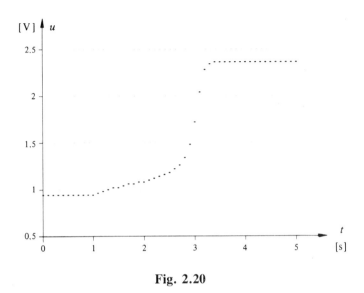

Fig. 2.20

2.1.13 Comment

A nonlinear circuit, to which a small signal is applied, normally behaves like a linear system. When increasing the signal amplitude, we expect the system to differ from a linear system *gradually*. This is the case with the transistor amplifier of figure 2.1. However, the nature of the circuit response in figure 2.10 may change *suddenly* when the amplitude is increased. The variation between figures 2.14 and 2.20 gives evidence of this effect.

2.1.14 Circuit with an Inductor

Let us replace the parallel capacitor in the circuit of figure 2.10 by a series inductor (fig. 2.21). Instead of the equation system (2.23) to (2.25), we get

$$e \;=\; u_1 + u_2 + u \tag{2.34}$$

$$i \;=\; Gu_1 \tag{2.35}$$

$$u_2 \;=\; L\,di/dt \tag{2.36}$$

$$i \;=\; g(u) \tag{2.37}$$

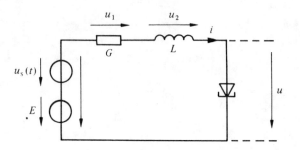

Fig. 2.21

The dc operating points are the same as those in the circuit of figure 2.10. By eliminating u_1 and u_2, we get

$$e = i/G + L\,di/dt + u \qquad (2.38)$$

$$i = g(u) \qquad (2.39)$$

It is tempting to eliminate u between (2.38) and (2.39), whereupon we get

$$L\,di/dt = e - g^{-1}(i) - i/G. \qquad (2.40)$$

However, this equation involves the inverse function of g, which does not exist. On the other hand, we can derive (2.39) and eliminate di/dt:

$$GL\,\frac{du}{dt} = \frac{1}{dg/du}[G(e-u)-g(u)] \qquad (2.41)$$

The similarity between (2.41) and (2.32) is remarkable. Nevertheless, the qualitative behavior of the solutions is entirely different. This is due to the incremental conductance dg/du, which can be positive, negative, or zero. Indeed, instead of relation (2.33), we have

$$du/dt > 0 \iff \begin{cases} G(e-u) > g(u) \quad \text{and} \quad dg/du > 0 \\ \text{or} \\ G(e-u) < g(u) \quad \text{and} \quad dg/du < 0 \end{cases} \qquad (2.42)$$

In figures 2.22 to 2.24, the direction of the solutions is marked by arrows on the characteristic curve of the diode for three different values of e. What distinguishes these figures from figures 2.17 to 2.19 is the change of direction at the extrema Q_1 and Q_2 of the characteristic.

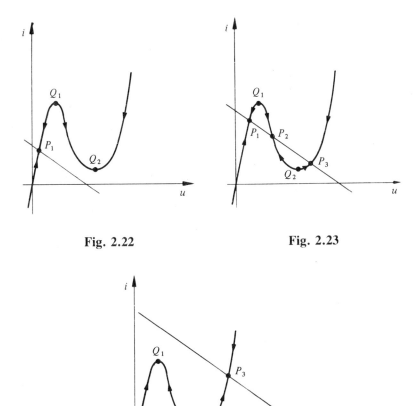

Fig. 2.22

Fig. 2.23

Fig. 2.24

Let us suppose that the constant source has a value corresponding to figure 2.23. If a solution starts from a point located to the left of Q_1, it tends toward P_1. If the initial point is between Q_1 and Q_2, it tends toward P_2, and if the starting point is to the right of Q_2, it tends toward P_3. Therefore, the three dc operating points are stable equilibria. Note that in the case of figure 2.10, P_2 was unstable (fig. 2.18).

Similarly, if the value of the constant source corresponds to figure 2.22 (2.24) and the initial point is located to the left of Q_1 (to the right of Q_2), the solution tends toward the stable equilibrium P_1 (P_3).

2.1.15 Impasse Points

Let us suppose that the source is constant, having a value which corresponds to figure 2.24. What happens if the initial point of a solution is located to the left of Q_1? The solution first reaches Q_1, but at this point it is unable to continue because the arrow points on both sides toward Q_1, although it is not a dc operating point. It is an *impasse point*.

We might object that it is impossible to implement the initial conditions to the left of Q_1. However, nothing prevents us from applying any given voltage source $u_s(t)$, e.g., that of figure 2.25. Let us assume that the value E of the constant source is that of figure 2.23 and the solution is initially located at the dc operating point P_1. When $u_s(t)$ increases from 0 to u_1, the solution follows, with some delay, the point of intersection $P_1(t)$ of the straight line corresponding to $E + u_s(t)$ and the characteristic curve of the diode. When the solution reaches Q_1, $P_1(t)$ has already disappeared, and hence we have the situation of figure 2.24. Again, it is impossible to continue, and Q_1 is an impasse point.

It is tempting to reject the circuit of figure 2.21 as the model for a physical device, since, in practice, a time evolution does not stop. We shall see in subsection 2.1.17 that the problem of impasse points is rather more subtle.

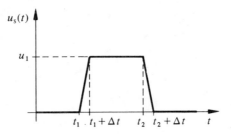

Fig. 2.25

2.1.16 Numerical Example

It is interesting to see how a numerical algorithm reacts at the impasse point. To do so, we replace the tunnel diode in the circuit of figure 2.21 by the same nonlinear resistor as in subsection 2.1.10, and we select $G = 0.30\ \Omega^{-1}$, $L = 1$ H. The voltage source has the form of figure 2.25, with $u_1 = 1$ V, $t_1 = 1$ s, $t_2 = 3.1$ s, $\Delta t = 0.1$ s. The voltage of the constant source is 9 V. The result obtained by SPICE is shown in figure 2.26. As soon as the solution reaches the impasse point $u = 1$ V, the program is unable to proceed with the calculation.

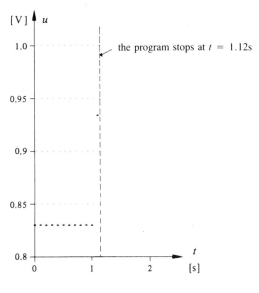

Fig. 2.26

Remember that in the presence of the parallel capacitor, in the circuit of figure 2.10, there was no impasse point. This result leads to the idea of adding a very small capacitor to the circuit of figure 2.21 in parallel with the nonlinear resistor. Figure 2.27 shows that the program no longer stops at $u = 1$ V. The path corresponding to the solution $u(t)$ of figure 2.27 is represented in figure 2.28. Observe that from point Q_1, with $u = 1$ V, the solution quickly moves toward point Q_3, where the current is the same as in Q_1.

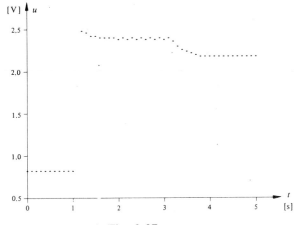

Fig. 2.27

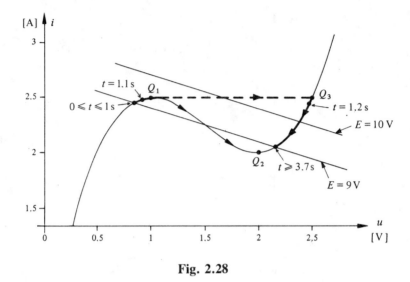

Fig. 2.28

2.1.17 Resolving the Impasse

We can say about figure 2.28 that, at the limit where the parallel capacitor tends toward zero, the solution jumps from point Q_1 to point Q_3. Consequently, the limit function $u(t)$ becomes discontinuous, while $i(t)$ remains continuous. Is this limit so-lution a real solution for the circuit of figure 2.21?

The answer depends on which viewpoint is adopted. If a circuit solution is required to be a continuous function of time, then this is obviously not a solution. On the other hand, if discontinuous functions are admissible, it actually is a solution of the circuit because, apart from the jump, the limit functions $u(t)$ and $i(t)$ satisfy the equation system (2.38) and (2.39).

Although we have this possibility of always finding a solution for the circuit of figure 2.21, we still refer to the points Q_1 and Q_2 as impasse points because it is indeed a local impasse from which the only escape is by way of a jump.

2.1.18 Comment

Despite the fact that the equation system (2.38) and (2.39) is continuous, it has a discontinuous solution. This phenomenon is due to the fact that g^{-1} is multivalued. Nevertheless, why should the jump occur exactly when the solution reaches Q_1?

There is no reason why it should. The equation system (2.38) and (2.39) also has a solution, with an orbit as shown in figure 2.29, with two jumps marked by dashed lines. Indeed, any given sequence of jumps can be imposed. This implies that, as soon as we allow discontinuous solutions for the equation system (2.38) and (2.39), the solution based on initial conditions is no longer unique. It is only when a capacitor is added in parallel with the tunnel diode that the uniqueness of the solution is restored. The response of a capacitor, however small it is, determines which solution among those of the circuit of figure 2.21 is appropriate. It is that solution which includes a jump only when it encounters an impasse point.

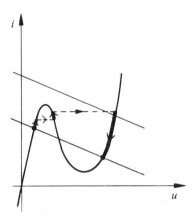

Fig. 2.29

2.1.19 Oscillator

Another choice of resistor and constant source values leads to the qualitative picture of the movement represented in figure 2.30. As before, there are two impasse

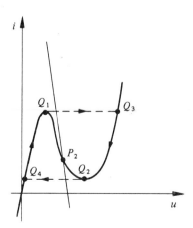

Fig. 2.30

points Q_1 and Q_2. However, there is only one dc operating point, P_2. Because it is unstable, the solution cannot tend toward this point. So, then, what is its asymptotic behavior?

Figure 2.30 shows that starting from an initial point to the left (or right) of P_2 always leads to $Q_1(Q_2)$. According to the rule given at the end of subsection 2.1.18, the solution jumps from $Q_1(Q_2)$ to $Q_3(Q_4)$. From $Q_3(Q_4)$, it proceeds toward $Q_2(Q_1)$, where it jumps to $Q_4(Q_3)$. Then, it heads toward $Q_1(Q_2)$ and the cycle starts again. Thus, the asymptotic movement is periodic; it is an *oscillator*, or (according to the terminology of vol. VIII, chap. 6) an *astable multivibrator*.

2.1.20 Numerical Example

By changing the conductance of the example in subsection 2.1.16 into 0.4444 Ω and connecting a 2.5 V constant source, we obtain the situation of figure 2.30. Let us assume that the circuit is initially located at P_2. Because it is an unstable equilibrium, a very small signal superposed to the constant source voltage is sufficient to keep the circuit well away from P_2 and to excite the oscillation. The response $u(t)$, calculated by SPICE, confirms this fact (fig. 2.32). The form of the source $u_s(t)$ is given in figure 2.31. The voltage of the constant source is 2.5 V. To make the numerical calculation possible (see subsect. 2.1.16), it has been necessary to introduce a 1 nF capacitor in parallel with the nonlinear resistor.

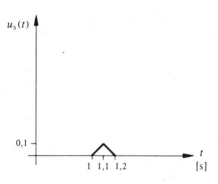

Fig. 2.31

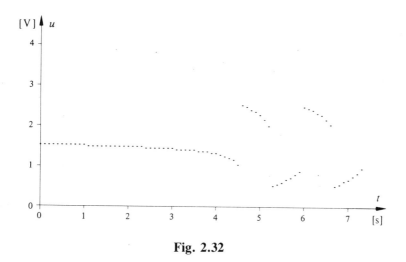

Fig. 2.32

2.2 EXISTENCE AND UNIQUENESS THEOREM FOR DIFFERENTIAL EQUATIONS IN NORMAL FORM

2.2.1 Introduction

The purpose of this section is to discuss the fundamental existence and uniqueness theorem for the solutions of a system of differential equations in *normal form*, that is, of the system:

$$
\begin{aligned}
dx_1/dt &= f_1(x_1, ..., x_n, t) \\
\vdots \qquad &\quad \vdots \\
dx_n/dt &= f_n(x_1, ..., x_n, t)
\end{aligned}
\tag{2.43}
$$

Hereafter, we shall write system (2.43) in abbreviated form as

$$
dx/dt = f(x, t)
\tag{2.44}
$$

where $x \in \mathbb{R}^n$ and $f : \mathbb{R}^{n+1} \to \mathbb{R}^n$.

The crucial hypothesis is that function f must be *Lipschitz* (definition 2.2.3). This condition can be locally or globally satisfied, and consequently we shall state a local theorem and a global theorem for the existence of solutions.

At first sight, the fact that differential equations are written in a normal form seems to be simply a hypothesis of convenience, without consequences. With the mild hypothesis that the functions f must be locally Lipschitz, we can ensure the existence of a solution starting from any initial condition. Consequently, the phenomenon of impasse points, which we discovered in the circuit of figure 2.21, is excluded. This shows that passing from the circuit equation system to a system of form (2.43) is far from being trivial. We shall return to this point in section 2.4.

2.2.2   Definition

A function $f: \mathbb{R}^m \to \mathbb{R}^n$ is *Lipschitz* on a set $D \subset \mathbb{R}^m$, if there is a constant K, such that

$$\| f(x) - f(y) \| \leqslant K \| x - y \| \tag{2.45}$$

for any pair of points $x, y \in D$. K is called a *Lipschitz constant* of f on D. The norm used in (2.45) is unimportant; normally for $z \in \mathbb{R}^r$, we use

$$\| z \| = \left| \sum_{i=1}^{r} z_i^2 \right|^{1/2} \tag{2.46}$$

which is the *Euclidean norm*.

In the context of differential equations, there are often functions which also depend on time. We do not require a condition of type (2.44) for time, but rather we require that (2.45) be verified for the other variables, with a constant K, which is not dependent on time. This leads to the following definition:
A function $f(x,t), f: \mathbb{R}^{m+1} \to \mathbb{R}^n$ is *Lipschitz with respect to x* in a range $D \subset \mathbb{R}^{m+1}$, if there is a constant K, such that

$$\| f(x, t) - f(y, t) \| \leqslant K \| x - y \| \tag{2.47}$$

for any pair of points $(x,t) \in D, (y,t) \in D$.

The function f is *locally Lipschitz with respect to x* at a point $(x_0, t_0) \in \mathbb{R}^{m+1}$, if there is a neighborhood V of (x_0, t_0), such that f is Lipschitz (with respect to x) on V.

The concept of neighborhood will be introduced in section 2.3. Let us simply note here that it suffices to take V as a set of points (x,t), such that $\| (x,t) - (x_0, t_0) \| \leqslant \epsilon$, where ϵ can be arbitrarily small.

2.2.3   Properties

The following properties are not difficult to prove. We refer to [4]:

- A function which is locally Lipschitz at a given point is continuous at that point.
- A function which has continuous partial derivatives at a given point is locally Lipschitz at that point.

These properties obviously can be generalized to functions which are not Lipschitz with respect to all the arguments.

2.2.4 Comment

According to subsection 2.2.3, the property providing for a function to be locally Lipschitz is more restrictive than continuity and less restrictive than differentiability.

A classical example of a function that is continuous, but not locally Lipschitz at 0, is given by

$$f(x) = \text{sgn}(x)\sqrt{|x|} \qquad (2.48)$$

which is represented in figure 2.33. If it were locally Lipschitz at 0, its graph ought to be located in the neighborhood V of 0, between the straight lines $y = \pm Kx$, for a certain K (fig. 2.34). This is obviously not the case for any finite K.

An example of a function having a discontinuous derivative at 0, but is nevertheless Lipschitz, is given by the piecewise linear function:

$$f(x) = \begin{cases} x & \text{for} \quad x < 0 \\ 2x & \text{for} \quad x > 0 \end{cases} \qquad (2.49)$$

which is represented in figure 2.35.

Clearly, the inequality:

$$|f(x) - f(y)| \leqslant 2|x - y| \qquad (2.50)$$

is verified at 0, and at any other point as well.

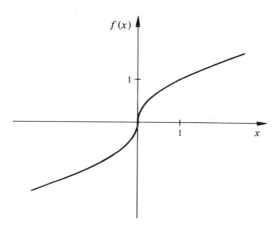

Fig. 2.33

84

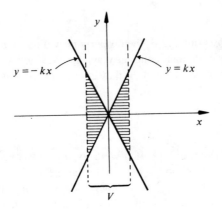

Fig. 2.34

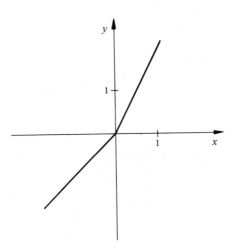

Fig. 2.35

2.2.5 Local Existence and Uniqueness Theorem

Suppose that $f:\mathbb{R}^{n+1} \to \mathbb{R}^n$ is a continuous function that is locally Lipschitz with respect to x at a point (x_0, t_0). Then, the system of differential equations:

$$dx/dt = f(x, t) \tag{2.51}$$

has one, and only one, solution $x(t)$ such that $x(t_0) = x_0$. This solution is defined within a time interval of nonzero length, which contains t_0. It has the following properties:

- It depends continuously on t, t_0, and x_0; the derivative with respect to t is continuous.
- If the function f still depends on a parameter λ, which means that if the system of differential equations is of the form:

$$dx/dt = f(x, t, \lambda)$$

(2.52)

where f is continuous in (x,t,λ) and locally Lipschitz with respect to x, then the solution passing through x_0 at the instant t_0 depends continuously on λ.
- If f is a smooth function, (definition 1.3.4), then the solution of (2.52) is a smooth function of t, t_0, x_0, and λ.

For the proof, see [4].

2.2.6 Comment

The continuity of f would suffice to ensure the local existence of the solution, but its uniqueness would no longer be ensured.

2.2.7 Examples

Let us consider the circuit of figure 2.36. Its state equation is

$$C \frac{du}{dt} = -(u/r)^{1/3}$$

(2.53)

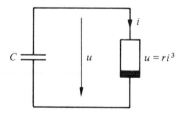

Fig. 2.36

Let us look for a solution which satisfies $u(0) = 0$. The trivial solution $u(t) \equiv 0$ fulfills this condition, but it is not the only one that does. Any function in the form:

$$u(t) = \begin{cases} a(t_0 - t)^{3/2} & \text{for } t < t_0 \\ 0 & \text{for } t_0 \leq t \end{cases} \tag{2.54}$$

with

$$a = r^{-1/2}(3C/2)^{-3/2}, \quad t_0 \leq 0 \tag{2.55}$$

is also a solution passing through 0 at the instant 0 (fig. 2.37), with t_0 *being any negative number or zero*. If $t_0 < 0$, it is identical to the trivial solution in the neighborhood of $t = 0$. On the other hand, for $t_0 = 0$, it is even locally different from the trivial solution. This is not in contradiction with theorem 2.2.5 because the right-hand side of (2.53) is not Lipschitz at $u = 0$.

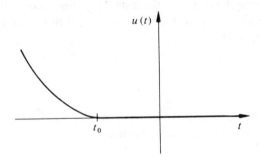

Fig. 2.37

Based on the initial condition $u(0) = 0$, the solution was not unique in the past. On the other hand, it is to be unique in the future, and thus the circuit is actually deterministic. However, it is easy to give an example where the situation is exactly the opposite.

The circuit of figure 2.38 differs from that of figure 2.36 because of the impedance converter inserted between the capacitor and the nonlinear resistor (vol. XIX, subsec. 9.3.3). This is equivalent to changing the sign for one of the two elements. Consequently, the state equation becomes

$$C\frac{du}{dt} = (u/r)^{1/3} \tag{2.56}$$

and, in addition to the trivial solution, we have

$$u(t) = \begin{cases} 0 & \text{for} \quad t \leq t_0 \\ a(t - t_0)^{3/2} & \text{for} \quad t_0 < t \end{cases} \tag{2.57}$$

which satisfy all $u(0) = 0$, **if t_0 is any positive number or zero** (fig. 2.39). Therefore, the circuit clearly is not deterministic and must be rejected as a model of a physical circuit.

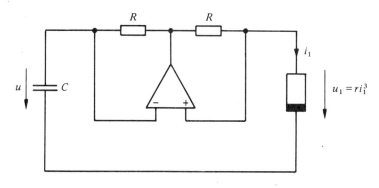

Fig. 2.38

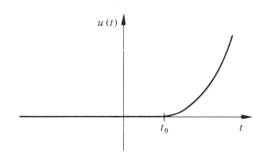

Fig. 2.39

2.2.8 Comment

Theorem 2.2.5 provides us with much information on the regularity of the solutions of a system of differential equations starting from initial conditions. On the other hand, it is quite vague regarding the time interval during which the solution exists. In fact, the existence of the solution for all of the times can only be obtained by imposing more restrictive conditions on the function f of (2.44).

2.2.9 Global Existence and Uniqueness Theorem

Suppose that $f: \mathbb{R}^{n+1} \to \mathbb{R}^n$ is a function which is continuous and Lipschitz with respect to x on the entire space $\mathbb{R}^{n+1}$. Let us assume that $f(0,t)$ is uniformly bounded in t. Then, the system of differential equations:

$$\mathrm{d}x/\mathrm{d}t = f(x,t) \tag{2.58}$$

has one, and only one, solution $x(t)$, which passes at a given instant, t_0, through a given point, $x(t_0) = x_0$. This solution is defined for any real t.

This theorem is a consequence of corollary 4.9 of [4].

2.2.10   Examples

Let us again consider the circuit of figure 2.36, but with an inductor instead of the capacitor (fig. 2.40). Its state equation is

$$L\,\mathrm{d}i/\mathrm{d}t = -ri^3 \tag{2.59}$$

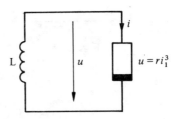

Fig. 2.40

The right-hand member, $f(i,t) = -ri^3$, is a locally Lipschitz function everywhere, since it is differentiable, with a continuous derivative. According to theorem 2.2.5, there is one, and only one, solution $i(t)$, which satisfies an arbitrary initial condition $i(t_0) = i_0$. However, its existence is only ensured for times which closely precede and follow t_0.

Can this solution be extended farther into the future or the past? In this simple case, we find the explicit solution:

$$i(t) = \pm (L/2r)^{1/2}\,(t-T)^{-1/2} \tag{2.60}$$

where T depends on the initial conditions in the following way (fig. 2.41):

$$T = t_0 - L/2ri_0^2 \tag{2.61}$$

The sign in (2.60) is that of i_0.

The solution (2.60) is only defined in the interval $t\epsilon(T,\infty)$, which includes, of course, t_0.

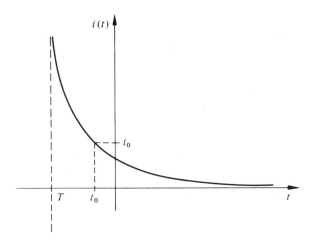

Fig. 2.41

Similarly, we find an example having solutions that cannot be indefinitely extended into the future. As was done in the circuit of figure 2.23, we introduce a negative impedance converter (fig. 2.42), which transforms (2.69) into

$$L\,di/dt = ri^3 \tag{2.62}$$

The solutions are in the form:

$$i(t) = \pm (L/2r)^{1/2}(T-t)^{-1/2} \tag{2.63}$$

where the sign is that of the initial current, and T is given by

$$T = t_0 + L/(2ri_0^2) \tag{2.64}$$

Function (2.63) tends toward infinity when t approaches T (fig. 2.43).

These two examples are not in contradiction with theorem 2.2.9 because function $f(i) = ri^3$ is not Lipschitz on the entire real axis. Indeed,

$$f(i_1) - f(i_2) = r(i_1 - i_2)(i_1^2 + i_1 i_2 + i_2^2) \tag{2.65}$$

The factor $i_1^2 + i_1 i_2 + i_2^2$ is not bounded if we vary i_1 and i_2 over the entire real axis. Consequently, although f is Lipschitz on each bounded interval, it is not Lipschitz on the entire $\mathbb{R}$.

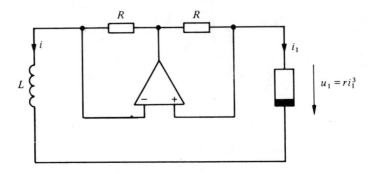

Fig. 2.42

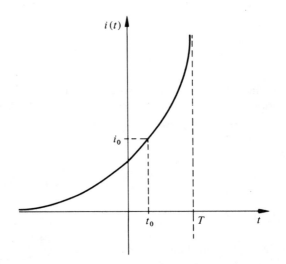

Fig. 2.43

2.2.11 Numerical Example

The circuit of figure 2.42 with $L = 1$ H, $R = 1$ Ω, $r = 1$ Ω/A is simulated by SPICE. The solution starting from the initial condition $i(0) = 0.5$A is represented in figure 2.44. According to (2.64), this solution must tend toward infinity when $t \rightarrow 2$ s. Indeed, the program is unable to continue after $t = 1.974$ s.

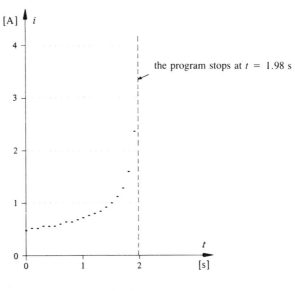

Fig. 2.44

2.2.12 Property

A piecewise linear continuous function $f:\mathbb{R}^n \to \mathbb{R}^m$ is globally Lipschitz.

We limit the proof to $n = m = 1$. In each linear domain, $f(x)$ is in the form $f(x) = ax + b$. If A is the largest of the numbers a, then

$$|df/dx| \leqslant A \tag{2.66}$$

Because

$$x) = \int_{x}^{y} (df/dx)(z)\,dz \tag{2.67}$$

we obtain

$$|f(y) - f'(x)| \leqslant A|y - x| \tag{2.68}$$

This proof can be generalized to higher dimensions.

2.2.13   Comment

A circuit such as that shown in figure 2.42, which includes a solution tending toward infinity within a finite period of time, is not a good model of a physical circuit. According to theorem 2.2.9, this phenomenon does not occur if the circuit equation system can be written in the form (2.58) and the right-hand member of (2.58) is a globally Lipschitz function.

If the elements of the circuit have piecewise linear characteristics, theorem 2.2.9 normally ensures the global existence of the solutions. However, if the characteristics are polynomial or exponential, which is the case with many ordinary models, we cannot expect to find a system of differential equations defined by a globally Lipschitz function. Does this then mean that the solutions of such a circuit necessarily have unique behavior? In particular, must they tend toward ∞ within a finite period of time? The answer is no. The circuit of figure 2.40 proves this to be so. In fact, what is required from a realistic model circuit is that, based on any initial conditions, there will be a solution which can be indefinitely continued *into the future*. The fact that the extension into the past exists, possibly, only for a limited period of time is irrelevant. In chapter 5, we will be able to prove, for a wide range of circuits, that any solution, if it exists, remains bounded in the future. To give some insight into this idea, we will now present the reasoning behind the circuit of figure 2.40. The energy stored in the inductor satisfies the relation:

$$\frac{d}{dt}\left(L\frac{i^2}{2}\right) = Li\frac{di}{dt} \tag{2.69}$$

$$= -ri^4 \leqslant 0 \tag{2.70}$$

A possible solution of (2.70) based on an initial current i_0 at the time t_0, therefore, would satisfy

$$|i(t)| \leqslant |i_0| \quad \text{for} \quad t \geqslant t_0 \tag{2.71}$$

Let us next consider the modified equation (2.59):

$$L\frac{di}{dt} = \begin{cases} -ri^3 & \text{for} & |i| \leqslant 2|i_0| \\ -8r|i_0|^3 & \text{for} & i > 2|i_0| \\ +8r|i_0|^3 & \text{for} & i < -2|i_0| \end{cases} \tag{2.72}$$

Because the right-hand member of (2.72) is globally Lipschitz, theorem 2.2.9 ensures the existence of a solution $i(t)$ of (2.72) with $i(t_0) = i_0$ for all the times. Then, (2.71) is verified for that solution by the preceding argument. Consequently, $i(t)$ is a solution of (2.59) for $t \geqslant t_0$.

This reasoning can be generalized, whereby we obtain the following theorem.

2.2.14 Global Existence and Uniqueness Theorem in the Future

Suppose that $f: \mathbb{R}^{n+1} \to \mathbb{R}^n$ is a continuous function and Lipschitz with respect to x on any domain of the form:

$$D = \{(x, t) \mid \|x\| \leq r\} \tag{2.73}$$

Let us assume that a solution $x(t)$ with $x(t_0) = x_0$ satisfies

$$\|x(t)\| \leq r \tag{2.74}$$

for any $t \geq t_0$ where it exists. The bound r can depend on x_0 and t_0. Then, there exists one, and only one, solution $x(t)$ defined for $[t_0, \infty)$ with $x(t_0) = x_0$.

2.3 SYSTEM OF CIRCUIT EQUATIONS AND CONFIGURATION SPACE

2.3.1 System of Circuit Equations

According to subsection 1.2.11, the system of circuit equations is composed of a maximal independent set of Kirchhoff equations and all the constitutive relations of the elements. In subsection 1.3.25, we listed the basic elements to be taken into account. Looking at that list, we can give a precise form to the system of equations.

A standard notation should be introduced. Suppose that b is the number of branches of the circuit. We designate u as the column matrix of all the branch voltages, and i as the column matrix of all the branch currents. The branches are numbered in the following order:

- the N_C capacitors;
- the N_L inductors;
- the N_S time-dependent independent sources;
- the N_R other resistors;
- the N_B linear two-ports.

Therefore, we can write

$$b = N_C + N_L + N_S + N_R + 2N_B \tag{2.75}$$

It is convenient to use a notation for the column matrices which groups together the voltages or currents of each type of element. We define

$$u_C = (u_1, ..., u_n)^T, \quad n = N_C \tag{2.76}$$

$$i_C = (i_1, ..., i_n)^T, \tag{2.77}$$

$$u_L = (u_{n+1}, ..., u_m)^T \quad m = N_C + N_L \tag{2.78}$$

and, similarly, we have i_L, u_S, i_S, u_R, i_R, u_B, and i_B. Furthermore, it is convenient to introduce the capacitor charges and the inductor fluxes:

$$q_C = (q_1, ..., q_n)^T, \qquad n = N_C \tag{2.79}$$

$$\varphi_L = (\varphi_{n+1}, . . . , \varphi_m)^T, \; m = N_C + N_L \tag{2.80}$$

In all cases, T designates the transposed matrix.

Using these notations, we can express the b independent Kirchhoff equations by means of two matrices, A and B, the elements of which are 0, $+1$, or -1:

$$Ai = 0 \tag{2.81}$$
$$Bu = 0 \tag{2.82}$$

and the constitutive relations of the elements, introduced in section 1.3, by means of $N_C + N_L + N_R$ functions $f_k:\mathbb{R}^2 \to \mathbb{R}$, of N_S functions $e_K:\mathbb{R} \to \mathbb{R}$ and N_B matrices G_k of dimension 2×2:

The capacitors are described by

$$\left. \begin{array}{l} f_k(u_k, q_k) = 0 \\ i_k = dq_k/dt \end{array} \right\} \quad k = 1, ..., N_C \tag{2.83} \tag{2.84}$$

The inductors are described by

$$\left. \begin{array}{l} f_k(\varphi_k, i_k) = 0 \\ u_k = d\varphi_k/dt \end{array} \right\} \quad k = N_C + 1, ..., N_C + N_L \tag{2.85} \tag{2.86}$$

The time-dependent sources are described by

or $$\left. \begin{array}{l} u_k - e_k(t) = 0, \\ i_k - e_k(t) = 0 \end{array} \right\} \quad k = N_C + N_L + 1, ..., N_C + N_L + N_S \tag{2.87} \tag{2.88}$$

The resistors are described by

$$f_k(u_k, i_k) = 0 \qquad k = N_C + N_L + N_S + 1, ..., N_C + N_L + N_S + N_R \tag{2.89}$$

The linear two-ports are described by

$$G_k \begin{pmatrix} u_k \\ u_{k+1} \end{pmatrix} + G_{k+1} \begin{pmatrix} i_k \\ i_{k+1} \end{pmatrix} = 0, \tag{2.90}$$

$$k = N_C + N_L + N_S + N_R + 1, \ N_C + N_L + N_S + N_R + 3, \ ..., \ N_C + N_L$$
$$+ N_S + N_R + 2N_B - 1$$

The equation system (2.81) to (2.90) is the *standard system of circuit equations*. There are $2b + N_C + N_L$ equations for the $2b + N_C + N_L$ unknowns $q_C(t)$, $\varphi_L(t)$, $u(t)$, and $i(t)$. This is the starting point of our subsequent studies.

2.3.2 Comment

The equation system (2.81) to (2.90) is a mixture of differential and nondifferential equations, as well as linear and nonlinear equations. It is in the form:

$$dq_C/dt = i_C \tag{2.91}$$
$$d\varphi_L/dt = u_L \tag{2.92}$$
$$0 = F(u, i, q_C, \varphi_L, t) \tag{2.93}$$

Thus, there are $N_C + N_L$ differential equations and $2b$ nondifferential equations.

It is useful to study the system of nondifferential equations (2.93) independently from (2.91) and (2.92). Because it includes more unknowns than equations, we expect to have an infinite number of solutions. The set of solutions normally forms a surface of dimension $N_C + N_L$ in $\mathbb{R}^{2b + N_C + N_L}$.

2.3.3 Example

Let us consider again the circuit of figure 2.21. All the currents and voltages are indicated and numbered according to the rule of subsection 2.3.1 in figure 2.45. We get $N_C = 0$, $N_L = 1$, $N_s = 1$, $N_R = 3$, $N_B = 0$.

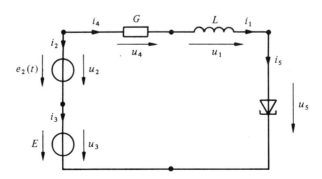

Fig. 2.45

A maximal independent set of Kirchhoff equations is

$$i_1 - i_5 = 0 \tag{2.94}$$
$$i_4 - i_1 = 0 \tag{2.95}$$
$$-i_2 - i_4 = 0 \tag{2.96}$$
$$-i_3 + i_2 = 0 \tag{2.97}$$
$$u_1 - u_2 - u_3 + u_4 + u_5 = 0 \tag{2.98}$$

The constitutive relations of the elements are

$$\varphi_1 - Li_1 = 0 \tag{2.99}$$
$$u_1 = d\varphi_1/dt \tag{2.100}$$
$$u_2 - e_2(t) = 0 \tag{2.101}$$
$$u_3 - E = 0 \tag{2.102}$$
$$Gu_4 - i_4 = 0 \tag{2.103}$$
$$g(u_5) - i_5 = 0 \tag{2.104}$$

In this example, the only differential equation is (2.100), and all the other equations constitute the system (2.93).

2.3.4 Definitions

An *operating point* of a circuit is a set $\xi = (u,i,q_C,\varphi_L)$ voltages, currents, charges, and fluxes, such that the circuit equations are satisfied, except for the differential equations (2.84) and (2.86). The set of all the operating points of a circuit is called the *configuration space*.

2.3.5 Comment

An operating point should not be confused with a solution of the circuit. An operating point is a set of values for u,i,q_C,φ_L, which corresponds to a point $\mathbb{R}^{2b + N_C + N_L}$. On the other hand, a solution is a set of functions $u(t)$, $i(t)$, $q_C(t)$, $\varphi_L(t)$, which corresponds to a curve in $\mathbb{R}^{2b + N_C + N_L}$ with time as a parameter.

At any instant, the solution is at an operating point. *Therefore, a solution can be considered as a movement in the configuration space.*

2.3.6 Example

Let us return to example 2.2.3. An operating point of the circuit is a point $(u_1, \varphi_1, i_1, u_2, \ldots, u_5, i_5)$ such that the equations (2.94) to (2.99) and (2.101) to (2.104) are satisfied. There are 10 constraints in $\mathbb{R}^{11}$. Consequently, the configuration space should be a surface of dimension $11 - 10 = 1$, i.e., a curve. It is difficult to visualize globally a curve in a space of dimension 11. What we can represent are the projections of the curve on different planes, as in descriptive geometry.

According to (2.94) and (2.97), the projections of the curve onto a plane that has two currents as coordinates are either the bisector of the first and third quadrants, or the bisector of the second and fourth quadrants. The projections onto the planes (φ_1, i_1), (u_2, i_2), $\ldots$, (u_5, i_5) are the characteristics of the elements. In particular, the projection onto the plane (u_2, i_2), which is actually the projection onto other planes with a coordinate u_2, is a straight line perpendicular to the axis of the u_2. This straight line moves as a function of time.

The projection of the curve onto the plane (u_1, φ_1) is the most interesting one. It involves the constitutive relations of all the elements. By combining (2.98) with (2.99), (2.101) to (2.103), and (2.95), we find

$$u_5 = -u_1 + e_2(t) + E - \varphi_1/GL \qquad (2.105)$$

When applying (2.105), (2.94), and (2.99), we obtain the equation that describes the projection of the curve onto the plane (u_1, φ_1):

$$g(-u_1 + e_2(t) + E - \varphi_1/GL) - \varphi_1/L = 0 \qquad (2.106)$$

This projection depends on time as well. It is represented in figure 2.46 for different values of $E + e_2(t)$.

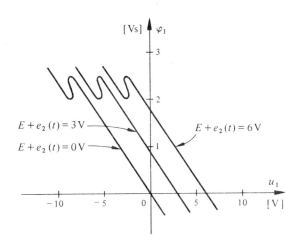

Fig. 2.46

Also note that (2.100) and (2.106) constitute a system of two equations for two unknowns, φ_1 and u_1. If we could resolve (2.106) for u_1, we would obtain a single differential equation in normal form. The figure shows, however, that u_1 is not a single-valued function of φ_1.

2.3.7 Comment

Because of the presence of time-dependent sources, the configuration space changes as a function of time. This renders less intuitive the picture of a solution as movement in the configuration space.

An analogy with mechanics can be of help to intuition. Let us consider a point mass that is forced to slide on a surface in the three-dimensional space under gravity (fig. 2.47). The surface corresponds to the configuration space of the circuit. In fact, it is also called configuration space in mechanics [5]. If we modify the surface by means of an external force, we obtain a time-dependent configuration space.

Although a configuration space varying with time, in fact, constitutes an intuitive object, it is still convenient to introduce a larger space, which remains fixed, by allowing the values of the time-dependent sources to remain free.

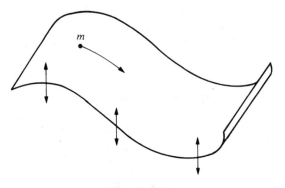

Fig. 2.47

2.3.8 Definitions

A *generalized operating point* is a set $\xi = (u, i, q_C, \varphi_L)$ of voltages, currents, charges, and fluxes, such that the circuit equations are satisfied, except for the differential equations (2.84) and (2.86), and the constitutive relations (2.87) and (2.88) of the time-dependent sources. The term "generalized" indicates that we abstract the special signals which are injected into the circuit by the sources. The set of the generalized operating points is called *generalized configuration space*.

We shall designate s as the set of the voltages of the time-dependent voltage sources and the currents of the time-dependent current sources.

A *dc operating point* is a generalized operating point, such that $i_C = u_L = s = 0$.

2.3.9   Property

In the absence of signals, a dc operating point is a solution of the circuit. More precisely, if during the time interval $[t_0, t_1]$ the voltages (or currents) of the time-dependent voltage (or current) sources are zero, the functions:

$$u(t) = \bar{u} \qquad (2.107)$$

$$i(t) = \bar{i} \qquad (2.108)$$

$$q_C(t) = \bar{q}_C \qquad (2.109)$$

$$\varphi_L(t) = \bar{\varphi}_L \qquad (2.110)$$

where $\bar{\xi} = (\bar{u}, \bar{i}, \bar{q}_C, \bar{\varphi}_L)$ is a dc operating point, constitute a solution of the circuit for $t_0 \leq t \leq t_1$.

Indeed, since by definition $i_C = u_L = 0$, equations (2.84) and (2.86) are satisfied by $q_C(t) = \bar{q}_C$ and $\varphi_L(t) = \bar{\varphi}_L$. The constitutive relations (2.87) and (2.88) are satisfied due to the absence of signals, and the other equations of the circuit are satisfied by any given generalized operating point.

2.3.10   Example

The circuit of figure 2.45 includes a time-dependent source. Therefore, the points of the generalized configuration space must satisfy one constraint less than is required of the nonautonomous configuration space. Because the latter space is a curve in $\mathbb{R}^{11}$, the generalized configuration space is a surface of dimension 2 in $\mathbb{R}^{11}$.

The dc operating points of this circuit have been previously studied in subsection 2.1.14.

2.3.11   Comment

With a simple count of equations and unknowns, we have concluded that the configuration space should be a surface or dimension $N_C + N_L$, and the generalized configuration space should be a surface of dimension $N_C + N_L + N_S$. While this idea is generally correct, there are numerous examples of systems of equations having sets of points that represent the solutions which cannot be called surfaces. We shall return to this fact in section 2.5. At present, we shall only discuss the case of linear circuits.

2.3.12 Configuration Space of a Linear Circuit

In the case of a linear circuit, all the equations are linear, most are homogeneous, and some are nonhomogeneous. Consequently, the (generalized) configuration space is, at any instant, a plane. Is said plane always of the same dimension $N_C + N_L$ ($N_C + N_L + N_S$)?

The answer to the above question is no. The system of circuit equations can be contradictory or linearly dependent, which leads to a (generalized) configuration plane having a dimension lower or higher than $N_C + N_L$ ($N_C + N_L + N_S$).

This concept deserves a deeper analysis. As shown by Boite and Neirynck (vol. IV, chap. 4), there are b independent Kirchhoff equations. Furthermore, the constitutive relations of the elements are independent as well, because each relation involves other variables. If the configuration plane is not of dimension $N_C + N_L$, there is contradiction or dependence between the Kirchhoff equations, on one hand, and the constitutive relations of the elements, on the other hand.

The simplest example which can be mentioned is the parallel connection of two independent voltage sources (fig. 2.48). In this case, $N_C + N_L = 0$ and because the circuit is linear, the configuration space should consist of a single point. The Kirchhoff equations are

$$u_1 - u_2 = 0 \tag{2.111}$$
$$i_1 + i_2 = 0 \tag{2.112}$$

and the constitutive relations of the sources are

$$u_1 = E_1 \tag{2.113}$$
$$u_2 = E_2 \tag{2.114}$$

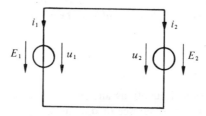

Fig. 2.48

If $E_1 \neq E_2$, (2.111) is in contradiction with (2.113) and (2.114), and thus the configuration space is empty. Conversely, if $E_1 = E_2$, then (2.111), (2.112), and (2.113) are dependent, and hence the configuration space is of dimension 1, since any current can flow in the two sources.

2.3.13 Definition

A linear circuit is *well posed* if its configuration space Λ is of dimension $N_C + N_L$ at any instant t. Otherwise, it is *ill posed*.

In the particular case of a linear resistive circuit, $N_C + N_L = 0$, and the circuit is well posed if, at any instant, it has exactly one solution.

2.3.14 Property

The generalized configuration space of a well posed linear circuit is of dimension $N_C + N_L + N_S$.

For the proof, we distinguish the amplitudes of the time-dependent sources s from the other variables of the circuit, which are designated by x. There are $2b - N_S$. The system of nondifferential circuit equations is in the form:

$$\begin{pmatrix} I & 0 \\ C & D \end{pmatrix} \begin{pmatrix} s \\ x \end{pmatrix} = \begin{pmatrix} e(t) \\ E \end{pmatrix} \tag{2.115}$$

where I is the unit matrix of dimension N_S; $e(t)$ is composed of signals from the time-dependent sources; E is a constant column matrix filled with zeros, except at the lines corresponding to constant sources.

The system (2.115) determines Λ_t. The space Λ is composed of the solutions of

$$(C, D) \begin{pmatrix} s \\ x \end{pmatrix} = E \tag{2.116}$$

According to the theorems of linear algebra, if Λ_t is of dimension $N_C + N_L$, the system (2.115) is of maximal rank, i.e., $2b$. Consequently, the system (2.116) is of maximal rank as well, and Λ is of dimension $N_C + N_L + N_S$.

2.3.15 Comments

Because the rank of (2.115) does not depend on its right-hand member, the dimension of Λ_t is the same for all the t.

For a linear circuit to be well posed, it is not enough that Λ be of dimension $N_C + N_L + N_S$. Indeed, two time-dependent voltage sources connected in parallel constitute a counter example.

2.3.16 Concept of Distance

It is indispensable to have a measure for the distance between the operating points. We need such a measure to prove that two solutions tend toward one another, that a solution remains bounded, or to give estimates of the exponential decay of transients, *et cetera*.

Most often, the distance between two points $x, y \in \mathbb{R}^n$ is defined by $\|x - y\|$ based on a *norm* $\|*\|$. In fact, we have already encountered the Euclidean norm in subsection 2.2.2, when defining a Lipschitz function. The corresponding distance is called *Euclidean distance*. There are lots of other norms. However, in a space of finite dimension, all the norms are equivalent [6], which means that all the qualitative concepts such as convergence, the fact of being bounded or not, the property of being a Lipschitz function, *et cetera* do not depend on the particular choice of the norm. To the contrary, the relative distances between different pairs of points can vary greatly.

If we introduce the Euclidean norm (2.46) in the space of all the variables, we must sum quantities of different dimensions, such as u_1^2 and i_1^2. Using constants with a dimension, we can bring all the terms of (2.46) back to the same dimension. Power would appear to be a reasonable choice for a common dimension. Therefore, we define

$$
\|(u, i, q_C, \varphi_L)\|_{R, L, C, \omega}^2 = \left[\frac{\|u\|^2}{R} + R \|i\|^2 + \frac{\|q_C\|^2}{C} \omega + \frac{\|\varphi_L\|^2}{L} \omega \right]
$$

(2.117)

where R, L, C, and ω, being *the normalization resistance, capacitance, inductance, and angular frequency*, are arbitrary positive constants and $\|*\|$ designates the Euclidean norm.

The normalization constants are valuable degrees of freedom when concrete bounds are to be deduced for example circuits. Each variable could also be provided with its own normalization constant, but we shall not do so here in order to avoid making the notation too heavy. For the same reason, we shall eliminate the indices R, L, C, and ω from the norm.

2.3.17 Geometric Concepts

Here, we shall review some geometric concepts of $\mathbb{R}^n$, each of which depends on the chosen norm.

A *solid sphere* of center x_0 and radius r is the set:

$$
B = \{x \,|\, \|x - x_0\| \leqslant r \}
$$

(2.118)

A *set* in $\mathbb{R}^n$ is *bounded* if it is contained in a solid sphere. A set $S \subset \mathbb{R}^n$ is *open* if, with each point $x_0 \in \Lambda$, a solid sphere centered at x_0 and having a positive radius is located entirely within S. A *neighborhood* of a point is an open set containing that point.

Suppose that Λ is a subset of $\mathbb{R}^n$. In our context, this will be the configuration space. A set $S \subset \Lambda$ is *open in* Λ if it is the intersection of Λ with an open set in $\mathbb{R}^n$. A neighborhood of a point $x \in \Lambda$ in Λ is an open set in Λ containing x.

A sequence $x_k \in \mathbb{R}^n$ *converges* toward $x \in \mathbb{R}^n$ if $\|x_k - x\|$ tends toward 0; x is called the *limit point* of the sequence. A set is *closed* if it contains all of its limit points. A set is *compact* if it is both closed and bounded.

2.3.18 Examples

The solid sphere B of (2.118) is not an open set because the points located on the limit sphere $\|x - x_0\| = r$ do not carry any sphere contained in B (fig. 2.49). Conversely, there is always a small solid sphere centered at an interior point of B which lies entirely within B. This shows that the set:

$$\widetilde{B} = \{x \,|\, \|x - x_0\| < r\} \tag{2.119}$$

is open. $\widetilde{B}$ is called an *open sphere*.

The solid sphere B is closed, whereas the open sphere $\widetilde{B}$ is not closed. Indeed, there are sequences of points of B which converge toward a point located on the limit sphere that belongs to B, but not to $\widetilde{B}$ (fig. 2.50). Since the sphere is bounded, it is compact. The **concepts** of closed set and open set are not complementary. Indeed, the entire space is both open and closed. Nevertheless, it is true that the complementary **set** of an open set is closed, and *vice versa*.

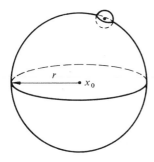

Fig. 2.49

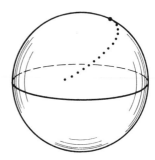

Fig. 2.50

In dimension 1, the solid sphere B of (2.118) is an interval of length $2r$ centered at x_0, which contains both its extreme points $x_0 - r$ and $x_0 + r$. We shall use the notation $[x_0 - r, x_0 + r]$ for this closed interval. The open sphere $\mathring{B}$ is the same interval, but without the extreme points. This interval is designated by $(x_0 - r, x_0 + r)$.

Figure 2.51 represents the neighborhood V of a point ξ on a surface $\Lambda \subset \mathbb{R}^3$, obtained by the intersection of an open sphere with Λ. This set is not open, but it is **open in** Λ.

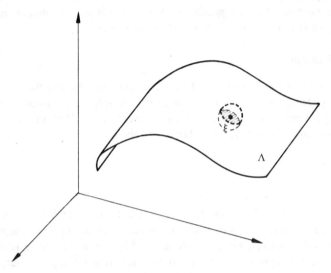

Fig. 2.51

2.3.19 Property

The configuration space, at any instant, as well as the generalized configuration space, are closed sets in $\mathbb{R}^{2b+N_C+N_L}$.

For the proof, let us first consider one of the equations defining the configuration space, for instance, a constitutive relation of a resistor in the form $f_k(u_k, i_k) = 0$. Suppose that $u^{(j)}$ is a sequence of voltages which converges toward $\bar{u}$, and $i^{(j)}$ is a sequence of currents which converges toward $\bar{i}$, such that $f_k(u_k^{(j)}, i_k^{(j)}) = 0$. Because, by hypothesis, all the functions intervening in the constitutive relations are continuous, the limit of $f_k(u_k^{(j)}, i_k^{(j)})$ is equal to $f_k(\bar{u}_k, \bar{i}_k)$, therefore, $f_k(\bar{u}_k, \bar{i}_k) = 0$. The same reasoning applies to all the other equations that determine the configuration space, whether generalized or not.

2.4 GLOBAL STATE EQUATIONS

2.4.1 Introduction

In this section, we shall study the transition from the system of circuit equations, the form of which is shown in (2.91) and (2.92), to a system of equations in normal form (2.43). The interest of this transition lies in the theorems 2.2.5, 2.2.9, and 2.2.14, which guarantee the existence, either local or global, and the uniqueness of the solutions starting from initial conditions. The solutions of a circuit will have these properties if the circuit equation can be transformed into a system in normal form. In particular, the impasse point phenomenon, observed in section 2.1, cannot occur.

The purpose of this section is not to provide a general and efficient method for deducing a system of differential equations in normal form from the system of circuit equations. What we are interested in is, in fact, its existence. The examples must also show the principle, rather than an economical derivation. Additionally, general analysis programs of circuits always use mixed systems of differential and nondifferential equations, and not a system in normal form.

2.4.2 Derivation

We obtain a system of equations in normal form if we eliminate from the equation system (2.91) to (2.93) all of the variables except for q_C and φ_L. Therefore, we must solve the nondifferential equations (2.93) in order to eliminate variables, which amounts to looking for a function p, such that

$$\begin{pmatrix} u \\ i \end{pmatrix} = p(q_C, \varphi_L, t) \tag{2.120}$$

for all the solutions of (2.93). In particular, we are interested in that part of p which provides i_C and u_L:

$$i_C = p_C(q_C, \varphi_L, t) \tag{2.121}$$
$$u_L = p_L(q_C, \varphi_L, t) \tag{2.122}$$

Using p_C and p_L, we can write

$$\frac{dq_C}{dt} = p_C(q_C, \varphi_L, t) \tag{2.123}$$

$$\frac{d\varphi_L}{dt} = p_L(q_C, \varphi_L, t) \tag{2.124}$$

which shows that the solutions of the circuit satisfy a system of differential equations in normal form.

Two questions arise. Do p_C and p_L always exist? Does any solution of the system (2.123) (2.124) always define a solution of the circuit?

Before dealing with these questions, we slightly modify the derivation of the equations (2.123) and (2.124). Instead of solving (2.93), we shall consider the reduced system by omitting the constitutive relations of the time-dependent sources. We write this reduced system as follows:

$$F(\tilde{u}, \tilde{i}, q_C, \varphi_L, s) = 0 \tag{2.125}$$

where u and i designate the voltages and currents other than the values s of the time dependent sources. Solving (2.125) amounts to representing the solution (2.125) in the form:

$$\begin{pmatrix} \tilde{u} \\ \tilde{i} \end{pmatrix} = f(q_C, \varphi_L, s) \tag{2.126}$$

By designating that part of f which provides i_C and u_L as

$$i_C = f_C(q_C, \varphi_L, s) \tag{2.127}$$

$$u_L = f_L(q_C, \varphi_L, s) \tag{2.128}$$

we obtain, using the constitutive relations of the time dependent sources,

$$\frac{dq_C}{dt} = f_C(q_C, \varphi_L, e(t)) \tag{2.129}$$

$$\frac{d\varphi_L}{dt} = f_L(q_C, \varphi_L, e(t)) \tag{2.130}$$

This system has the same form as (2.123) and (2.124). Its advantage is that it involves a function f which does not depend on time.

2.4.3 Definition

The system of differential equations (2.129) and (2.130) is a *system of global state equations* of the circuit if both of the following conditions are satisfied:

- If $\xi(t) = (u(t), i(t), q_C(t), \varphi_L(t))$ is a solution of the circuit, then $(q_C(t), \varphi_L(t))$ is a solution of (2.129) and (2.130).
- If $(q_C(t), \varphi_L(t))$ is a solution of (2.129) and (2.130), then there exists exactly one solution $\xi(t)$ of the circuit with these charges and fluxes.

In this case, q_C and φ_L are *global state variables*.

2.4.4 Example

Let us try to find state equations for the circuit of figure 2.52, the only linear element of which is an inductor with the characteristic of figure 1.36 or 1.37.

Now, i_1 and u_2 should be expressed as functions of q_1, φ_2, and u_3. Since the only nondifferential equations involving q_1 and φ_2 are the constitutive relations of the capacitor and the inductor,

$$u_1 = q_1/C_1 \tag{2.131}$$

$$i_2 = g_2(\varphi_2) \tag{2.132}$$

then i_1 and u_2 should be expressed as functions of u_1, i_2, and u_3. We can easily find that

$$u_1 + u_2 - u_3 + R_4 i_1 = 0 \tag{2.133}$$

$$u_2 - R_5 (i_1 - i_2) = 0 \tag{2.134}$$

and, therefore,

$$i_1 = \frac{1}{R_4 + R_5} (-u_1 + R_5 i_2 + u_3) \tag{2.135}$$

$$u_2 = \frac{R_5}{R_4 + R_5} (-u_1 - R_4 i_2 + u_3) \tag{2.136}$$

Consequently,

$$\frac{d}{dt} \begin{pmatrix} q_1 \\ \varphi_2 \end{pmatrix} = \frac{1}{R_4 + R_5} \begin{pmatrix} -1 & R_5 & 1 \\ -R_5 & -R_4 R_5 & R_5 \end{pmatrix} \begin{pmatrix} q_1/c_1 \\ q(\varphi_2) \\ e_3(t) \end{pmatrix} \tag{2.137}$$

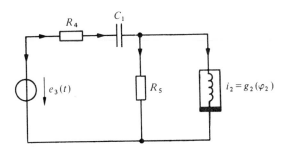

Fig. 2.52

From these calculations, it follows that a solution of the circuit satisfies (2.137). Nonetheless, it is not difficult to obtain from u_1, i_2, u_3, and therefore from q_1, φ_2, u_3, all of the currents and voltages for the circuit. Given a solution of (2.137), therefore, we can construct a solution of the circuit.

2.4.5 Geometric Interpretation

The function f of (2.126) plays a crucial role for the state equations. If the solutions of (2.125) are apt to be represented in that form, we obtain global state equations. We may ask, however, in what case does this happen?

The geometric concepts of section 2.3 make things somewhat more clear. The solutions of (2.125) are the generalized operating points and the set of solutions forms the generalized configuration space Λ. The function f of (2.126) exists if, for each combination of values q_C, φ_L, and s, there is exactly one generalized operating point $\xi = (\tilde{u}, \tilde{i}, q_C, \varphi_L, s)$. In other words, the projection of Λ onto the space $\mathbb{R}^{NC+NL+NS}$ of the coordinates q_C, φ_L, and s is a bijective function.

In figure 2.53, we show the example of a surface Λ such that this projection is not bijective. Indeed, the points $\bar{\xi}_1$ and $\bar{\xi}_2$ have the same coordinates q, φ, and s.

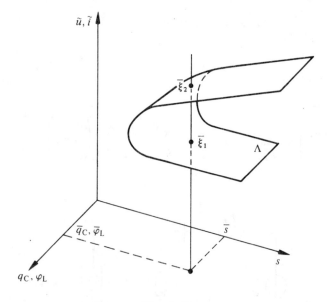

Fig. 2.53

2.4.6 Definition

Suppose that $\pi:\mathbb{R}^{2b+NC+NL} \to \mathbb{R}^{NC+NL+NS}$ is the projection defined by

$$\pi : (\tilde{u}, \tilde{i}, q_C, \varphi_L, s) \to (q_C, \varphi_L, s) \tag{2.138}$$

and π/Λ is the restriction of π to the generalized configuration space Λ. The variables q_C, φ_L, and s are the *global parameters* of Λ, if π/Λ is a bijective function. In that case, $(\pi/\Lambda)^{-1}$ is a *global parameterization of Λ by q_C, φ_L, and s*.

2.4.7 Proposition

If the variables q_C, φ_L, and s characterize the global parameters of the generalized configuration space Λ, then q_C and φ_L are global state variables of the circuit. The global state equations are obtained from the global parameters of Λ according to (2.126) to (2.130).

For the proof, we first observe that the derivation of (2.129) and (2.130) proves that any solution of the circuit satisfies this system.

Conversely, let us suppose that $(q_C(t), \varphi_L(t))$ is a solution of the system (2.129) and (2.130). That is,

$$\xi(t) = (f(q_C(t), \varphi_L(t), e(t)), q_C(t), \varphi_L(t), e(t)) \tag{2.139}$$

Then, through (2.127) and (2.128), $\xi(t)$ satisfies (2.91) and (2.92). Furthermore, because any point $\xi = (f(q_C, \varphi_L, s), q_C, \varphi_L, s)$ is located on Λ, all the nondifferential equations are satisfied, except possibly the constitutive relations of the time-dependent sources. Finally, replacing s by $e(t)$ ensures as well the constitutive relations of the time-dependent sources.

2.4.8 Comment

The situation described in proposition 2.4.7 has a simple geometric interpretation. A solution $\xi(t)$ of the circuit is a movement on the surface Λ (fig. 2.54). Its projection in the space of the coordinates q_C, φ_L, and s satisfies the state equations (2.129) and (2.130). Conversely, if a solution $(q_C(t), \varphi_L(t))$ of the state equations is given, then its image by inverse projection on Λ satisfies the entire system of equations of the circuit.

2.4.9 Decomposition of the Parameterization

The question of whether q_C, φ_L, and s are global parameters of Λ is far from being trivial. For small circuits, we can try to obtain the function f of (2.126) explicitly. Such was the method applied to example 2.4.4. However, for large circuits, the difficulty is insurmountable and the problem must be decomposed.

Let us detail system (2.125):

$$F_C(u_C, q_C) = 0 \tag{2.140}$$

$$F_L(\varphi_L, i_L) = 0 \tag{2.141}$$

$$F_{RB}(u, i) = 0 \tag{2.142}$$

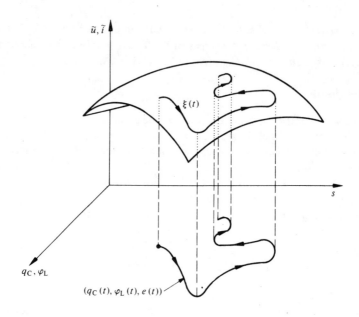

Fig. 2.54

The equations (2.140) are the constitutive relations of the capacitors, equations (2.141) are those of the inductors, and the system (2.142) includes the Kirchhoff equations as well as the constitutive relations of the resistors and the linear two-ports. Because q_C and φ_L are not involved in (2.142), we obtain the following lemma.

2.4.10 Lemma

The generalized configuration space Λ has the global parameters of q_C, φ_L, and s if and only if the three following conditions are satisfied:

- The capacitors are charge controlled:

$$u_C = h_C(q_C) \tag{2.143}$$

- The inductors are flux controlled:

$$i_L = g_L(\varphi_L) \tag{2.144}$$

- The solutions of (2.142) have the global parameters u_C, i_L, and s; in other words, there is a function $\psi \colon \mathbb{R}^{NC+NL+NS} \to \mathbb{R}^{2b}$ such that all the solutions of (2.142) are of the form:

$$\begin{pmatrix} u \\ i \end{pmatrix} = \psi(u_C, i_L, s) \tag{2.145}$$

2.4.11   Corollary

Assuming that the three conditions of lemma 2.4.10 are satisfied, let us designate those parts of ψ which provide i_C and u_L by

$$i_C = \psi_C(u_C, i_L, s) \tag{2.146}$$

$$u_L = \varphi_L(u_C, i_L, s) \tag{2.147}$$

Then, the system (2.129) and (2.130) is a system of global state equations. It is written in the form:

$$\frac{dq_C}{dt} = \psi_C(h_C(q_C), g_L(\varphi_L), e(t)) \tag{2.148}$$

$$\frac{d\varphi_L}{dt} = \psi_L(h_C(q_C), g_L(\varphi_L), e(t)) \tag{2.149}$$

2.4.12   Associated Resistive Circuit

The equation system (2.142) describes the resistive part of the circuit, except for the constitutive relations of the time-dependent sources. We can augment it to a complete equation system of a resistive circuit. Indeed, we simply add the equations:

$$u_k = E_k \tag{2.150}$$

for the branches which carry a capacitor or a time-dependent voltage source, and

$$i_k = I_k \tag{2.151}$$

for the branches which carry an inductor or a time-dependent current source. The resulting system describes the *associated resistive circuit*, and the function ψ of (2.145) expresses the solution of this circuit as a function of the values E_k and I_k of the sources.

2.4.13   Definition

The *associated resistive circuit for a circuit* is obtained by replacing the capacitors and the time-dependent voltage sources by constant voltage sources, and the inductors and time-dependent current sources by constant current sources (fig. 2.55). In order to distinguish these from other possible constant sources which may exist in the original circuit, we call them *substitution sources*.

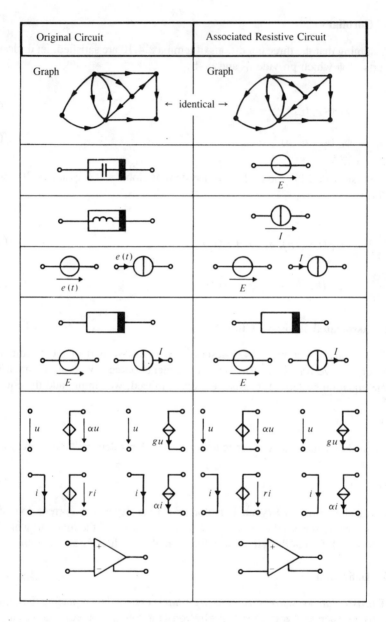

Fig. 2.55

2.4.14   Example

Let us consider the Wien-bridge oscillator of figure 2.56 (vol. VIII, subsec. 5.3.5). The associated resistive circuit is shown in figure 2.57. According to (2.146), the currents i_1 and i_2 must be expressed as a function of the voltages E_1 and E_2. If the two-port were linear, the admittance matrix should be calculated.

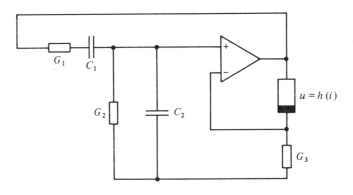

Fig. 2.56

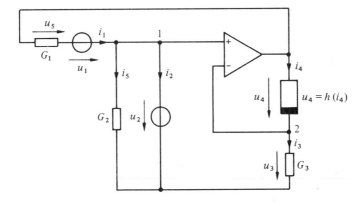

Fig. 2.57

Taking into account that the input voltage of the operational amplifier is zero, we obtain the following Kirchhoff voltage laws:

$$u_1 - u_4 + u_5 = 0 \tag{2.152}$$

$$u_2 - u_3 = 0 \tag{2.153}$$

Because the input currents of the operational amplifier are zero, the Kirchhoff current laws at nodes 1 and 2 are

$$i_1 - i_2 - i_5 = 0 \tag{2.154}$$

$$-i_3 + i_4 = 0 \tag{2.155}$$

Combining these equations with the constitutive relations of the elements, we find

$$i_1 - i_2 - G_2 E_2 = 0 \tag{2.156}$$

$$E_1 + i_1/G_1 - h(G_3 E_2) = 0 \tag{2.157}$$

from which we get

$$i_1 = -G_1 E_1 + G_1 h(G_3 E_2) \tag{2.158}$$

$$i_2 = -G_1 E_1 - G_2 E_2 + G_1 h(G_3 E_2) \tag{2.159}$$

The expressions i_3, i_4, i_5, u_3, u_4, u_5, as functions of E_1 and E_2, can be easily obtained from (2.158) and (2.159).

The state equations resulting from (2.158) and (2.159) are

$$dq_1/dt = -\frac{G_1}{C_1} q_1 + G_1 h\left(G_3 \frac{q_2}{C_2}\right) \tag{2.160}$$

$$dq_2/dt = -\frac{G_1}{C_1} q_1 - \frac{G_2}{C_2} q_2 + G_1 h\left(G_3 \frac{q_2}{C_2}\right) \tag{2.161}$$

2.4.15   Theorem: Existence of Global State Equations

Consider a circuit with charge-controlled capacitors and flux-controlled inductors. Let us assume that the associated resistive circuit has a solution for any choice of values for the substitution sources.

Suppose that ψ is the function which expresses the solution of the associated resistive circuit as a function of the values of the substitution sources. Suppose that q_C and φ_L are respectively given by (2.146), (2.147), (2.143), and (2.144). Then, (2.148) and (2.149) constitute a system of global state equations of the original circuit.

This theorem results from corollary 2.4.11 and definition 2.4.13.

2.4.16   Theorem: Existence and Uniqueness of the Solutions

Let us assume that the hypotheses of theorem 2.4.15 are satisfied. Thus,

- If the circuit is piecewise linear, then for any choice of initial charges q_0, initial fluxes φ_0, and initial time t_0, there is exactly one solution $\xi(t)$ of the circuit with $q_C(t_0) = q_0$ and $\varphi_L(t_0) = \varphi_0$. This function $\xi(t)$ is defined for all times, and it has a continuous derivative.

- If the circuit is smooth, and the function ψ is smooth as well, then the conclusion is the same, except that the solution might only exist during a finite interval around t_0. Nonetheless, wherever $\xi(t)$ exists, the function is smooth. If we can show, as an additional condition, that $\xi(t)$ is bounded for $t \geqslant t_0$, then $\xi(t)$ exists for all $t \geqslant t_0$.

This theorem results from theorem 2.4.15 as well as theorems 2.2.5, 2.2.9, and 2.2.14.

2.4.17   Comments

With theorem 2.4.16, we have been able to reduce the general problem of existence and uniqueness of the solution for a time-dependent nonlinear circuit to the simpler problem of existence, uniqueness, and regularity of the solution for a resistive, autonomous, nonlinear circuit. The latter will be discussed in chapter 3.

We could still wonder why it would not be possible to describe the solutions of a circuit by means of other systems of state equations, using other state variables. Indeed, there often are several state variables from which to choose. However, this is only the case when additional conditions are satisfied. For instance, u_C and i_L can be global state variables only when the capacitors are both charge and voltage controlled, and the inductors are flux and current controlled. Such is the case particularly with linear capacitors and inductors.

Therefore, if there are state equations, they also exist in the variables q_C and φ_L. The reason for this is that the system of circuit equations makes use of the time derivatives q_C and φ_L only. It follows that it is not possible to get stronger results regarding the existence and uniqueness of the solutions by using other state variables. This subject is treated more rigorously in [7].

2.5 LOCAL STATE EQUATIONS

2.5.1 Introduction

There actually exist circuits that do not have any system of global state equations, while still having solutions for almost all of the initial conditions. This suggests that we should extend the concept of state equation, while limiting its validity to the restricted domain of the configuration space.

In the first part of this section, the global concepts of section 2.4 will be transformed into local concepts. Then, a criterion for the existence of local state equations with respect to linear circuits will be set forth.

2.5.2 Example

Let us consider again the circuit of figure 2.45 and try to derive the global state equations (2.148) and (2.149). Therefore, u_1 must be expressed as a function of φ_1 and u_2, starting from equations (2.94) to (2.99) and (2.102) to (2.104). We start from (2.98):

$$u_1 = u_2 + u_3 - u_4 - u_5 \qquad (2.162)$$

Using (2.101) to (2.102), we get

$$u_1 = u_2 + E - i_4/G - g^{-1}(i_5) \qquad (2.163)$$

Then, through (2.93), (2.94), and (2.98), we get

$$u_1 = u_2 + E - \varphi_1/LG - g^{-1}(\varphi_1/L) \qquad (2.164)$$

The corresponding state equation is

$$\frac{d\varphi_1}{dt} = u_s(t) + E - \varphi_1/LG - g^{-1}(\varphi_1/L) \qquad (2.165)$$

Except for a factor L, we again obtained equation (2.38). As we previously observed in section 2.1, g is not invertible, thereby equation (2.164) does not define a global parameterization of Λ, and thus (2.165) is not a global state equation.

However, g is locally invertible in almost all of the operating points. Indeed, the restriction of g to a small enough neighborhood W of a point $(\bar{u}, \bar{i})$ of its graph is a bijective function (fig. 2.58). Only points Q_1 and Q_2 do not have such a neighborhood.

We conclude that the solutions of the circuit satisfy (2.165), unless they are located at Q_1 or Q_2.

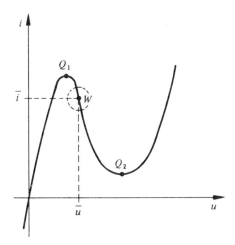

Fig. 2.58

2.5.3 Definitions

Consider a circuit with a generalized configuration space Λ, and a generalized operating point $\bar{\xi} \in \Lambda$. A system of differential equations of the form (2.129) and (2.130) is a *system of local state equations at* $\bar{\xi}$, if there is a neighborhood W of $\bar{\xi}$ in Λ, and a neighborhood V of (q_C, φ_L, s) in $\mathbb{R}^{NC+NL+NS}$, such that both of the following conditions are satisfied:

- If $\xi(t) = (u(t), i(t), q_C(t), \varphi_L(t))$ is a solution of the circuit which is located, during the time interval $t \in (a,b)$, in W, then $(q_C(t), \varphi_L(t))$ is a solution of (2.129) and (2.130), such that $(q_C(t), \varphi_L(t), e(t)) \in V$ for $t \in (a,b)$.
- If $(q_C(t), \varphi_L(t))$ is a solution of (2.129) and (2.130) such that $(q_C(t), \varphi_L(t), e(t)) \in V$ for (a,b), then there exists exactly one solution $\xi(t)$ of the circuit with these charges and fluxes, such that $\xi(t) \in W$ for $t \in (a,b)$.

In that case, q_C and φ_L are *local state variables at* $\bar{\xi}$.

2.5.4 Geometric Representation

In figure 2.59, we have represented the point $\bar{\xi} \in \Lambda$, the neighborhoods W and V, and a solution $\xi(t)$ with its projection $(q_C(t), \varphi_L(t), e(t))$. Note that the inverse projection of $(q_C(t), \varphi_L(t), e(t))$ on Λ is not unique. It is only by limiting ourselves to W that we can define $\xi(t)$ by inverse projection.

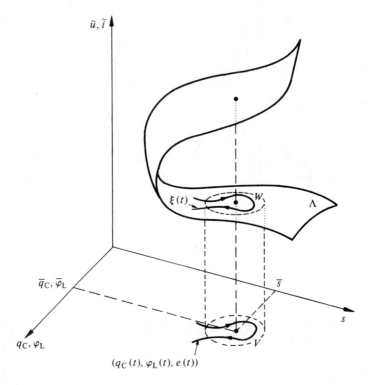

Fig. 2.59

2.5.5   Definition

Suppose that π is the projection defined by (2.138). Let $\xi \in \Lambda$ be a generalized operating point. The variables q_C, φ_L, and s are *local parameters of* Λ *at* $\bar{\xi}$, if there exists a neighborhood W of $\bar{\xi}$ in Λ, such that the restriction of π to W, πW is an injective function. In that case,

$$(\pi/W)^{-1} : V \rightarrow W \tag{2.166}$$

where $V = \pi(W)$ is the *local parameterization of* Λ *by* q_C, φ_L, *and* s *at* $\bar{\xi}$.

2.5.6   Comment

If the variables q_C, φ_L, and s are local parameters of Λ at $\bar{\xi}$, then close to $\bar{\xi}$ is the image of a small piece of $\mathbb{R}^{N_C + N_L + N_S}$. If the parameter $(\pi/W)^{-1}$ has a minimum of regularity, this image can be viewed as a deformed version of the piece in the plane

(q_C, φ_L, s). If this is the case at all the points of Λ, we say that Λ is a *surface of dimension* $N_C + N_L + N_S$. In mathematical texts, the term *differential manifold* [8] is used when the local parameter has continuous partial derivatives at any point. A discussion of the configuration space viewed as a differential manifold is given in [9]. For the study of the state equations, however, the concept of differential manifold is too general. Indeed, although a differential manifold has a set of local parameters at any point, nothing guarantees that these parameters are q_C, φ_L, and s. As for global state equations, it is possible to consider local state equations in other variables; but if such equations exist, they also exist at the variables q_C, φ_L, and s. Therefore, $\dot{q}_C$, φ_L, and s must be local parameters.

The following example illustrates the fact that it is not always possible to characterize the parameters of a surface by the same variables. The sphere of figure 2.60 is indeed a surface of dimension 2 in the space of dimension 3. At all points, except on the equator, it has local parameters characterized by the variables x and y. On the equator, we are forced to change to other parameters. At these points, the pairs x, z or y,z will do. We can easily see that none of the three pairs of variables provides a local parameter at all the points of the sphere. Thus, if the sphere were the generalized configuration space of a circuit, the variables q_C, φ_L, and s would not be able to characterize the local parameters at all points, although it is indeed a surface of the suitable dimension.

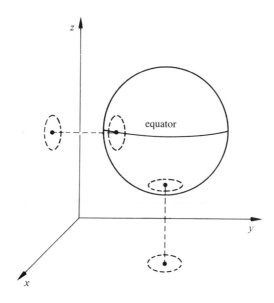

Fig. 2.60

2.5.7 Proposition

If the variables q_C, φ_L, and s are the local parameters of the generalized configuration space Λ at $\bar{\xi}$, then q_C and φ_L are local state variables at $\bar{\xi}$. The state equations are obtained from the local parameterization at $\bar{\xi}$ according to (2.126) to (2.130).

The proof for this proposition consists of a transposition from global to local of the proof for proposition 2.4.7

2.5.8 Definitions

A capacitor is *locally charge controlled* at a point $(\bar{u},\bar{q})$ of its characteristic if there exists a neighborhood W of $(\bar{u},\bar{q})$ such that the following condition is satisfied. If (u_1,q) and (u_2,q) are two points of the characteristic, and if $(u_1,q) \in W$, then $(u_2,q) \notin W$ (fig. 2.61).

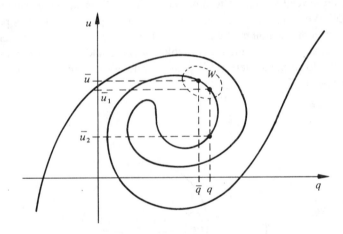

Fig. 2.61

We define analogously the concepts of *locally voltage-controlled capacitor*, *locally flux-* or *current-controlled inductor*, and *locally voltage-* or *current-controlled resistor*.

2.5.9 Lemma

The generalized configuration space Λ has the local parameters q_C, φ_L, and s at $\bar{\xi} = (\bar{u}, \bar{i}, \bar{q}_C, \bar{\varphi}_L)$ if and only if the three following conditions are satisfied:

- The capacitors are locally charge controlled at $(\bar{u}_C, \bar{q}_C)$.
- The inductors are locally flux controlled at $(\bar{\varphi}_L, \bar{i}_L)$.

- The associated resistive circuit has exactly one solution in a neighborhood of $\bar{\xi}$. More precisely, there exists a neighborhood $W \subset \mathbb{R}^{2b}$ of $(\bar{u}, \bar{\imath})$ and a neighborhood $V \subset \mathbb{R}^{N_C + N_L + N_S}$ of $(\bar{u}_C, \bar{\imath}_L, \bar{s})$, such that the associated resistive circuit has exactly one solution in W if the set of the values E and I of the substitution sources is located in V.

2.5.10 Proposition

Suppose that $\bar{\xi} \in \Lambda$ is a generalized operating point. If the hypotheses of lemma 2.5.9 are verified, then q_C and φ_L are local state variables at $\bar{\xi}$. If ψ_C, ψ_L, h_C, and g_L are the functions defined by (2.146), (2.156), (2.143), and (2.144), respectively, but the existence of which is only guaranteed here in a neighborhood of $\bar{\xi}$, then (2.148) and (2.149) comprise a system of local state equations at $\bar{\xi}$.

2.5.11 Comment

Up to this point, we have simply transposed the expansions of section 2.4 from global to local. In this way, we have been able to reduce the problem of the local existence of the state equations to the simpler problem of the existence and local uniqueness of the solution of the associated resistive circuit.

We can now proceed one step further. The existence and uniqueness of the solution of the associated resistive circuit can be established from its linearization. The mathematical tool allowing this transition is the *implicit function theorem*.

2.5.12 Implicit Function Theorem

The implicit function theorem addresses the question of whether a system of equations:

$$
\begin{aligned}
f_1(x_1, ..., x_n, y_1, ..., y_m) &= 0 \\
\vdots \qquad\qquad\qquad \vdots & \\
f_m(x_1, ..., x_n, y_1, ..., y_m) &= 0
\end{aligned}
\tag{2.167}
$$

locally defines $y_1, \ldots, y_m$ as a function of $x_1, \ldots x_n$. We write (2.167) in the abbreviated form:

$$
f(x, y) = 0
\tag{2.168}
$$

Suppose that $f:\mathbb{R}^{n+m} \to \mathbb{R}^m$ is a smooth function and $x = a$, $y = b$ is a solution of (2.168). If the *Jacobian matrix*:

$$\frac{\partial f}{\partial y} = \begin{pmatrix} \dfrac{\partial f_1}{\partial y_1} & \cdots & \dfrac{\partial f_1}{\partial y_m} \\ \vdots & & \vdots \\ \dfrac{\partial f_m}{\partial y_1} & \cdots & \dfrac{\partial f_m}{\partial y_m} \end{pmatrix} \tag{2.169}$$

is not singular at (a,b), that is, if its determinant, the *Jacobian*, is nonzero at (a,b), then there exist a neighborhood V of a in $\mathbb{R}^n$, a neighborhood of W of (a,b) in $\mathbb{R}^{n+m}$, and a smooth function $V \to \mathbb{R}^m$ such that

$$f(x, g(x)) = 0 \tag{2.170}$$

for any $x \epsilon V$, and all the solutions of (2.168) with $(x,y)\epsilon W$ are of that form.

For the proof, refer to [10].

2.5.13 Example

The simplest case of the implicit function theorem is $n = m = 1$. Suppose a resistor is defined by a constitutive relation:

$$f(u, i) = 0 \tag{2.171}$$

which is represented by a curve in the plane (u,i). Let $(\bar{u},\bar{\imath})$ be a point of that characteristic. If $(\partial f/\partial u)(\bar{u},\bar{\imath}) \neq 0$, then u, in a neighborhood of $(\bar{u},\bar{\imath})$, is a function of i. In other words, the resistor is locally current controlled at $(\bar{u},\bar{\imath})$.

2.5.14 Linearized System of Equations

We can reformulate the theorem 2.5.12 by linearizing the system of equations (2.168) around (a,b). As a first approximation, we get

$$f(x, y) \cong f(a, b) + \frac{\partial f}{\partial x}(a, b) \cdot \Delta x + \frac{\partial f}{\partial y}(a, b) \cdot \Delta y \tag{2.172}$$

where $\partial f/\partial y$ is the Jacobian matrix (2.169) and $\partial f/\partial x$ is the corresponding matrix of the partial derivatives of f with respect to x, $\Delta x = (x_1 - a_1, \ldots, x_n - a_n)$ and $\Delta y = (y_1 - b_1, \ldots, y_m - b_m)$.

If $f(x,y) = f(a,b) = 0$, it follows that

$$\frac{\partial f}{\partial x}(a, b) \cdot \Delta x + \frac{\partial f}{\partial y}(a, b) \cdot \Delta y \cong 0 \tag{2.173}$$

If we set the equality in (2.173), we obtain the *equation system (2.168) linearized around (a,b)*.

Let us express Δy as a function of Δx from the linearized system:

$$\Delta y = - \left[\frac{\partial f}{\partial y}(a, b) \right]^{-1} \left[\frac{\partial f}{\partial x}(a, b) \right] \Delta x \tag{2.174}$$

The right-hand member of (2.174) is well defined if the Jacobian matrix $\partial f/\partial y$ is not singular at (a,b). Exactly the same condition has been formulated in theorem 2.5.12. This fact is expressed by the following corollary.

2.5.15 Corollary

The system of equations:

$$f(x, y) = 0 \tag{2.175}$$

with $x \in \mathbb{R}^n$, $y \in \mathbb{R}^m$, $f:\mathbb{R}^{n + m} \to \mathbb{R}^m$ locally defines y as a function of x around its solution (a,b), if the linearized equation:

$$\frac{\partial f}{\partial x}(a, b) \cdot \Delta x + \frac{\partial f}{\partial y}(a, b) \cdot \Delta y = 0 \tag{2.176}$$

defines $\Delta y = y - b$ as a function of $\Delta x = x - a$.

More precisely, if (2.176) has exactly one solution Δy for any Δx, then the conclusions of theorem 2.5.12 are verified.

2.5.16 Comment

According to the theorems of linear algebra, the equation system (2.176) has exactly one solution Δy either for all the Δx, or for none of the Δx. For this reason, we shall say hereafter that the system (2.176) has exactly one solution Δy, without referring to Δx.

2.5.17   Linearized Resistive Circuit

We intend to apply corollary 2.5.15 to the system of equations (2.142), which describes the associated resistive circuit. We set $x = (u_C, i_L, s)$ and $y = (u, i)$.

Let us assume that $\bar\xi = (\bar u, \bar i, \bar q_C, \bar\varphi_L)$ is a generalized operating point of the original circuit. Then, $(\bar u, \bar i)$ is a solution of the associated resistive circuit. According to corollary 2.5.15, the latter has exactly one solution for each x close to $\bar x$, if the system (2.142) linearized around (x,y) thusly

$$\frac{\partial F_{RB}}{\partial x}(\bar x, \bar y)\cdot \Delta x + \frac{\partial F_{RB}}{\partial y}(\bar x, \bar y)\cdot \Delta y = 0 \qquad (2.177)$$

has exactly one solution Δy.

Let us examine the linear system on an equation by equation basis. The linear equations do not change when passing from (2.142) to (2.177). Such is the case in particular for the Kirchhoff equations and the constitutive relations of linear two-ports. The only equations which may be nonlinear are the constitutive relations of the resistors. Let us assume that the branch k carries a resistor having a constitutive relation, which is part of (2.142), in the form:

$$f_k(u_k, i_k) = 0 \qquad (2.178)$$

The linearized equation around $(\bar u_k, \bar i_k)$, which is part of (2.177), is

$$\frac{\partial f_k}{\partial u_k}(\bar u_k, \bar i_k)\cdot \Delta u_k + \frac{\partial f_k}{\partial i_k}(\bar u_k, \bar i_k)\cdot \Delta i_k = 0 \qquad (2.179)$$

This is the constitutive relation of a linear resistor in the incremental variables $\Delta u_k = u_k - \bar u_k$ and $\Delta i_k = i_k - \bar i_k$. It follows that (2.179) constitutes the system of equations of a linear resistive circuit.

2.5.18   Definitions

Let a resistor be defined by the constitutive relation $f(u,i) = 0$, and let $(\bar u, \bar i)$ be a point of its characteristic. The *linearized resistor around* $(\bar u, \bar i)$ is the linear resistor of value:

$$r(\bar u, \bar i) = -\frac{\partial f/\partial i}{\partial f/\partial u}(\bar u, \bar i) \qquad (2.180)$$

The value $r(\bar u, \bar i)$ is called the *differential resistance at* $(\bar u, \bar i)$. The inverse value is the *differential conductance at* $(\bar u, \bar i)$.

2.5.19 Properties

If the constitutive relation of a resistor is of the form $u = h(i)$, then the differential resistance at $(\bar{u}, \bar{i})$ is

$$r(\bar{u}, \bar{i}) = \frac{dh}{di}(\bar{i}) \tag{2.181}$$

If the constitutive relation of a resistor is of the form $i = g(u)$, then the differential conductance at $(\bar{u}, \bar{i})$ is

$$g(\bar{u}, \bar{i}) = \frac{dg}{du}(\bar{u}) \tag{2.182}$$

Indeed, (2.181) and (2.182) result from (2.180) when setting $f(u,i) = u - h(i)$ and $f(u,i) = i - g(u)$, respectively.

2.5.20 Comments

The differential conductance of a tunnel diode can be positive, negative, or zero. As shown in subsection 2.1.9, it plays a crucial role for the stability of the dc operating points of the circuit in figure 2.10.

It follows from the constitutive relation (1.12) that the differential resistance of a voltage source is zero everywhere. A linearized voltage source around any given operating point is a short circuit. Analogously, a linearized current source is always an open circuit.

The concept of linearized resistor is not very useful for piecewise linear resistors. Indeed, if $(\bar{u}, \bar{i})$ is located inside a linear domain, the linearized resistor, up to a translation of the characteristic, is simply the linear resistor which corresponds to the linear domain. However, if $(\bar{u}, \bar{i})$ is located on the boundary between the two linear domains, the linearization of the constitutive relation does not exist.

Therefore, the concept of linearized resistor and linearized circuit will be restricted to smooth circuits in the following.

2.5.21 Definition

Let $\xi \in \Lambda$ be a generalized operating point of a circuit. The *associated linearized resistive circuit at* ξ is obtained from the associated resistive circuit by replacing each nonlinear resistor with the corresponding linearized resistor at ξ. The complete transformation of the original circuit into the associated resistive circuit is shown in figure 2.62.

126

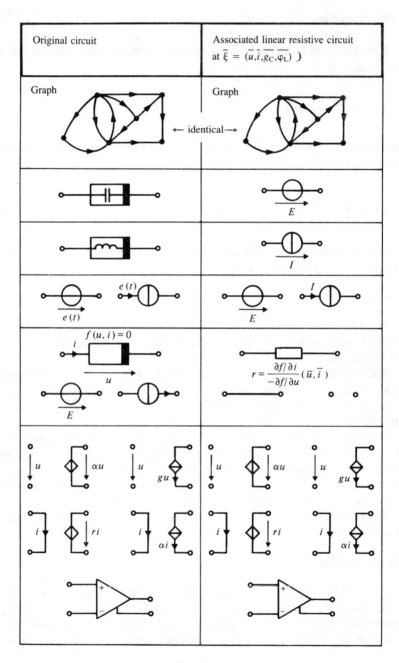

Fig. 2.62

2.5.22 Theorem: Existence of Local State Equations

Let $\bar{\xi}$ be a generalized operating point of a ***smooth*** circuit. Let us assume that

- the capacitors are locally charge controlled around $\bar{\xi}$;
- the inductors are locally flux controlled around $\bar{\xi}$; and
- the associated linear resistive circuit at $\bar{\xi}$ (2.62) is well posed.

Then q_C and φ_L are local state variables at $\bar{\xi}$. The local state equations are obtained, like the global state equations (theorem 2.4.15), from the functions ψ_C, ψ_L, h_C, q_L. The only difference is that these functions possibly exist only in a neighborhood of $\bar{\xi}$.

This theorem is a consequence of proposition 2.5.10 and corollary 2.5.15.

2.5.23 Theorem: Existence and Uniqueness of the Solutions

Let us assume that the hypotheses of theorem 2.5.22 are satisfied.

Then, there exists a neighborhood W of $\bar{\xi} = (\bar{u}, \bar{\imath}, q_C, \varphi_L)$ in Λ, and a neighborhood V of $(\bar{q}_C, \bar{\varphi}_L, \bar{s})$ in $\mathbb{R}^{NC+NL+NS}$ with the following properties. For any choice of initial charges q_0, initial fluxes φ_0, and initial time t_0 such that $(q_0, \varphi_0, e(t_0)) \in V$, there is a solution $\xi(t) = (u(t), i(t), q_C(t), \varphi_L(t))$ of the circuit, such that $q_C(t_0) = q_0$, and $\xi(t) \in W$ for t sufficiently close to t_0. This solution is smooth and it exists in a time interval around t_0 of nonzero length. It is unique, at least as long as it does not leave W.

Indeed, the existence of local state equations of the form (2.148) and (2.149) is guaranteed by theorem 2.5.22. The functions h_C and g_L involved in (2.148) and (2.149) are smooth because the circuit is smooth, and the functions ψ_C and ψ_L are smooth due to theorem 2.5.12. Thus, the local existence and uniqueness of the solution results from theorem 2.2.5.

2.5.24 Comment

If the hypotheses of theorem 2.5.22 are satisfied at any generalized operating point and if the solutions are bounded in the future, then they exist for any time $t \geq t_0$. We will not give a proof of this theorem because the circuits that have solutions bounded in the future and configuration spaces which are characterized by the local parameters q_C, φ_L, and s, without characterizing the global parameters by the same variables, are not common.

The crucial condition for the existence and local uniqueness of the solution of a circuit, according to theorem 2.5.23, is that the associated linear resistive circuit should be well posed. If it is ill posed at a point $\bar{\xi} \in \Lambda$, the local state equations at that point might not exist, and a solution with initial conditions close to this point might not exist or not be unique.

However, there is nothing stated about this in theorem 2.5.23. The following examples, in fact, show various possible behaviors for the solutions.

2.5.25 Example

Let us return to the circuit of figure 2.21. The associated linear resistive circuit is represented in figure 2.63. This circuit is evidently well posed, provided that the differential conductance g of the tunnel diode is nonzero.

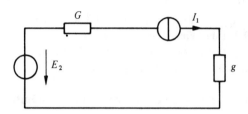

Fig. 2.63

The only critical points are the points Q_1 and Q_2 of the characteristic of the tunnel diode (fig. 2.58), where the tangent is horizontal. At Q_1 and Q_2, the tunnel diode becomes an open circuit, the series connection of which with the current source in figure 2.63 is contradictory. Thus, the circuit of figure 2.63 is ill posed at the generalized operating points having a projection in the plane (u,i) of Q_1 or Q_2. We have shown in section 2.1 that they actually are impasse points.

The criterion of theorem 2.5.23 also shows how to eliminate the impasse points. We only need add a voltage source in parallel with the conductance g, and the circuit will be well posed, even when $g = 0$, because the current of the current source will now be able to flow through this voltage source. In the original circuit, this amounts to adding a capacitor in parallel with the tunnel diode, and this was indeed our solution in the numerical example 2.1.16.

2.5.26 Example

Another type of example is given by the circuits of figures 2.36 and 2.38. Let $\bar{\xi}$ be the operating point where the capacitor voltage is zero. The differential resistance at $\bar{\xi}$ is also zero, and the current through the differential resistance in the associated linear resistive circuit at $\bar{\xi}$ is indeterminate. Therefore the associated linear resistive circuit at $\bar{\xi}$ is ill posed.

Nevertheless, the associated nonlinear resistive circuit has a single solution for any value E of the voltage source which replaces the capacitor. Thus, according to theorem 2.4.15, there is not only a local state equation at $\bar{\xi}$, but also a global state equation. Indeed, it is equation (2.53) for the circuit of figure 2.36 and equation (2.56) for the circuit of figure 2.38, which is valid up to a factor C.

As shown in subsection 2.2.27, there are two solutions which pass through $\bar{\xi}$. This fact is not in contradiction with theorem 2.4.16 because the function ψ is not differentiable at $\bar{\xi}$. Indeed, the right-hand members of (2.53) and (2.56) are both precisely the function ψ_C of (2.146). They are not differentiable at 0.

2.5.27   Example

A circuit which has a system of global state equations has, *a fortiori*, a system of local state equations at every generalized operating point. Is the converse true? The following example shows that such is not the case. Let us consider the resistive two-port defined by

$$i_1 - I e^{u_1/U} \cos(u_2/U) = 0 \qquad (2.183)$$
$$i_2 - I e^{u_2/U} \sin(u_2/U) = 0 \qquad (2.184)$$

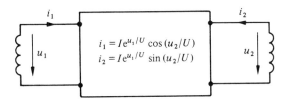

Fig. 2.64

which is closed on two positive linear inductors (fig. 2.64).

In order to obtain the global state equations, we must express u_1 and u_2 as functions of i_1 and i_2 from (2.183) and (2.184). However, u_1 and u_2 cannot be **global** functions of i_1 and i_2 because, when the value of u_2 is increased by $2\pi U$, we obtain the same i_1 and i_2 in (2.183) and (2.184). Consequently, there is no system of global state equations.

On the contrary, it is possible to solve (2.183) and (2.184) **locally** for u_1 and u_2. This fact is guaranteed by theorem 2.5.12. Indeed, the Jacobian matrix for the variables u_1 and u_2 is

$$J = (I/U) e^{u_1/U} \begin{pmatrix} \cos(u_2/U) & -\sin(u_2/U) \\ \sin(u_2/U) & \cos(u_2/U) \end{pmatrix} \qquad (2.185)$$

where the determinant:

$$\det J = (I/U)^2 e^{2u_1/U} \qquad (2.186)$$

is everywhere positive.

Thus, according to proposition 2.5.10, there are local state equations at any point, and because theorem 2.5.12 also allows us to conclude that u_1 and u_2 are smooth functions of i_1 and i_2, the solution of the circuit exists and it is unique for any initial condition $\varphi_1(t_0)$ and $\varphi_2(t_0)$.

2.5.28 Example

The circuit of figure 2.65 is a low-pass filter of degree 3 (vol. XIX). The capacitors C_1, C_2, and C_3 form a loop. Thus, the associated linear resistive circuit includes a loop of voltage sources. Therefore, it is ill posed everywhere. Does this mean that there is no solution, or no solution determined by its initial charges and fluxes? Since the circuit being considered is linear, this cannot be the case. Let us try to find the state equations.

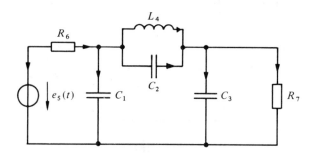

Fig. 2.65

The generalized configuration space Λ should be a plane of dimension $N_C + N_L + N_S = 5$ in the space $\mathbb{R}^{18}$ of the variables q_1, q_2, q_3, φ_4 and u_k, i_k, $k = 1, \ldots , 7$. We should characterize the parameters of Λ by q_1, q_2, q_3, φ_4, and u_5 to obtain the state equations. However, the three capacitors form a loop, resulting in the Kirchhoff equation:

$$u_1 - u_2 - u_3 = 0 \tag{2.187}$$

which implies

$$q_1/C_1 - q_2/C_2 - q_3/C_3 = 0 \tag{2.188}$$

For this reason, the three charges are not independent, and only two of them can be used as parameters for Λ.

Let us try to characterize the parameters of Λ by q_1, q_2, φ_4, and u_5. We obtain

$$u_1 = q_1/C_1 \tag{2.189}$$

$$u_2 = q_2/C_2 \tag{2.190}$$

$$u_4 = u_2 \tag{2.191}$$

$$u_3 = q_1/C_1 - q_2/C_2 = u_7 \qquad (2.192)$$

$$q_3 = \frac{C_3}{C_1} q_1 - \frac{C_3}{C_2} q_2 \qquad (2.193)$$

$$u_6 = u_5 - q_1/C_1 \qquad (2.194)$$

$$i_4 = \varphi_4/L_4 \qquad (2.195)$$

$$i_5 = -i_6 = -\frac{u_5}{R_6} + \frac{q_1}{R_6 C_1} \qquad (2.196)$$

$$i_7 = \frac{q_1}{R_7 C_1} - \frac{q_2}{R_7 C_2} \qquad (2.197)$$

Nonetheless, i_1, i_2, and i_3 cannot be expressed by q_1, q_2, φ_4, and u_5.

Let us add i_3 as a fifth parameter. Then, the characterization of the parameters of Λ is completed by

$$i_1 = -i_3 - i_7 + i_6 \qquad (2.198)$$

$$i_1 = -i_3 - \frac{q_1}{C_1}\left(\frac{1}{R_6} + \frac{1}{R_7}\right) + \frac{q_2}{R_7 C_2} + \frac{u_5}{R_6} \qquad (2.199)$$

$$i_2 = i_3 + \frac{q_1}{R_7 C_1} - \frac{q_2}{R_7 C_2} - \frac{\varphi_4}{L_4} \qquad (2.200)$$

The differential equations:

$$i_k = dq_k/dt \,, \quad k = 1, 2, 3 \qquad (2.201)$$

combined with (2.193), give

$$i_3 = \frac{C_3}{C_1} i_1 - \frac{C_3}{C_2} i_2 \qquad (2.202)$$

By replacing (2.202) in (2.199) and (2.200), and by solving for i_1 and i_2, we find i_1, i_2, and u_4 as functions of q_1, q_2, φ_4, and u_5, which leads to the state equations:

$$\frac{dq_1}{dt} = \frac{1}{C^2}\left[-\left(\frac{C_2+C_3}{R_6} + \frac{C_2}{R_7}\right) q_1 + \frac{C_1}{R_7} q_2 - \frac{C_1 C_3}{L_4} \varphi_4 + \frac{C_1(C_2+C_3)}{R_6} e_5(t)\right]$$

(2.203)

$$\frac{dq_2}{dt} = \frac{1}{C^2}\left[\left(\frac{C_2}{R_7} - \frac{C_2 C_3}{R_6 C_1}\right) q_1 - \frac{C_1}{R_7} q_2 - \frac{(C_1+C_3)C_2}{L_4} \varphi_4 + \frac{C_2 C_3}{R_6} e_5(t)\right]$$

(2.204)

$$d\varphi_4/dt = q_2/C_2$$ (2.205)

with

$$C^2 = C_1 C_2 + C_1 C_3 + C_2 C_3$$ (2.206)

We have just shown that all the solutions of the circuit of figure 2.65 satisfy the system of differential equations in normal form, (2.203) to (2.205). Conversely, starting from a solution of (2.203) to (2.205), we obtain a solution of the circuit by setting $i_1 = dq_1/dt$, $i_2 = dq_2/dt$, $u_4 = d\varphi_4/dt$, by determining i_3 through (2.202), and by expressing all the other variables through q_1, q_2, i_3, φ_4, and u_5.

In conclusion, we have just shown that the system (2.203) to (2.205) is equivalent to the complete system of circuit equations. Therefore, it deserves to be called a system of global state equations, although it does not comply with definition 2.4.3, since it fails to include the variable q_3.

2.5.29 Comments

If a circuit includes a loop of capacitors and voltage sources or a cut set of inductors and current sources, the same phenomenon occurs as in example 2.5.28. We obtain a system of equations in normal form for a subset of the variables q_C and φ_L. Therefore, definitions 2.4.3 and 2.5.3 should be generalized accordingly.

Instead of doing so, we shall introduce a transformation in section 4.3, which enables us to eliminate the loops of capacitors and voltage sources as well as the cut sets of inductors and current sources. If we apply this transformation to the circuit of figure 2.65, for instance, equations (2.203) to (2.205) become the global state equations of the resulting circuit, according to definition 2.4.3.

What is the geometric interpretation of the phenomenon identified by example 2.5.28? Although the generalized configuration space of the circuit in figure 2.65 is a plane of dimension 5, the solutions remain within a plane $\tilde{\Lambda} \subset \Lambda$ of dimension 4, defined by relation (2.202). If we choose the initial conditions outside $\tilde{\Lambda}$, obviously the corresponding solution of the circuit does not exist. Conversely, if we choose the initial conditions in $\tilde{\Lambda}$, the solution exists and it is unique.

2.5.30 Summary

In this chapter, the existence and uniqueness of solutions of the circuit based on initial conditions have been discussed.These properties should not be confused with the existence and uniqueness of an asymptotic steady-state solution for $t = \infty$, which will be treated in chapter 5.

A set of values for the voltages, currents, charges, and fluxes qualifies as an initial condition for a solution if and only if it satisfies the nondifferential equations of the circuit, i.e., if it is an operating point.The set of operating points is called the configuration space. A solution is a movement in the configuration space.

If we do not consider the constitutive relations of the time-dependent sources, we obtain the generalized configuration space Λ, which is larger than the configuration space, except when the circuit is autonomous. A solution is also a movement in Λ. The advantage of the generalized configuration space is that it does not depend on time, as opposed to the configuration space.

If q_C, φ_L, and s, being the charges of the capacitors, the fluxes of the inductors, and the voltages (or currents) of the time-dependent voltage (or current) sources, are global parameters of Λ, then the complete system of circuit equations is equivalent to a system of differential equations in normal form for q_C, φ_L, and s, which is called the system of global state equations. It follows that, for arbitrary initial conditions $q_C(t_0)$ and $\varphi_L(t_0)$, there exists exactly one solution of the circuit. We have no information regarding the time interval of its existence. However, if the circuit is piecewise linear, the solutions exist up to $t \mapsto \pm\infty$, and if the solutions of a circuit remain bounded in the future, they exist up to $t \mapsto +\infty$.

The space Λ has global parameters q_C, φ_L, and s if the capacitors are charge controlled, the inductors are flux controlled, and the associated resistive circuit has exactly one solution. We obtain the global state equations by resolving the associated resistive circuit.

If Λ has local parameters characterized by q_C, φ_L, and s in a neighborhood of a generalized operating point $\bar{\xi}$, the complete system of circuit equations is locally equivalent to a system of differential equations in normal form, i.e., the system of local state equations at $\bar{\xi}$. It follows that there exists exactly one solution of the circuit for each initial condition close to $\bar{\xi}$. In particular, $\bar{\xi}$ cannot be an impasse point.

The space Λ has local parameters q_C, φ_L, and s around $\bar{\xi}$ if the capacitors are locally charge controlled at $\bar{\xi}$, the inductors are locally flux controlled at $\bar{\xi}$, and the associated linear resistive circuit at $\bar{\xi}$ is well posed. Consequently, if we want to find the singular points of the solutions of a circuit, we must search among the generalized operating points in which a capacitor is not locally charge controlled, an inductor is not locally flux controlled, or the associated linear resistive circuit is ill posed.

Chapter 3

Resistive Circuits

3.1 TOOLS OF GRAPH THEORY

3.1.1 Introduction

Although circuits without inductors and capacitors constitute a very special class, their study is fundamental to the theory of nonlinear circuits. We have already seen, in chapter 2, that the criteria of existence and uniqueness of a solution starting from initial conditions, refer to a resistive circuit. Similarly, we shall see in chapter 5 that the stability theorems involve quantities which again relate to a resistive circuit associated with the original circuit.

Before coming to the heart of the matter, we must introduce two theorems which belong to graph theory. The first one is *Tellegen's theorem* (vol. IV, subsec. 1.2.15), which plays a central role in linear as well as nonlinear circuit theory. The second one is the *colored branch theorem* [11], the usefulness of which for circuits has been realized only recently [12].

3.1.2 Tellegen's Theorem

Let us consider an oriented graph. We assume that two real numbers—the current and the voltage—are associated with each branch in such a way as to satisfy the Kirchhoff lemmas stated in subsection 1.2.10. If $u_k(i_k)$ is the voltage (current) of the kth branch, then

$$\sum_{k=1}^{b} u_k \, i_k = 0 \tag{3.1}$$

The proof of Tellegen's theorem is given in volume IV (subsec. 1.3.15).

136

3.1.3 Note

Tellegen's theorem does not refer in any way to the nature of the circuit elements. The only thing that is required is a set of currents and voltages which satisfy Kirchhoff's laws. These currents and voltages can be values taken by a solution of a circuit at a given instant, or they can be currents of one solution and voltages of another, currents of a circuit and voltages of another circuit having the same graph, *et cetera*. In short, the strength of this theorem is its generality, and subsequently we will have the opportunity to take advantage of it.

3.1.4 Definition

Let $\xi(t) = (u(t), i(t), q_C(t), \varphi_L(t))$ be a solution of a circuit. The *instantaneous power absorbed by the kth branch* of this circuit at *the instant t* is defined by

$$p_k(t) = u_k(t) i_k(t) \tag{3.2}$$

3.1.5 Corollary: Power Balance

Let $\xi(t) = (u(t), i(t), q_C(t), \varphi_L(t))$ be the solution of a circuit. Then,

$$\sum_{k=1}^{b} u_k(t) i_k(t) = 0 \tag{3.3}$$

This means that the total power absorbed by all the branches is zero at any instant.

3.1.6 Definition

Let $\xi(t)$ and $\tilde{\xi}(t)$ be two solutions of a circuit. The *incremental power absorbed by the kth branch at the instant t* is defined by

$$p_{\Delta k}(t) = \Delta u_k(t) \Delta i_k(t) \tag{3.4}$$
$$= [u_k(t) - \tilde{u}_k(t)][i_k(t) - \tilde{i}_k(t)] \tag{3.5}$$

3.1.7 Comment

The difference between the two solutions:

$$\Delta\xi(t) = (\Delta u(t), \Delta i(t), \Delta q_C(t), \Delta\varphi_L(t)) \tag{3.6}$$

is a solution of the circuit if and only if the circuit is linear. However, $\Delta u(t)$ and $\Delta i(t)$ always satisfy the Kirchhoff equations because they are linear equations, even when the circuit is nonlinear. This fact allows for a second application of Tellegen's theorem.

3.1.8   Corollary: Incremental Power Balance

Let $\xi(t) = (u(t), i(t), q_C(t), \varphi_L(t))$ and $\tilde{\xi}(t) = (\tilde{u}(t), \tilde{i}(t), \tilde{q}_C(t), \tilde{\varphi}_L(t))$ be two solutions of a circuit. Then,

$$\sum_{k=1}^{b} \Delta u_k(t)\, \Delta i_k(t) = 0 \tag{3.7}$$

where

$$\Delta u_k = u_k - \tilde{u}_k \tag{3.8}$$
$$\Delta i_k = i_k - \tilde{i}_k \tag{3.9}$$

This means that the total incremental power absorbed by the set of all branches is zero.

3.1.9   Definitions

Let us assume that two oriented branches and one loop containing these branches are given in a graph. Additionally, an arbitrary orientation is given to the loop. The two branches have the *same orientation within the loop* if both of them are in the same or opposite orientation as the loop. This concept depends on the loop, but not on the orientation given to it. The occurrence of two branches having the same orientation within a cut set is defined analogously. A graph is *partially oriented* if one part of its branches is oriented and the other part is not. The extreme cases of the nonoriented graph and the oriented graph are special cases of partially oriented graphs.

A *loop (cut set)* in a partially oriented graph is *uniform* if all the oriented branches of the loop (cut set) have the same orientation within the loop (cut set).

3.1.10   Example

The graph of figure 3.1 is partially oriented. The branches a and c have the same orientation within the loop $aecd$. Because the other branches are not oriented, the loop is uniform. Conversely, a and c are of opposite orientation with respect to the loop

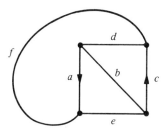

Fig. 3.1

afcb, and that loop is not uniform. In the cut set *abcf*, *a* and *c* are of opposite orientation, whereas in the cut set *aecd*, *a* has the same orientation as *c*. The branches *b*, *d*, *e*, and *f* simultaneously form a loop and a cut set. Since none of these branches is oriented, they constitute a uniform loop and a uniform cut set.

3.1.11   Colored-Branch Theorem

Consider a partially oriented graph. Let us assume that the oriented branches of the graph are green colored and each nonoriented branch is either red or blue colored. Then, each green branch is part of either

- a uniform loop composed of green and red branches only; or
- a uniform cut set composed of green and blue branches only.

3.1.12   Comments

Of course, the green, red, and blue colors do not play any special role in graph theory. In fact, we could have simply spoken of a decomposition of the set of branches into three disjoined subsets. Colors are used for the sake of mnemonics only.

Nothing has been assumed regarding the way in which the branches are colored or oriented. Thus, there are many different ways to apply this theorem to the circuits. This is similar to Tellegen's theorem, where nothing has been assumed regarding the elements located on the branches.

3.1.13   Proof of the Colored-Branch Theorem

The proof of the colored-branch theorem consists of an algorithm designed to find a uniform loop of green and red branches containing the green branch v [12]. When the algorithm does not reach its goal, we obtain a uniform cut set of green and blue branches. An example will be given in subsection 2.1.14. It is recommended that it be studied simultaneously.

At each step j of the algorithm, a set B_j of branches and a set N_j of nodes are constructed. The set B_1 includes only the v branch, and the single node contained in N_1 is the one toward which v is directed.

The transition from B_j to B_{j+1}, and from N_j to N_{j+1} is carried out as follows. To B_j, we add a green branch or a red branch, which is connected between a node of N_j and a node that does not belong to N_j. The latter is added to N_j in order to form N_{j+1}. In the case of a green branch, it is required to be oriented from the node of N_j toward the node not belonging to N_j. Note that B_j includes j branches, and N_j includes j nodes.

It may be impossible to find any branch to add to B_j. In such a case, we say that the algorithm stops by failure. On the other hand, as soon as the other node of v is part of N_j, we stop the algorithm, and we say that it ended with success.

First, note that the algorithm must necessarily end because N_j includes j nodes and the graph has only a finite number of nodes. If we do not stop upon a failure, the second node of v must eventually be part of N_j. Second, note that success and failure evidently exclude one another. Thus, the theorem is proved if we can show that v is part of a uniform loop of green and red branches in the case of success, and part of a uniform cut set of green and blue branches in the case of failure.

Let us assume that, after J steps, the algorithm has led to a failure. Then, all the branches connected between a node of N_J and a node not belonging to N_J are either blue or green, which, in the latter case, are oriented toward N_J. The set of these branches constitutes a cut set, which separates the nodes of N_J from the nodes not belonging to N_J. Branch v is also part of it and oriented toward N_J. Therefore, it is a uniform cut set of green and blue branches including N_J.

Let us assume that, after J steps, the algorithm has led to a success. Then, the last branch is connected to the second node of v and, in the case of a green branch, it is oriented toward that node. Let p be the other node of that branch. Then, at a certain step of the algorithm, a branch w has been added, which is connected to p and, in the case of a green branch, it is oriented toward p. Likewise, there will be a branch connected to the other node of w, *et cetera*. In this way, by following the algorithm in reverse, we obtain the uniform loop of the green and red branches which includes v.

Note as well that in the case where the algorithm stops by failure, the set of branches which separates the nodes of N_J from the nodes not belonging to N_J may be a union of more than one cut set. Branch v is part of one of these cut sets and the theorem is also verified in this case.

3.1.14 Example

The algorithm of subsection 3.1.12 is illustrated by means of the graph in figure 3.2, where the green branches are drawn in solid lines, the red branches in dashed lines, and the blue branches in dotted lines. Figure 3.3 shows the various steps of the algorithm. It stops by failure at the seventh step. The corresponding cut set is represented in figure 3.4. The two nodes not belonging to B_7 are marked by an x. If the branch a were oriented in the opposite direction, the algorithm would have ended up with success at the eighth step (fig. 3.5). The corresponding loop is evident.

Note that the sets B_j and N_j are not unique. Indeed, in figure 3.3, we could have added one of the green branches instead of the red branch in order to pass from B_2, N_2 to B_3, N_3. It follows that the loop or the cut set, the existence of which is guaranteed by the theorem, is also not unique. Nonetheless, it follows from proof 3.1.12 that the outcome of the algorithm does not depend on the particular choice of the B_j, N_j.

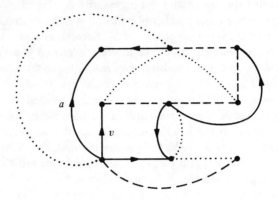

Fig. 3.2

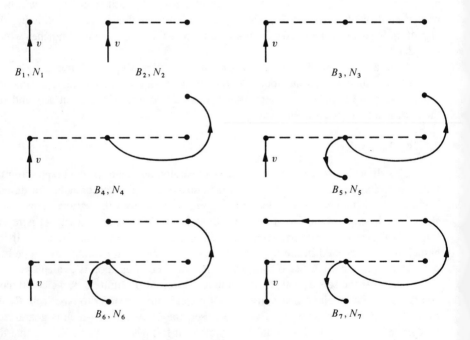

Fig. 3.3

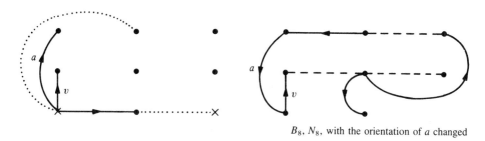

Fig. 3.4

B_8, N_8, with the orientation of a changed

Fig. 3.5

3.1.15 Corollary

Let us consider a branch v of a nonoriented graph. The other branches are arbitrarily grouped into two disjointed sets R and B. Then, either v forms a loop with branches of R, or v forms a cut set with branches of B, but not both.

Indeed, this results directly from the colored-branch theorem if v is colored green, the branches of R are red, and the branches of B are blue.

3.1.16 Comment

In the remainder of this section, we shall show some elementary applications of the colored-branch theorem within the framework of graph theory. It would be possible to obtain the same results with other methods. In fact, the importance of the colored-branch theory will only become evident when it is combined with Tellegen's theorem.

3.1.17 Reminder

A *tree* is a subgraph that includes all the nodes, but no loop, and which is composed of as many connected parts as the complete graph (vol. IV, subsec. 4.1.1). Therefore, it includes a maximal set of branches without loops. A *cotree* is the subgraph with the complementary set of branches of a tree.

3.1.18 Property

The two following conditions are necessary and sufficient for a set of branches of a graph to be a tree: the set of branches includes no loop; the complementary set includes no cut set.

In order to prove that the condition is necessary, we first note that, by definition, there is no loop in a tree. Let v be a branch of the cotree. Then, v must form a loop with branches of the tree only. If this were not the case, we could join v to the tree, and we would still get a set of branches without loops, which is in contradiction with definition 3.1.17.

Consequently, corollary 3.1.15 allows us to conclude that there exist no cut sets of branches of the cotree passing through v. It is sufficient to group the branches of the tree in R and the branches of the cotree—except for v—in B. Since v is an arbitrary branch of the cotree, there is no cut set in the cotree.

By the same reasoning, the condition is sufficient as well. Indeed, a set of branches without a loop is a tree if increasing it generates a loop. If we take a branch v of the complementary set, according to corollary 3.1.15, there is a loop passing through v, and through no other branch of the complement, because there is no cut set of branches of the complement. Therefore, no branch can be added to the original set without producing a loop, thus proving that it is a tree.

3.1.19 Property

Let us assume that a graph G is transformed into a graph $\tilde{G}$ by short-circuiting (removing) a branch g. Let S be a set of at least two branches of G, and let $g \in S$. Then, S is a tree (cotree) of G if and only if $S-\{g\}$ is a tree (cotree) of $\tilde{G}$.

Indeed, when g is short-circuited, the loops of G are transformed into loops of $\tilde{G}$, and all the loops of $\tilde{G}$ are obtained in this way. A cut set of G is either eliminated or left invariant in the reduction from G to $\tilde{G}$, depending on whether g is part of the cut set. If S is a tree of G, then there is no loop in S and no cut set in $G-S$, according to property 3.1.18. Then $S-\{g\}$ does not include any loop, and $G-(S-\{g\})$ does not include any cut set. Therefore, $S-\{g\}$ is a tree of $\tilde{G}$.

By the same argument, but in reverse order, we prove that S is a tree of G if $\tilde{S}-\{g\}$ is a tree of $\tilde{G}$.

If we remove a branch, the proof is analogous.

3.1.20 Property

Let S_1 and S_2 be two disjointed sets of branches of a graph such that S_1 does not include any loop, and S_2 does not include any cut set. Then, there exists a tree which includes all the branches of S_1, and no branch of S_2.

Indeed, the following algorithm enables us to find this tree. To S_1, we add a branch t which does not form any loop with S_1 and does not belong to S_2. Let $S'_1 = S_1 \cup \{t\}$. Then, we add to S'_1 a branch which does not form any loop with S'_1, et cetera. Because there is a finite number of branches in the complement of $S_1 \cup S_2$, the algorithm must stop. Let S be the set of branches obtained upon stopping. Then, $S_1 \subset S \subset \tilde{S}_2$ where $\tilde{S}_2$ is the complement of S_2. If $S = \tilde{S}_2$, then S is a tree because it does not include any loop and its complement does not include any cut set. If $S \subset \tilde{S}_2$, then S is a tree because it is a maximal set of branches without any loop.

3.2 PASSIVITY AND LOCAL PASSIVITY

3.2.1 Introduction

According to (3.3), the sum of the powers absorbed by all the branches is zero. Consequently, at any given instant, some power is absorbed by certain branches and supplied by others. Their distribution within the circuit generally changes as a function of time and depends on the solution. However, certain elements can only absorb power; they are the *passive resistors*. Others, the *eventually passive resistors*, always absorb power as soon as the current or the voltage exceeds a certain value. If a circuit is composed of eventually passive resistors, we will be able to show that there is at least one solution.

It might even be possible that the incremental power absorbed by an element will always be positive, irrespective of the two solutions to which it is related. This is roughly the case of the *locally passive resistors*. This property, combined with (3.7), will enable us to prove the uniqueness of the solution of the resistive circuit.

3.2.2 Definitions

A resistor is *passive* if, for any pair (u,i) of current and voltage which satisfies its constitutive relation, we have

$$u \cdot i \geq 0 \tag{3.11}$$

If such is not the case, the resistor is *active*. If the inequality (3.11) is strict, except for the case $u=i=0$, then the resistor is *strictly passive*. If there exists a constant $K>0$, such that

$$u \cdot i \geq 0 \quad \text{if} \quad ||(u, i)|| > K \tag{3.12}$$

then the resistor is *eventually passive*. If the first inequality of (3.12) is strict, we have a *strictly eventually passive* resistor.

3.2.3 Comment

The norm involved in (3.12) is defined according to subsection 2.3.16. As for other concepts, the asymptotic passivity does not depend on the particular form of the norm.

3.2.4 Property

The instantaneous power that can be produced by an eventually passive resistor is bounded.

Indeed, if a resistor is eventually passive, we have, according to (3.12),

$$u^2 / R_0 + R_0 i^2 \leqslant K^2 \tag{3.13}$$

as soon as $u \cdot i < 0$. Here, R_0 is the normalization resistance of the norm. Consequently, the power, $-u \cdot i$, that can be produced by the resistor is bounded by

$$-u \cdot i = -\frac{1}{2} \left(\frac{u}{\sqrt{R_0}} + \sqrt{R_0}\, i \right)^2 + \frac{1}{2} \left(\frac{u^2}{R_0} + R_0 i^2 \right) \leqslant K^2 \tag{3.14}$$

3.2.5 Restrictions on the Characteristics

The allowable ranges for the characteristics of the different types of resistors are cross-hatched in figures 3.6 to 3.9. We observe, for instance, that a continuous characteristic of a passive resistor must pass through the point $u = i = 0$ if it does not remain fully within a single quadrant. Figure 3.10 shows the relations among the different concepts.

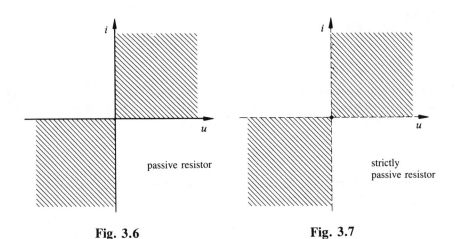

Fig. 3.6   Fig. 3.7

3.2.6 Examples

Let us consider the examples of resistors from section 1.3. The positive-valued linear resistor, the exponential diode, the tunnel diode, the thyristor with disconnected gate, and the piecewise linear diode are all strictly passive resistors. Conversely, the ideal diode is passive without being strictly passive. The voltage and current sources do not agree with any of the four concepts of passivity.

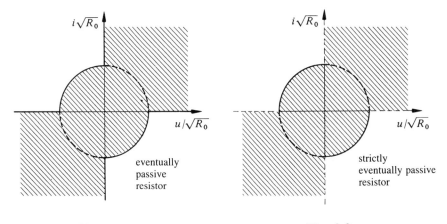

Fig. 3.8　　　　　　　　　　Fig. 3.9

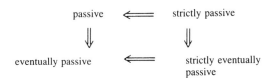

Fig. 3.10

The exponential *diode* that is *polarized* by a voltage source, according to figure 3.11, again constitutes a resistor, the characteristic of which is shown in figure 3.12. It is active, but it remains strictly eventually passive. It is interesting to note that the voltage source, which is not passive at all, becomes eventually passive in series with the diode. This is because the currents opposed to the source voltage are limited by I_s, the inverse current of the diode. Consequently, the active power that can be produced by the source is limited to EI_s.

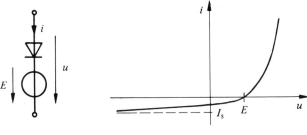

Fig. 3.11　　　　　　　　Fig. 3.12

146

If the diode is polarized in the opposite direction (fig. 3.13), the result is still strictly eventually passive (fig. 3.14). In this case, the source produces power when the current flows in the direction of the diode. This power is not bounded, but when the current is sufficiently large, it is exceeded by the power dissipated in the diode. For the same reason, the resistor composed of an exponential diode in parallel with a current source, in the opposite direction of the diode (fig. 3.15), is strictly eventually passive (fig. 3.16).

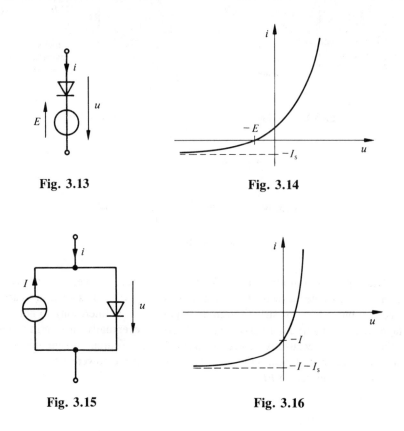

Fig. 3.13

Fig. 3.14

Fig. 3.15

Fig. 3.16

On the contrary, if the current of the source is oriented the same as the diode (fig. 3.17), and if it is larger than the inverse current of the diode, then the composite resistor is no longer eventually passive (fig. 3.18). Indeed, in the presence of a voltage u of opposite orientation with respect to the diode, the source produces the power $u \cdot I$, while the power dissipated by the diode is limited to $u \cdot I_s$.

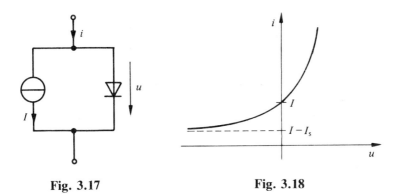

Fig. 3.17   Fig. 3.18

3.2.7 Relative Passivity

Looking at figures 3.6 to 3.9, we note that the origin of the plane (u,i) plays a special role in the concept of passivity. A generalization of this concept is obtained by shifting the cross-hatched areas of figures 3.6 to 3.9. The properties of circuits with passive resistors are generalized as well to circuits composed of resistors that are passive with respect to the origin of the shifted plane (u,i).

3.2.8 Definition

A resistor is *passive with respect to a point* (u_0,i_0) *of its characteristic* if, for any other point (u,i) of its characteristic, we have

$$\Delta u \cdot \Delta i = (u - u_0)(i - i_0) \geqslant 0 \tag{3.15}$$

It is *strictly passive with respect to* (u_0,i_0) if (3.15) is a strict inequality, unless $u = u_0$ and $i = i_0$.

3.2.9 Comment

We could also introduce the concept of relative asymptotic passivity. In most cases, however, it would coincide with the concept of asymptotic passivity of subsection 3.2.2.

The allowable ranges for the characteristics of passive and strictly passive resistors with respect to (u_0,i_0) are cross-hatched in figures 3.19 and 3.20, respectively.

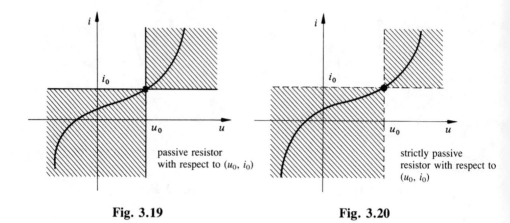

Fig. 3.19 Fig. 3.20

3.2.10 Examples

The positive-valued linear resistor, as well as the exponential diode and the piecewise linear diode, are all strictly passive with respect to any point of their characteristic.

It should be noted that the constant independent sources are passive relative to any point of their characteristic, but they are not passive at all according to definition 3.2.2.

In figure 3.21, we have indicated with a bold line the points of the characteristic of a tunnel diode at which the diode is passive.

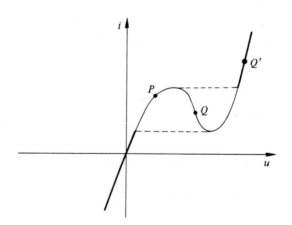

Fig. 3.21

3.2.11 Local Passivity

Both passivity and relative passivity are global concepts by virtue of the fact that they involve the entire characteristic of the resistor. For a resistor not to be passive with respect to a point P of its characteristic, it suffices to have another point Q such that (3.15) is not satisfied.

In the case of eventually passive resistors, the points sufficiently distant from the reference point satisfy (3.15). At the other extreme, the points sufficiently close to the reference point can be required to satisfy (3.15). This is the concept of local passivity.

3.2.12 Definition

A *resistor* is *locally passive at a point* (u_0, i_0) *of its characteristic* if, for any other point of the characteristic in a neighborhood of (u_0, i_0) inequality (3.15) is satisfied. More precisely, there exists a $\epsilon > 0$, such that if

$$||(u - u_0, i - i_0)|| < \epsilon \qquad (3.16)$$

and if (u, i) belongs to the characteristic, we have (3.15). If such is not the case, the resistor is *locally active at* (u_0, i_0). If a resistor is locally passive at every point of its characteristic, we simply call it *locally passive*.

If the inequality (3.15) is strict, except for the point (u_0, i_0), we have a *strictly locally passive resistor* at (u_0, i_0). If such is the case at every point, it is a *strictly locally passive resistor*.

3.2.13 Property

Let us assume that a smooth resistor is voltage controlled in a neighborhood of the point (u_0, i_0) of its characteristic, i.e., that its constitutive relation can be written in the form $i = g(u)$. If it is locally passive, then $(dg/du)(u_0) \geq 0$. Conversely, if $(dg/du)(u_0) > 0$, then the resistor is strictly locally passive at (u_0, i_0).

If the resistor is current controlled in a neighborhood of (u_0, i_0) with the constitutive relation $u = h(i)$, and if it is locally passive, then $(dh/di)(u_0) \geq 0$. Conversely, if $(dh/di)(i_0) > 0$, then the resistor is strictly locally passive at (u_0, i_0).

Indeed, if $i = g(u)$, the derivative $(dg/du)(u_0)$ is the limit of $[g(u) - g(u_0)]/(u - u_0) = (i - i_0)/(u - u_0) = (i - i_0)(u - u_0)/(u - u_0)^2$. Since this quantity is non-negative when the resistor is locally passive, so is the limit. Conversely, if $(dg/du)(u_0) > 0$, then the derivative is positive within a whole neighborhood of u_0 by reason of continuity, and, therefore, $g(u) - g(u_0)$ is positive for $u > u_0$ and negative for $u < u_0$ in this neighborhood, which implies strict local passivity. The proof for the locally current-controlled resistors is identical.

3.2.14 Examples

The positive linear resistor, the exponential diode, and the piecewise linear diode are all strictly locally passive. The constant independent sources and the ideal diode are locally passive only.

The tunnel diode is locally active as well in certain points. In figure 3.22, we have indicated with a bold line the set of points at which it is locally passive. It is interesting to compare figures 3.22 and 3.21.

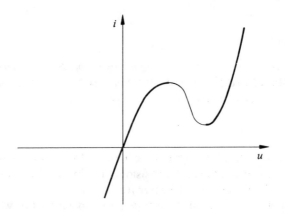

Fig. 3.22

3.2.15 Comments

At first glance, it appears that the (strict) passivity at one point is equivalent to a non-negative (positive) derivative at that point. However, a resistor with the constitutive relation $i = g(u) = u^3$ with $\alpha<0(\alpha>0)$ constitutes a counter example. Indeed, if $\alpha<0$, the resistor is locally active at $i=u=0$ although $(dg/du)(0)=0$. If $\alpha>0$, the resistor is strictly locally passive at $u=i=0$, although $(dg/du)(0)=0$. Therefore, property 3.2.13 is all that can be said about it. Only the condition that $dg/du \geqslant 0$ at any point of its characteristic is equivalent to the local passivity at any point.

The range excluded for the locally passive characteristic with respect to (u_0,i_0) is cross-hatched in figure 3.23.

It is obvious that a resistor which is (strictly) passive with respect to any point of its characteristic is also (strictly) locally passive. However, the converse is not true. Indeed, figure 3.24 represents the characteristic of a resistor which is not passive with respect to any point of its characteristic, while being strictly locally passive.

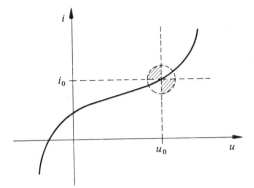

Fig. 3.23

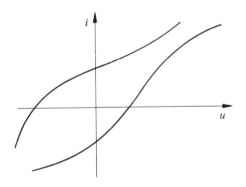

Fig. 3.24

If a resistor is voltage controlled or current controlled, then the concepts of local passivity and passivity with respect to any point become equivalent. If, in addition, the local passivity is strict, the resistor is controlled by the other variable as well. Thus, it has a constitutive relation that is both of form (1.5) and form (1.6), with strictly increasing bijective functions g and h, and non-negative derivatives. The differential resistance and conductance are non-negative wherever they exist.

3.3 EXISTENCE AND UNIQUENESS OF THE SOLUTION OF A LINEAR RESISTIVE CIRCUIT

3.3.1 Introduction

In chapter 2, we discussed the existence and uniqueness of the solution of a nonlinear circuit. We have seen that the associated linear resistive circuit plays a central role in this problem.

In electronics, the dc operating point is a fundamental concept. As shown in subsection 2.3.9, a dc operating point corresponds to a constant solution, i.e., an equilibrium solution of the circuit. It is important to know whether a circuit has either none, exactly one, or several dc operating points. Because the dc operating point is a solution of the nondifferential circuit equations, with $i_C = u_L = u_S = 0$, it is the solution of the nonlinear resistive circuit obtained by removing the capacitors, short-circuiting the inductors, and cancelling the time-dependent sources. The problem of the existence and uniqueness of the solution of a nonlinear resistive circuit will be treated in section 3.4. We shall see that it is closely related to the existence and uniqueness of the linear resistive circuit obtained by replacing the nonlinear resistors with linear resistors.

This is sufficient motivation to devote considerable effort to the study of linear resistive circuits.

The equations of a resistive circuit, be it linear or not, involve voltages and currents at the same instant only. Therefore, we can study the circuit separately at each instant, even when time-dependent sources are present. We need only consider the same circuit, wherein the time-dependent sources are replaced by autonomous sources, and study only those properties which do not depend on the particular values of the sources.

In short, in this section we shall study circuits composed of linear resistors, constant independent sources, controlled sources, and ideal operational amplifiers. We shall investigate whether the circuit is well posed, i.e., whether it has exactly one solution.

3.3.2 System of Equations of a Linear Resistive Circuit

The standard system of equations (2.81) to 2.90) of a linear resistive circuit is given in the form:

$$\eta = H\xi \tag{3.17}$$

where $\xi = (u,i)^T, \eta$ includes the values of the independent sources and the zeros, and H is a matrix of dimension $2b \times 2b$.

A theorem of linear algebra states that the solution of (3.17) exists, and that it is unique if and only if H is not singular, i.e., if and only if $\det H \neq 0$. This allows us to state immediately the following property.

3.3.3 Property

Whether a linear resistive circuit is well posed does not depend on the values of the independent sources. In particular, a linear resistive circuit is well posed if and only if the circuit obtained by short-circuiting the voltage sources and removing the current sources has only the zero solution.

3.3.4 Comments

There is a pitfall in property 3.3.3. Let us consider the circuit of figure 3.25. By short-circuiting the voltage sources, we obtain the circuit of figure 3.26, which manifestly has only the zero solution if $R_2 \neq 0, R_3 \neq 0$. We conclude that the circuit of figure 3.25 is well posed. However, when $E_1 = E_2$, it obviously has no solution.

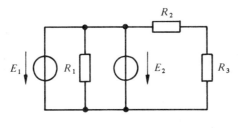

Fig. 3.25

This apparent contradiction arises from the fact that by short-circuiting both sources, we obtain, in addition to the two resistors connected in parallel, a short circuit closed on itself (fig. 3.27). An arbitrary current i in the short circuit is compatible with the circuit equations, a fact which shows that the circuit of figure 3.27, as well as the circuit of figure 3.25, are ill posed.

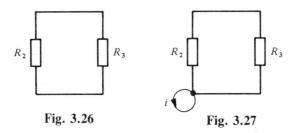

Fig. 3.26   **Fig. 3.27**

154

The same phenomenon occurs when a cut set of current sources is present (fig. 3.28). When these sources are removed, the resulting circuit is no longer connected (fig. 3.29). Nevertheless, when taken as a circuit composed of the resistors R_1 to R_4, it has only the zero solution. More correctly, though, there is still an open circuit, which is shown by a dotted line in figure 3.29. Without violating the circuit equations, the open circuit can carry an arbitrary voltage, and thus we conclude that the circuits of figures 3.28 and 3.29 are ill posed.

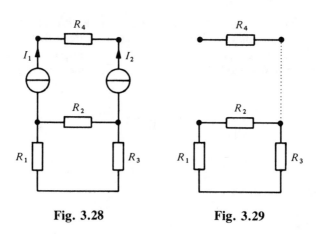

Fig. 3.28 **Fig. 3.29**

Now, we shall show that when controlled sources and operational amplifiers are absent, and if all the resistors are positive and finite, only loops of voltage sources and cut sets of current sources can produce an ill-posed circuit.

3.3.5 Theorem

Suppose that a circuit is composed of

- positive linear resistors;
- voltage and current sources.

It is well posed if and only if there are no loops of voltage sources, nor any cut sets of current sources.

Indeed, it is clear that when loops of voltage sources or cut sets of current sources are present, there is no solution, except for certain combinations of values of these sources. Thus, according to property 3.3.3, the circuit is ill posed.

In order to prove that the absence of these loops and cut sets is sufficient for the circuit to be well posed, we set the sources to zero and show that the resulting circuit, which includes positive resistors, open circuits, and short circuits, has only the solution $u = i = 0$.

Let us assume that there is no loop of short circuits, nor any cut set of open circuits, and suppose (u,i) to be a solution. According to the power conservation law (3.3), we have

$$\sum_{k=1}^{b} u_k i_k = 0 \qquad\qquad (3.18)$$

In (3.18), the short and open circuits do not contribute to the sum, and the terms corresponding to the resistors cannot be negative. Consequently, all the terms are zero. With regard to the resistors, this is only possible if the voltage and the current are simultaneously zero.

What remains to be proved is that the currents of the short circuits and the voltages of the open circuits are zero as well. To this end, let us apply the colored-branch theorem 3.1.11 twice. First, the circuit graph is oriented and colored in the following way:

- the short circuits are green colored and oriented in the direction of the current which is part of the solution. If this current is zero, we arbitrarily orient the short circuit;
- no branch is red colored;
- the resistors and open circuits are blue colored.

Because there is no loop exclusively composed of short circuits, theorem 3.1.11 guarantees the existence of a cut set such that all the short circuits have the same orientation within the cut set. According to the Kirchhoff lemma, we have

$$\sum_{coupé} i_k = 0 \qquad\qquad (3.19)$$

where the sum ranges over the branches of the cut set. We have $i_k = 0$ if the branch carries an open circuit or a resistor, and $i_k \geqslant 0$ in the case of a short circuit, as a result of the chosen orientation. Consequently, $i_k = 0$ for the short circuits as well. The second application of the colored-branch theorem is the dual of the first one:

- the open circuits are green colored and oriented in the direction of the voltage, which is part of the solution. If said voltage is zero, we arbitrarily orient the open circuit;
- the resistors and short circuits are red colored;
- no branch is blue colored.

The reasoning which permits us to conclude that the voltages of the open circuits are zero is also dual to the first application of theorem 3.1.11.

Thus, we have proved that the absence of loops composed of short circuits and cut sets composed of open circuits guarantees that the original circuit is well posed.

3.3.6 Corollary: Local Existence of the Solution of a Nonlinear RLC Circuit

Suppose a smooth circuit composed of linear and nonlinear resistors, capacitors, and inductors and of time-dependent independent sources. Let us assume that the capacitors (inductors) are charge (flux) controlled and the resistors are locally passive. Suppose ξ_0 to be an operating point at the instant t_0 such that there is no loop exclusively composed of voltage sources, capacitors, and resistors with a differential resistance which is zero at ξ_0, nor any cut sets composed of current sources, inductors, and resistors with an infinite differential resistance at ξ_0. Then, there exists one, and only one, solution $\xi(t)$ of the circuit with $\xi(t_0) = \xi_0$.

This theorem results from theorems 3.3.5 and 2.5.23.

3.3.7 Example

The circuit of figure 3.30 is a frequency multiplier. If the model for the diodes is that of the exponential diode, their differential resistance is positive everywhere. Because the voltage source and the capacitors do not form a loop and the inductors do not form a cut set, the hypotheses of corollary 3.3.6 are satisfied; hence, through any operating point, there is one, and only one, solution of the circuit passing at a given instant.

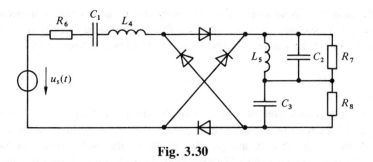

Fig. 3.30

If the circuit includes piecewise linear diodes, it is piecewise linear. The associated resistive circuit is piecewise linear as well. In each linear domain, it is well posed for the same reasons as in the case of exponential diodes. Thus, for any choice of charges and initial fluxes within a linear domain, there exists exactly one solution of the circuit. It can be extended toward the future or the past, provided that it does not leave the linear domain. In fact, we could show by another argument that it can even be extended beyond the linear domain boundaries.

If the model of the diodes is that of the ideal diode, we should distinguish the different linear domains. When two diodes are conducting and two diodes are blocked, nothing is changed with regard to the above arguments, and the local existence and uniqueness of the solutions around such operating points are guaranteed. On the contrary, when all four diodes are conducting (blocked), they form a loop (cut set), and thus the resistive circuit is ill posed. At such an operating point, the solution either does not exist, or it is not unique.

We can conclude that in the domains where all four diodes are either passing or blocked, the parasitic resistances of the diodes play a decisive role.

3.3.8 Example

The parallel connection of a positive resistor and a negative resistor (fig. 3.31) leads to the four equations:

$$i_1 + i_2 = 0 \tag{3.20}$$

$$u_1 - u_2 = 0 \tag{3.21}$$

$$u_1 - R_1 i_1 = 0 \tag{3.22}$$

$$u_2 + R_2 i_2 = 0 \tag{3.23}$$

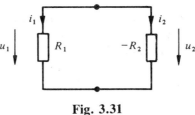

Fig. 3.31

Equations (3.20) to (3.23) only have the trivial solution, except for the case $R_1 = R_2$, where $i_1 = -i_2 = i, u_1 = u_2 = R_1 i$ is a nontrivial solution for any constant $i \neq 0$.

3.3.9 Comment

Example 3.3.8 shows two things. First, theorem 3.3.5 would be wrong if we allowed negative resistors. Second, a circuit can be well posed or ill posed, depending on the value of its elements. In our case, it is ill posed by reason of exception only. Indeed, R_1 should be exactly equal to R_2. On the other hand, when a circuit includes positive resistors and independent sources only, theorem 3.3.5 tells us that it is the connection of the elements, not their value, which determines whether a circuit is well posed.

The negative resistor can be implemented by a negative impedance converter (vol. XIX, subsec. 9.3.3), which produces the circuit of figure 3.32, composed of four positive resistors and an operational amplifier. The negative resistor, the terminals

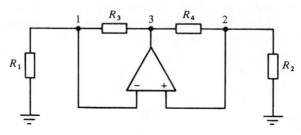

Fig. 3.32

of which are node 1 and ground, corresponds to $-R_2$ in figure 3.31. Its value is $-R_2 R_3 / R_4$. This example suggests that the problem of whether a resistive linear circuit is well posed, therefore, is much more difficult to solve in the presence of operational amplifiers or controlled sources. The remainder of this section is devoted to this problem.

3.3.10 Circuits with Nullators and Norators

Let us consider circuits composed of positive linear resistors, independent sources, ideal operational amplifiers, and controlled sources. As shown in subsection 1.3.23, in principle, it is sufficient to limit the elementary linear two-ports to the operational amplifier. In this section, we still prefer to decompose it into a nullator-norator pair, according to fig. 1.13. As an example, the circuit of figure 3.32 is represented according to figure 3.33.

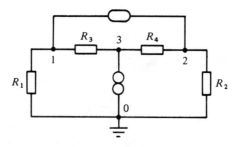

Fig. 3.33

The nullator is defined by two constitutive relations, whereas the norator has none. Consequently, if a circuit includes more norators than nullators, its equation system has more unknowns than equations, and it is necessarily ill posed. Conversely, if there are more nullators than norators, the number of equations exceeds the number of unknowns, and the circuit is ill posed, unless the equations in excess are linear combinations of the other equations. In order to avoid such an anomaly, in this section we will study circuits composed of positive linear resistors, independent sources, nullators, and norators, the latter two being in equal number.

3.3.11 Classification

We classify circuits composed of linear resistors, voltage and current sources, and nullators and norators into the following three classes:

- Circuits that are well posed for any choice of finite positive values for the resistors; these shall be called *always well-posed circuits.*
- Circuits that are well posed or ill posed, depending on the choice of finite positive values for the resistors; these shall be called *sometimes ill-posed circuits.*
- Circuits that are ill posed for any choice of finite positive values for the resistors; these shall be called *always ill-posed circuits.*

It is possible to state topological criteria that allow us to decide the class to which a given circuit belongs. This is the subject of the remainder of this section.

3.3.12 Examples

If the circuit does not include any nullators nor any norators, it is, according to theorem 3.3.5, always well posed or always ill posed.

The circuit of figure 3.33 is an example of a sometimes ill-posed circuit.

3.3.13 Property

Suppose that we have a circuit which is composed of resistors, independent sources, and an equal number of nullators and norators. If there exists a loop exclusively composed of nullators and voltage sources, or a cut set exclusively composed of nullators and current sources, the circuit is always ill posed. Likewise, if there exists a loop of norators and voltage sources, or a loop of norators and current sources, the circuit is always ill posed.

Indeed, if there is a loop of nullators and voltage sources, then the sum of the source voltages in the loop should be zero. If, according to the constitutive relations of the sources, this is not the case, the circuit has no solution. If such is the case, then the Kirchhoff equations and the constitutive relations are linearly dependent. In both cases, $\det H = 0$ in (3.17), and thus the circuit is ill posed. If there is a cut set of nullators and current sources, then we follow an analogous reasoning.

If there is a loop of norators and voltage sources, the solution, if it exists, cannot be unique. Indeed, suppose (u,i) to be a solution of the circuit. Let us add an arbitrary loop current I to the current. More precisely, let us add a current I to the currents of the branches that are part of the loop, in the same direction with respect to the loop. The other currents remain unchanged. The new set i' of currents still satisfies Kirchhoff's current laws. Because no branch of the loop involves any current in the constitutive relation, if there is any, (u,i) is a solution of the circuit as well. The proof for a cut set of norators and current sources is analogous.

3.3.14 Property

Suppose that we have a well posed circuit which is composed of resistors, independent sources, and nullators and norators. Let us consider an arbitrary resistor. We obtain a new well-posed circuit by replacing the resistor by either a short circuit or an open circuit.

For the proof, we consider the constitutive relation of branch k carrying a resistor R_k in the system (3.17). The corresponding line of H is of the following form:

$$H = \begin{pmatrix} & u_k & & i_k & \\ & \vdots & & \vdots & \\ 0 & 0 \ldots 0 & 1 & 0 \ldots 0 & -R_k & 0 \ldots 0 \\ & \vdots & & \vdots & \end{pmatrix} \tag{3.24}$$

The expansion of the determinant of H, according to this line, includes two terms: the first with respect to the column related to u_k; the other with respect to the column related to i_k. The corresponding minors are the determinants of matrix H for the circuits respectively obtained by short-circuiting branch k and by removing branch k. Because, by hypothesis, $\det H \neq 0$, at least one minor must be nonzero, and thus the corresponding circuit is well posed.

3.3.15 Comments

Possibly, both operations on the resistors may lead to a well posed circuit.

By replacing the resistors one by one using open circuits or short circuits, we obtain a circuit composed of nullators and norators only. According to proposition 3.3.14, it is possible to carry out these operations in such a way that a circuit which is initially well posed always remains well posed. According to property 3.3.13, a circuit composed of nullators, norators, and independent sources is ill posed as soon as there is a loop (cut set) composed of nullators and voltage (current) sources only, or of norators and voltage (current) sources only. Property 3.3.14, in this case, implies that the initial circuit is already ill posed.

As we shall see, these conditions on the norator and nullator connections are not only sufficient but also necessary for the circuit to be always ill posed. Additionally, we shall give the structure of the original circuit for this case.

3.3.16 Definition

Let us consider the graph of a circuit composed of resistors, voltage and current sources, and nullators and norators. Two *trees* of this graph are *conjugate*; if one includes all the norators and no nullator, then the other includes all the nullators and no norator; both include all the voltage sources and no current source, and they both include the same resistors.

3.3.17 Example

Figure 3.34 shows a pair of conjugate trees for the circuit of figure 3.33.

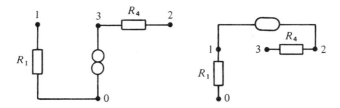

Fig. 3.34

3.3.18 Comments

A pair of conjugate trees is determined by the set of resistors for both trees.

If we replace a tree (cotree) resistor or a voltage (current) source by a short circuit (open circuit) in a pair of conjugate trees, then again we obtain a pair of conjugate trees. This is due to property 3.1.19. By fully completing this operation, we obtain a circuit composed of nullators and norators only, where the nullators simultaneously form a tree and a cotree; the same property is true for the norators. The concept of conjugate trees coincides with that of *complementary trees* in the absence of resistors and independent sources.

3.3.19 Analysis in the Presence of a Pair of Conjugate Trees

If there exists a pair of conjugate trees, the system of the $2b$ equations (3.17) is easily transformed into an equivalent system of r equations, where r is the number of resistors.

Let us first consider the tree with the norators and the fundamental set of loops (vol. IV, subsec. 4.2.10), generated by the cotree branches. The current of a tree branch is the superposition of currents for these loops, for which we write

$$i_{TR} = Ci_{CO} + Di_{SC} \tag{3.25}$$

where i_{TR}, i_{CO}, and i_{SC} are the column matrices, respectively composed of the currents of the tree resistors, the cotree resistors, and the current sources. We have taken into account that the loops generated by the nullators have zero current. The matrices C and D are submatrices of the matrix associated with the fundamental set of loops generated by the tree (vol. IV, subsec. 4.2.11).

Let us now consider the tree with the nullators and the fundamental system of the cut sets generated by the tree branches. By using a reasoning that is dual to the preceding one, we obtain

$$u_{CO} = Eu_{TR} + Fu_{ST} \tag{3.26}$$

where u_{CO}, u_{TR}, and u_{ST} are the column matrices, respectively composed of the voltages of the cotree resistors, the tree resistors, and the voltage sources. The matrices E and F are submatrices of the matrix associated with the fundamental set of cut sets generated by the tree (vol. IV, subsec. 4.2.19).

Let $R(G)$ be the diagonal matrix composed of the tree (cotree) resistances (conductances). Then, we obtain the equivalent system of (3.17):

$$\begin{pmatrix} u_{TR} \\ i_{CO} \end{pmatrix} = \begin{pmatrix} R & 0 \\ 0 & G \end{pmatrix} \left[\begin{pmatrix} 0 & C \\ E & 0 \end{pmatrix} \begin{pmatrix} u_{TR} \\ i_{CO} \end{pmatrix} + \begin{pmatrix} 0 & D \\ F & 0 \end{pmatrix} \begin{pmatrix} u_{ST} \\ i_{SC} \end{pmatrix} \right] \tag{3.27}$$

Let $x = (u_{TR}, i_{CO})^T, y = (-RDi_{SC}, -GFu_{ST})^T,$

$$Q = \begin{pmatrix} R & 0 \\ 0 & G \end{pmatrix} \tag{3.28}$$

and

$$M = \begin{pmatrix} 0 & C \\ E & 0 \end{pmatrix} \tag{3.29}$$

Then, (3.27) becomes

$$x = QMx - y \tag{3.30}$$

or

$$y = (QM - 1)x \tag{3.31}$$

Note that matrix M depends on the topology of the circuit only, and Q depends on the values of the resistors only.

We deduce from (3.31) that the circuit is ill posed if and only if the determinant of $QM-1$ is zero. This condition is equivalent to matrix QM having an eigenvalue 1.

3.3.20 Example

We illustrate the method of subsection 3.3.19 by the circuit of figure 3.35, which is that of figure 3.33 with two additional sources. The branches carrying the resistors have been oriented and numbered, and the resistors for a pair of conjugate trees have been cross-hatched. Then, equation (3.25) becomes

$$\begin{pmatrix} i_1 \\ i_4 \end{pmatrix} = \begin{pmatrix} 0 & 1 \\ 1 & 0 \end{pmatrix} \begin{pmatrix} i_2 \\ i_3 \end{pmatrix} + \begin{pmatrix} 0 \\ -1 \end{pmatrix} I \tag{3.32}$$

equation (3.26) becomes

$$\begin{pmatrix} u_2 \\ u_3 \end{pmatrix} = \begin{pmatrix} 1 & 0 \\ 0 & 1 \end{pmatrix} \begin{pmatrix} u_1 \\ u_4 \end{pmatrix} + \begin{pmatrix} 1 \\ 0 \end{pmatrix} E \tag{3.33}$$

and equation (3.31) becomes

$$\begin{pmatrix} 0 \\ R_4 I \\ -E/R_2 \\ 0 \end{pmatrix} = \begin{pmatrix} -1 & 0 & 0 & R_1 \\ 0 & -1 & R_4 & 0 \\ 1/R_2 & 0 & -1 & 0 \\ 0 & 1/R_3 & 0 & -1 \end{pmatrix} \begin{pmatrix} u_1 \\ u_4 \\ i_2 \\ i_3 \end{pmatrix} \tag{3.34}$$

The determinant of the matrix for the right-hand side of (3.34) is $\det(QM-1)=1-R_1R_4/R_2R_3$. We again obtain the result of subsection 3.3.9, according to which the circuit is ill posed if $R_1=R_2R_3/R_4$.

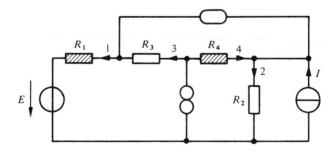

Fig. 3.35

3.3.21 Theorem: Criterion for Always Ill-Posed Circuits

Suppose that we have a circuit which is composed of linear resistors, voltage and current sources, and an equal number of nullators and norators. It is always ill posed if and only if there does not exist any pair of conjugate trees.

For the proof, we first assume that the circuit is well posed for a given choice of resistor values. As already noted in subsection 3.3.15, by short-circuiting certain resistors and all voltage sources, and by removing the other resistors and all current sources, we obtain a circuit exclusively composed of nullators and norators, without any loop or cut set composed of nullators only, nor of norators only. Thus, according to property 3.1.18, the nullators and norators form complementary trees of this reduced circuit. Accordingly, property 3.1.19 shows that the set of resistors which have been short-circuited along with the voltage sources forms a tree with the nullators as well as the norators. Thus, this set of resistors defines a pair of conjugate trees. This shows that if a circuit is not always ill posed, there exists a pair of conjugate trees.

Now let us assume that there exists a pair of conjugate trees, and, for a certain choice of resistor values, the circuit is ill posed. Consequently, the matrix QM of (3.30) has the eigenvalue 1. Let us multiply all the resistances of the tree by a factor λ and divide the resistances of the cotree by the same factor. This amounts to multiplying Q, and therefore QM, by λ. By choosing λ different from the inverse values of all the eigenvalues of QM, all the eigenvalues of λQM are different from 1, and the circuit with the modified resistors is well posed. This shows that a circuit which has at least one pair of conjugate trees is not always ill posed.

3.3.22 Corollary

Consider a circuit composed of linear resistors, voltage and current sources, and an equal number of nullators and norators. Let us assume that there exists a pair of conjugate trees. Let us consider a choice of resistor values for which the circuit is well posed. If we replace a tree (cotree) resistor by a voltage (current) source without changing the values of the other resistors, then the resulting circuit is still well posed.

Indeed, in the proof of theorem 3.3.21, we have identified the tree (cotree) branches by the fact that the circuit remains well posed if we short-circuit (remove) those branches. Regarding whether the circuit is well posed or ill posed, nothing changes, according to property 3.3.3, when the short circuit (open circuit) is replaced by a voltage (current) source.

3.3.23 Example

The simplest circuit that is always ill posed is an operational amplifier without feedback (fig. 3.36). Because resistor R_3 constitutes a loop and a cut set with the norator, this circuit cannot have any pair of conjugate trees. Only if a feedback resistor is connected between node 3 and node 1 or 2 would there no longer be a cut set

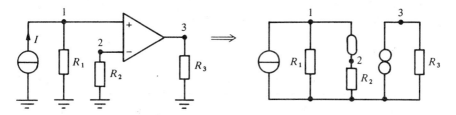

Fig. 3.36

composed of resistor R_3 and the norator; consequently, R_3 may be part of the cotree of a pair of conjugate trees.

According to the criterion of subsection 3.3.21, the circuit of figure 3.36 is always ill posed. This result is easily verified. If a solution exists, it cannot be unique because an arbitrary current can always be added in the R_3 norator loop.

3.3.24 Always Well-Posed Circuits

Now, we should distinguish between sometimes ill-posed circuits and always well-posed circuits. Therefore, regarding criterion 3.3.21, we assume the existence of at least one pair of conjugate trees and we can use the equations (3.30).

Let us assume that the circuit is ill posed and the sources are zero. Thus, we have $y = 0$ in (3.31), and there must be a nontrivial solution of this homogeneous system. We can always orient the resistors in such a way that their currents and voltages are non-negative. For such an orientation, all the elements of x and Mx are non-negative.

3.3.25 Definition

Suppose x to be a column matrix. We say that $x > 0 (x \geq 0)$ if $x_k > 0 (x_k \geq 0)$ for all the elements of x.

3.3.26 Property

Let us assume that a circuit has a pair of conjugate trees, the resistors can be oriented, and a column matrix $x > 0$ such that $Mx > 0$ can be found. Then, the circuit is sometimes ill posed.

Indeed, let us assume that we have $x > 0$ with $Mx > 0$. For each branch k of the tree, we define $R_k = x_k/(Mx)_k$, and if k is a branch of the cotree, we set $R_k = (Mx)_k/x_k$. With these resistances, we have for all k:

$$(QMx)_k = (x_k/(Mx)_k)(Mx)_k = x_k \tag{3.35}$$

which proves that x is an eigenvector of QM having the eigenvalue 1. Consequently, the circuit with this choice of resistances is ill posed.

3.3.27 Comment

In consideration of the property 3.3.26, we will look for some $x>0$ such that $Mx>0$. As we shall see, their existence only depends on whether it is possible to orient the graph in a certain way. More precisely, we should not uniformly orient loops and cut sets that are exclusively composed of resistors and nullators. Such an objective is inferred from observing a simple fact: if all the voltages within a loop are either oriented in the same direction or zero, then all the voltages of this loop must be zero. Likewise, if all the currents within a cut set are either oriented in the same direction or zero, then all the currents of this cut set must be zero.

3.3.28 Definition

Consider a circuit composed of resistors, voltage and current sources, and nullators and norators. An *orientation of the resistors* is *uniform* if the two following conditions are satisfied:

- each resistor is part of a uniform loop exclusively composed of resistors, voltage sources, and norators;
- each resistor is part of a uniform cut set exclusively composed of resistors, current sources, and norators.

This concept will be generalized in subsection 3.3.34.

3.3.29 Property

Suppose we have a circuit composed of resistors, voltage and current sources, and norators and nullators. An orientation of the resistors is uniform if and only if the two following conditions are satisfied:

- no resistor is part of a uniform loop exclusively composed of resistors, voltage sources, and nullators;
- no resistor is part of a uniform cut set exclusively composed of resistors, current sources, and nullators.

The proof consists of two applications of the colored-branch theorem 3.1.11. For the first application, we use the following colors: green for the resistors, red for the voltage sources and nullators, and blue for the norators and current sources. If it is possible to orient the green branches in such a way that every green branch is part of a uniform cut set of green and blue branches, according to the second condition of definition 3.3.28, then there is no uniform loop of green and red branches which includes at least one resistor, according to the first condition of property 3.3.29.

Conversely, the absence of red and green uniform loops with resistors implies the existence of a uniform green and blue cut set for each green branch. This proves the equivalence of the first condition of subsection 3.3.29 with the second condition of subsection 3.3.28. The other half of the proof is obtained from an analogous argument by exchanging the colors of the nullators and norators.

3.3.30 Property

Consider a circuit composed of resistors, voltage and current sources, and nullators and norators. The circuit is sometimes ill posed if its graph satisfies the two following conditions:

- there exists a pair of conjugate trees;
- there exists a uniform orientation of the resistors.

For the proof, we will choose a pair of conjugate trees. Our purpose is to identify an $x > 0$ such that $Mx > 0$, and to apply property 3.3.26.

Let us assume that the resistors are oriented according to the conditions of the theorem. An arbitrary orientation is chosen for the other branches, which will be used as a reference for the currents and voltages. Let us consider a loop L, where the resistors are uniformly oriented. Suppose $I > 0$ to be a loop current. More precisely, suppose $i^{(L)}$ to be defined by

$$
i_k^{(L)} = \begin{cases} +I & \text{if the branch } k \text{ is a resistor of } L, \text{ or another branch of } L \\ & \text{oriented as the resistors;} \\ -I & \text{if branch } k \text{ belongs to } L, \text{ with an orientation opposite to} \\ & \text{that of the resistors;} \\ 0 & \text{for the branches that do not belong to } L. \end{cases} \tag{3.36}
$$

Then $i^{(L)}$ complies with Kirchhoff's current laws. Consequently, we can write (3.25):

$$
i_{TR}^{(L)} = C i_{CO}^{(L)} \tag{3.37}
$$

where $i_{TR}^{(L)}(i_{CO}^{(L)})$ are the currents of the tree (cotree) resistors that are part of $i^{(L)}$. Note that there is no current source on L, and, therefore, the second term of (3.25) disappears. The components of $i_{TR}^{(L)}$ and $i_{CO}^{(L)}$, because they are resistors, are equal to $+I$ or 0.

Now, let us define

$$
i = \sum_L i^{(L)} \tag{3.38}
$$

where the sum extends over all the loops composed of resistors, voltage sources, and norators, where the resistors are uniformly oriented. If

$$i_{TR} = \sum_L i_{TR}^{(L)} \tag{3.39}$$

and

$$i_{CO} = \sum_L i_{CO}^{(L)} \tag{3.40}$$

then we have $i_{TR} \geq I$ and $i_{CO} \geq I$ because each resistor is part of at least one of these loops, and we also have

$$i_{TR} = Ci_{CO} \tag{3.41}$$

In an analogous way, for an arbitrary cut set voltage $U > 0$, we obtain column matrices of voltages $u_{TR} \geq U$ and $u_{CO} \geq U$, such that

$$u_{CO} = Du_{TR} \tag{3.42}$$

By setting $x = (u_{TR}, i_{CO})^T$, we find $x > 0$ with $Mx > 0$.

3.3.31 Example

The orientation of the resistors of the circuit in figure 3.35 is uniform. Therefore, according to proposition 3.3.30, there should be a choice of values $R_1, \ldots R_4$ and a solution of the corresponding circuit with $E = I = 0$, which is nonzero on the resistors. Indeed, we can choose $u_1 > 0, u_4 > 0, i_2 > 0, i_3 > 0$ as arbitrary, and a solution will be constructed by setting $i_1 = i_3, i_4 = i_2, u_2 = u_1, u_3 = u_4, R_1 = u_1/u_3, R_2 = u_1/i_2, R_3 = u_4/i_3, R_4 = u_4/i_3$.

3.3.32 Counterexample

Property 3.3.30 provides conditions which are sufficient for a circuit to be ill posed. Are these conditions necessary as well?

Such is not the case. The circuit of figure 3.37 constitutes a counter example. Because resistor R_5 forms a loop with the nullator by itself, it is impossible to orient it, unless this loop is uniform. Thus, according to property 3.3.29, there is no uniform resistor orientation.

Nonetheless, except for resistor R_5, the circuits of figures 3.35 and 3.37 are identical when the sources are set to zero. Because the circuit of figure 3.35 is ill posed if $R_1 R_4 = R_2 R_3$, there is a nonzero solution of the circuit with zero sources, if R_1 to R_4 satisfy this relation. The same voltages and currents constitute a solution of the circuit of figure 3.37 with a zero source and the same resistors R_1 to R_4, if we

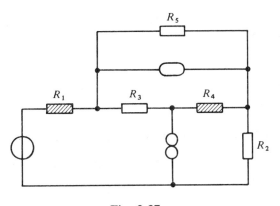

Fig. 3.37

add a voltage and a current that are zero for R_5. Consequently, the circuit of figure 3.37 is sometimes ill posed.

3.3.33 Comment

The nonzero solution with vanishing sources is nonzero on any resistor in the case of example 3.3.31, whereas it is zero on R_5 in the case of the counter example 3.3.32. Similarly, in the former case it is possible to orient the resistors uniformly, whereas in the latter case we can satisfy the two conditions of definition 3.3.28 for resistors R_1 to R_4 only. This suggests that the concept of uniform orientation should be extended to the partial orientations of the resistors.

3.3.34 Definition

Consider a circuit composed of resistors, voltage and current sources, and nullators and norators. A *partial orientation of the resistors* is *uniform* if the two following conditions are satisfied:

- each *oriented* resistor is part of a uniform loop exclusively composed of *oriented* resistors, voltage sources, and norators;
- each *oriented* resistor is part of a uniform cut set exclusively composed of *oriented* resistors, current sources, and norators.

3.3.35 Property

Consider a circuit composed of resistors, voltage and current sources, and norators and nullators. A partial orientation of the resistors is uniform if and only if the two following conditions are satisfied:

- no oriented resistor is part of a uniform loop exclusively composed of resistors, voltage sources, and nullators;
- no oriented resistor is part of a uniform cut set exclusively composed of resistors, current sources, and nullators.

The proof is identical to that of property 3.3.29, if the unoriented resistors are treated like nullators.

3.3.36   Theorem: Criterion for Sometimes Ill-Posed Circuits

Consider a circuit composed of resistors, voltage and current sources, and nullators and norators. The circuit is sometimes ill posed if and only if its graph satisfies the two following conditions:

- there exists a pair of conjugate trees;
- there exists a uniform partial orientation of the resistors, wherein at least one resistor is oriented.

We prove that these two conditions are sufficient by the same arguments as applied to the proof of property 3.3.30, if we join the unoriented resistors to the nullators.

In order to prove that both conditions are necessary, we assume that the circuit is sometimes ill posed. According to theorem 3.3.21, there exists a pair of conjugate trees. What remains to be proved is the second condition.

The uniform partial orientation of the resistors is obtained in the following way. There exists a choice of values for the resistors such that the circuit is ill posed. Let us consider the circuit with these resistors and vanishing sources. Because the circuit is ill posed, it has a nonzero solution. We orient each resistor in the direction of the voltage and the current associated with it by way of this solution. If the voltage and the current are zero, the resistor is not oriented.

In order to prove that this orientation is uniform, we consider a uniform loop of resistors, voltage sources, and nullators. In the Kirchhoff equation of the corresponding voltages, all the voltages are either of the same sign or zero. Indeed, the voltages of the oriented resistors have the same orientation within the loop because the loop is uniform; the unoriented resistors and the nullators have a zero voltage, and because the sources vanish, the voltage sources have a zero voltage as well. Because the sum of the voltages vanishes, all the loop voltages must vanish, which implies that no resistor within the loop can be oriented. Analogously, we reach the same conclusion for uniform cut sets composed of resistors, current sources, and nullators. It follows from property 3.3.35 that the partial orientation of the resistors is uniform.

Finally, we must establish that at least one resistor is oriented. Were this not the case, we would have a nonzero solution of the circuits with vanishing sources, all the resistor voltages and currents of which are zero. However, in the presence of a pair of conjugate trees, all the currents are linear combinations of the currents flowing in the cotree resistors and current sources. Were these currents zero, all current would vanish. By dual reasoning, all the voltages should vanish as well. Thus, the solution considered would be zero, contrary to what was assumed.

3.3.37 Comment

The possible presence of unoriented resistors makes it a little more difficult to apply the criterion of theorem 3.3.36 than that of property 3.3.30. If we try to orient the resistors of a given circuit one by one, we have to consider three alternatives for each resistor: the orientation in one direction; the orientation in the opposite direction; no orientation.

For certain resistors, it is easy to see that they cannot be oriented. This is the case for the set S_0 of the resistors which form a loop or a cut set with the nullators. For instance, resistor R_5 of the circuit in figure 3.37 belongs to S_0. Then, this is the case of the set S_1 of the resistors which form a loop or a cut set with the nullators and resistors of S_0, *et cetera*.

3.3.38 Example

We intend to show that the circuit of figure 3.38 is always well posed. The tree resistors for a pair of conjugate trees have been cross-hatched. The orientation of the resistors in figure 3.38 indicates an attempt to satisfy the conditions of theorem 3.3.36. We start by giving an arbitrary orientation to branch 1. Then, the orientation of branch 2 is subject to the requirement of avoiding a uniform cut set of branches 1, 2, and 7. The orientation of branch 3 avoids a uniform loop 2, 3, 7. The direction of branch 4 is compulsory; otherwise, cut set 3, 4, 7, 8 would be uniform. Branch 5 is oriented in view of loop 4, 5, 8, and branch 6 is oriented in view of cut set 5, 6, 8. However; this is not sufficient; loop 1, 9, 6, 8, 7 is uniform. Thus, this attempt has failed.

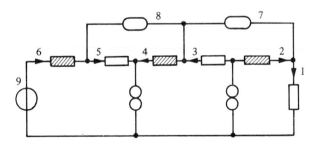

Fig. 3.38

For a uniform partial resistor orientation, the only possibility which is not excluded by the preceding reasoning is when branch 1 is unoriented. Then, neither can branch 2 be oriented; otherwise, the cut set of branches 1, 2, and 7 would necessarily become uniform. Continuing this type of reasoning, we notice that those cut sets and loops which previously caused the resistors to be oriented are now preventing them from being oriented. Eventually, no resistor is oriented.

We conclude that there is no uniform partial orientation such that at least one resistor is oriented. According to theorem 3.3.36, the circuit of figure 3.38 is always well posed.

3.4 EXISTENCE AND UNIQUENESS OF THE SOLUTION OF A NONLINEAR RESISTIVE CIRCUIT

3.4.1 Introduction

Nonlinear resistive circuits can have either no solution, one solution, or several solutions. While the absence of any solutions indicates a defective model, this is not the case for the presence of several solutions. Indeed, the presence of multiple dc operating points reflects reality and is even indispensable for semiconductor memory circuits. Nevertheless, it is important to be able to exclude multiple solutions whenever they are not desired. This is particularly the case for circuits that must amplify, limit, or transform the form of a signal, or for circuits of combinatorial logic.

In this section, as in the preceding one, we first deal with circuits having neither controlled sources nor operational amplifiers. Then, we will study circuits composed of *locally passive* resistors, current and voltage sources, and operational amplifiers. The restriction to locally passive resistors is justified by the fact that locally active resistors, such as the tunnel diode, can be synthesized by means of locally passive resistors and controlled sources or operational amplifiers. Doing so will enable us to state topological criteria rather than inequalities which involve the resistor characteristics.

Nonetheless, the locally active resistors will be taken into consideration in the study of circuits having neither controlled sources nor operational amplifiers. For this reason, the range of application for these two types of theorems somewhat overlap.

A collection of previous research papers on this subject has been published in [13]. Unlike the present book, most of the material is concerned with equations rather than the circuits themselves. A recent paper [14] provides topological criteria that are slightly different from those presented in this section.

3.4.2 Theorem: Uniqueness for Circuits Composed of Locally Passive Resistors

Suppose we have a circuit composed of

- voltage-controlled locally passive resistors, including the current sources;
- current-controlled locally passive resistors, including the voltage sources.

If there is at least one voltage-controlled resistor in each loop and there is at least one current-controlled resistor in each cut set, then the circuit has at most one solution.

The proof follows the method for the proof of theorem 3.3.36. Let us assume that we have two solutions $(u^{(1)}, i^{(1)})$ and $(u^{(2)}, i^{(2)})$. Let us define the increments $\Delta u = u^{(2)} - u^{(1)}$ and $\Delta i = i^{(2)} - i^{(1)}$. According to the incremental power balance 3.1.8, we have

$$\sum_k \Delta u_k \Delta i_k = 0 \tag{3.43}$$

Because all the resistors are locally passive, we have $\Delta u_k \Delta i_k \geqslant 0$ for all k. Then, (3.43) can be satisfied if and only if $\Delta u_k \Delta i_k = 0$ for all k. If branch k carries a voltage-controlled (current-controlled) resistor, then $\Delta u_k = 0 (\Delta i_k = 0)$ implies that $\Delta i_k = 0$ $(\Delta u_k = 0)$. Consequently, we can have $\Delta u_k \neq 0 (\Delta i_k \neq 0)$ only if the resistor is not current-controlled (voltage-controlled). This case shall be excluded by two applications of the colored-branch theorem, as discussed in subsection 3.3.5.

First, we color green the resistors which are not current controlled, and we orient them in the same way as their Δu_k. Then, we color red the current-controlled resistors. By hypothesis, there is no cut set of green branches. It follows that there exists a uniform loop for each green branch. The Δu_k corresponding to the resistors of this loop, which are (are not) current controlled, are zero (non-negative). Because their sum is zero, all the Δu_k of the loop must be zero.

We conclude that $\Delta u_k = 0$ for all the k; then, by a second application of the colored-branch theorem, we find $\Delta i_k = 0$ for all the k.

3.4.3 Comments

There are many resistors which are voltage- and current-controlled at the same time. Thus, they are strictly locally passive. If a circuit is composed of strictly locally passive resistors and sources, the absence of voltage source loops and current source cut sets, therefore, according to theorem 3.4.2, is sufficient to guarantee the uniqueness of the solution. This corollary is very similar to that of theorem 3.3.5.

The condition of local passivity is essential for the uniqueness of the solution. Indeed, the tunnel diode circuits, which were considered in section 2.1, frequently have several dc operating points.

Even for such a restricted class of circuits as considered by theorem 3.4.2, the *existence* of the solution is not always guaranteed. For example, if we connect a Zener diode in parallel with a voltage source larger than U_z (fig. 3.39), the resulting circuit has no solution. This shows that the circuit is ill modeled, and that, even with a more adequate model, it will exhibit singular behavior. In practice, either the voltage drops together with a large power dissipation in the internal resistance of the physical source, or the Zener diode is destroyed.

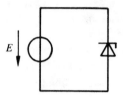

Fig. 3.39

The question of existence will be treated within the broader context of eventually passive resistors. Before addressing this question, we still need an analysis theorem, which will be given in 3.4.7 without proof.

3.4.4 Definition

Let $f:\mathbb{R}^n \to \mathbb{R}^n$ be a differentiable function. Suppose that $K>0$ and $y \in R^n$. Let us consider all the x with $f(x)=y$ and $\|x\| \leq K$. Let us assume that there are none for $\|x\| = K$, and that those with $\|x\| < K$ satisfy

$$[\det(\partial f/\partial x)](x) \neq 0 \tag{3.44}$$

Then, the *degree of f at y with respect to region* $\|x\| < K$ is defined by

$$d(f, y, K) = \sum_{f(x)=y,\ \|x\| < K} \mathrm{sgn}\ \{[\det(\partial f/\partial x)](x)\} \tag{3.45}$$

3.4.5 Example

The solutions of the resistive circuit in figure 2.8 are given, according to (2.21) and (2.22), by the intersections of the curve $i = g(u)$ and the straight line $i = G(E-u)$ in the plane (u,i)(fig. 2.9). Equivalently, these intersections are given by the solutions of the equation $f(u)=0$, where

$$f(u) = g(u) - G(E - u) \tag{3.46}$$

Suppose U_0 to be sufficiently large, such that all the solutions of $f(u)=0$ are located within $|u| < U_0$. We have

$$df/du = dg/du + G \tag{3.47}$$

Then, $df/du > 0(<0)$ if the slope of the tangent to the curve is larger (smaller) than the slope of the straight line, and $df/du = 0$ if the tangent to the curve coincides with the straight line. Except for the latter case, we can apply definition 3.4.4.

In the example of figure 2.9, we have $df/du > 0$ at P_1 and P_3, and $df/du < 0$ at P_2. Consequently,

$$d(f, 0, U_0) = 1 \qquad (3.48)$$

If there is only one intersection between the curve and the straight line, (3.48) remains valid.

3.4.6 Comment

Since (3.48) is verified in all cases where definition 3.4.4 is applicable, it is natural to require (3.48) as well for the limit cases $P_1 = P_2$ and $P_2 = P_3$, where $df/du = 0$. As a general rule, it is possible to extend definition 3.4.4 to the case where det $(\partial f/\partial x) = 0$, and even to functions that are continuous only, but not necessarily differentiable. By way of this extension, the degree $d(f,y,K)$ becomes a function having a value which can change, when we vary y and f, if and only if there is an x with $f(x) = y$ that crosses sphere $\|x\| = K$.

The degree theory exceeds the scope of this text, and we refer the reader to [15]. Here, we mention a theorem which precisely expresses that, by a continuous change of f, $d(f,y,K)$ does not change its value. For this purpose, we shall consider a family of functions f_λ, where λ varies between 0 and 1. This amounts to studying a function $F:[0,1] \times \mathbb{R}^n \to \mathbb{R}^n$ such that $F(\lambda,x) = f_\lambda(x)$.

We define the degree so as to obtain information about the number of solutions of $f(x) = y$. We could have used this number directly as the degree. However, it does not remain constant, neither under a variation of y, nor under a variation of f. Consequently, such a definition would be of no use.

If $d(f,y,K) = 1$, the equation $f(x) = y$ has at least one solution in the solid sphere $\|x\| \leq K$. This clearly results from (3.45) in the case where the formula is applicable. The degree theory allows us to reach the same conclusion in the general case. We shall use this result to prove the existence of the solution of a resistive circuit.

More precisely, if $d(f,y,K) = 1$, and if formula (3.45) is applicable, then the equation $f(x) = y$ has an odd number of solutions in the solid sphere $\|x\| \leq K$. By interpreting a solution of $f(x) = y$ with $\partial f/\partial x = 0$ as several coalesced solutions, we say that $f(x) = y$ has an odd number of solutions in $\|x\| \leq K$ if $d(f,y,K) = 1$.

In example 3.4.5, this interpretation actually coincides with our intuition. Indeed, if we increase parameter E in figure 2.9 until the straight line $i = G(E - u)$ becomes tangent to the curve $i = g(u)$, the intersections P_1 and P_2 become closer to one another and coalesce in the limit case. Thus, the tangent point, where $\partial f/\partial u = 0$ is considered to comprise two solutions of $f(u) = 0$.

176

3.4.7 Theorem

Suppose that $F:[0,1]\times \mathbb{R}^n \to \mathbb{R}^n$ is a continuous function, that $y \in R^n$, and that $K>0$. Let us consider the functions $f_\lambda:\mathbb{R}^n \to \mathbb{R}^n$, given by $F_\lambda(x)=F(\lambda,x)$, for $0\leqslant\lambda\leqslant1$. Let us assume that there is no x with $\|x\|=K$ and $f_\lambda(x)=y$ for any $\lambda \in [0,1]$. Then,

$$d(f_0,y,K) = d(f_1,y,K) \tag{3.49}$$

For the proof, refer to [15].

3.4.8 Application

The idea to apply theorem 3.4.7 to resistive circuits comes from Chua and Wang [16]. The function F is defined in different ways, depending on the desired result. We start directly from the standard equation system for a nonlinear circuit. We write it in the form:

$$f_1(u,i) = 0 \tag{3.50}$$

where $f_1:\mathbb{R}^{2b} \to \mathbb{R}^{2b}$. Then we modify the nonlinear resistors, depending on a parameter λ, in the following way.

- If the resistor of branch k is voltage controlled, with the constitutive relation $i_k=g_k(u_k)$, then the modified resistor is defined by the constitutive relation:

$$i_k = \lambda g_k(u_k) + (1-\lambda)u_k/R \tag{3.51}$$

  where R is an arbitrary positive resistance.
- If the resistor of branch k is current controlled, with the constitutive relation $u_k=h_k(i_k)$, then the modified resistor is defined by the constitutive relation:

$$u_k = \lambda h_k(i_k) + (1-\lambda)R i_k \tag{3.52}$$

We write the system of equations for the modified nonlinear circuit in the form:

$$f_\lambda(u,i) = 0 \tag{3.53}$$

where $f_\lambda:\mathbb{R}^{2b} \to \mathbb{R}^{2b}$. The equation:

$$f_0(u,i) = 0 \tag{3.54}$$

describes the linear resistive circuit obtained by replacing all the nonlinear resistors with linear resistors of value R.

Function F is defined by $F(\lambda,u,i)=f_\lambda(u,i)$. Because the right-hand sides (3.51) and (3.52) are continuous functions of λ, u_k, and i_k, function F is also continuous. If we can find a constant K such that the modified circuit has no solution with $\|(u,i)\|\geq K$ for any λ with $0\leq\lambda\leq 1$, we can apply theorem 3.4.7. If the linear circuit of the case $\lambda=0$ is well posed, we may conclude that the original circuit has an odd number of solutions.

3.4.9 Comments

We require the resistors to be either voltage controlled or current controlled. We could generalize the arguments to the case of resistors with a parametric representation [17].

The current of the resistor defined by (3.51) is located between the current of the original nonlinear resistor and that of a linear resistor of value R (fig. 3.40). By a dual reasoning, if the resistor is defined by (3.52), then the voltage is located between the voltage of the original nonlinear resistor and that of a linear resistor of value R (fig. 3.41).

The crucial point in application 3.4.8 is checking to see that the solutions of the modified circuit remain bounded as a function of λ. For this purpose, we use the power balance (3.3). Were the circuit exclusively composed of strictly eventually passive resistors, beyond a certain value for voltages and currents, all the resistors would then absorb power. However, this is not compatible with the power balance, and hence we conclude that he solutions must include voltages and currents below this limit.

We will provide for a class of slightly larger resistors, which specifically includes the independent sources. The conditions to be imposed on their connection will first be worked out through examples.

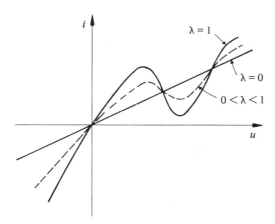

Fig. 3.40

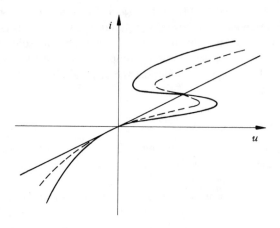

Fig. 3.41

3.4.10 Examples

The circuits of figures 3.42 to 3.44 are composed of the same elements, but they are connected in a different way. In order to determine whether there exists a solution for a current of an arbitrary source I, we may provisionally imagine that voltage u is imposed, and we determine the set of the values of I by varying u from $-\infty$ to $+\infty$.

If $u \rightarrow -\infty$ in the circuit of figure 3.42, then D_1 and D_2 are blocked, and $I \cong u/R \rightarrow -\infty$. If $u \rightarrow +\infty$, then D_1 and D_2 are conducting and $I = u/R + 2g(u) \rightarrow +\infty$. Because I is a continuous function of u, all the values of I are taken by varying u from $-\infty$ to $+\infty$. Consequently, the circuit of figure 3.42 always has a solution.

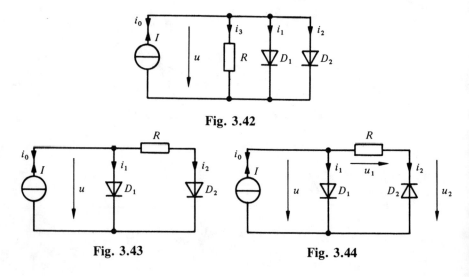

Fig. 3.42

Fig. 3.43   **Fig. 3.44**

If $u \rightarrow -\infty$ in the circuit of figure 3.43, D_1 and D_2 are blocked, and $I \rightarrow -2I_s$. When $u \rightarrow +\infty$, D_1 and D_2 are conducting and $I > g(u)$, therefore $I \rightarrow +\infty$. It is not difficult to see that only the values $I > -2I_s$ are taken when u varies between $-\infty$ and $+\infty$. Consequently, the circuit of figure 3.43 has no solution if $I < -2I_s$.

When $u \rightarrow -\infty$ in the circuit of figure 3.44, D_1 is blocked and D_2 is conducting. Because $u = u_1 + u_2$, at least one of the two voltages u_1 and u_2 tends toward $-\infty$. As a consequence, $I \rightarrow -\infty$. Conversely, if $u \rightarrow +\infty$, D_1 is conducting and D_2 is blocked, which leads to $I > g(u) \rightarrow +\infty$. All the values of I are taken by varying u between $-\infty$ and $+\infty$. Thus, the circuit of figure 3.44 always has a solution.

What constitutes the difference between the circuit of figure 3.43 and the other two circuits? The diodes and the current source form a cut set in the circuit of figure 3.43, the Kirchhoff equation of which is $i_1 + i_2 + i_0 = 0$. The admissible currents are $i_1 > -I_s$, $i_2 > -I_s$, and $i_0 = -I_s$. If $-2I_s - I > 0$, the Kirchhoff equation cannot be satisfied.

Conversely, in the circuit of figure 3.42, resistor R is added to the cut set, the equation of which becomes $i_0 + i_1 + i_2 + i_3 = 0$. This equation can always be satisfied because all the currents are admissible for the resistor.

The cut set of the circuit of figure 3.44 is composed of the same elements as in figure 3.43, but the diode D_2 is turned around. Thus, we have $i_2 < +I_s$ instead of $i_2 > -I_s$, and the equation $i_0 + i_1 + i_2 = 0$ can always be satisfied.

Let us return to the circuit of figure 3.39. It has no solution for a reason dual to that for the circuit of figure 3.43. Indeed, there is a loop of elements having a set of admissible voltages bounded on the same side.

3.4.11 Definitions

A resistor is *weakly active* if there are two constants $U_0 > 0$ and $I_0 > 0$ such that for any point (u,i) of the characteristic such that $u < -U_0(u > +U_0)$, we have $i \leq +I_0(i \geq -I_0)$, and for any point with $i < -I_0(i > +I_0)$, we have $u \leq +U_0(u \geq -U_0)$. The admissible range for the characteristic of plane (u,i) is cross-hatched in figure 3.45.

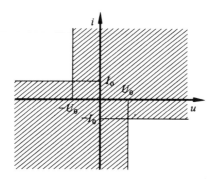

Fig. 3.45

180

A resistor is *strongly passive at* $+\infty$ if it satisfies

$$u \to +\infty \iff i \to +\infty \tag{3.55}$$

and a resistor is *strongly passive at* $-\infty$ if it satisfies

$$u \to -\infty \iff i \to -\infty \tag{3.56}$$

3.4.12 Property

A weakly active resistor having a voltage u (or a current i), can produce at most the power $|u| I_0 (|i| U_0)$.

3.4.13 Comments

An eventually passive resistor is weakly active. Hence, according to property 3.2.4, it can only produce a finite amount of power, whereas the power produced by a weakly active resistor can linearly tend toward infinity with u or i. The most important example of such a resistor, which is weakly active without being eventually passive, is the independent source.

On the contrary, the concept of strong passivity at infinity is more restrictive than that of eventual passivity. Indeed, an eventually passive resistor must have only a positive (negative) current when $u \to +\infty (u \to -\infty)$, and a positive (negative) voltage when $i \to +\infty (i \to -\infty)$.

We distinguish strong passivity from $+\infty$ and $-\infty$ to adapt the resistors which only satisfy (3.55) and (3.56), but not both. The exponential diode is an example. It is obvious that the distinction between $+\infty$ and $-\infty$ depends on the orientation of the resistor.

3.4.14 Theorem: Existence of a Solution

Suppose that a circuit is composed of weakly active resistors, which are either voltage controlled, with all the voltages admissible, or current controlled, with all the currents admissible. Let us assume that the following conditions concerning their connection are satisfied.

- In each loop that is exclusively composed of resistors, which are current controlled, but not voltage controlled, at least one resistor is strongly passive at $+\infty$, and at least one—possibly the same one—is strongly passive at $-\infty$. The orientations are relative to some loop orientation.
- In each cut set that is exclusively composed of resistors, which are voltage controlled, but not current controlled, at least one resistor is strongly passive at $+\infty$, and at least one—possibly the same one—is strongly passive at $-\infty$. The orientations are relative to some cut set orientation.

Thus, the circuit has an odd number of solutions. In particular, a solution always exists.

3.4.15 Comment

The two conditions imposed on the connection in theorem 3.4.14 exclude voltage source loops and current source cut sets.

3.4.16 Proof

We will follow the procedure of subsection 3.4.8 for the proof. All the resistors are modified according to (3.51) or (3.52), depending on whether they are voltage controlled or current controlled. In the limit case $\lambda = 0$, we obtain a circuit that is exclusively composed of linear resistors of value R. According to theorem 3.3.5, this circuit is always well posed. It has only the zero solution.

Now, we shall show that the circuit has no solution if $\|(u,i)\| \geqslant K$ for an adequate K that does not depend on λ. To this end, we will introduce various constants which are linked to the characteristics of the resistors. According to definition 3.4.11, two constants, U_{0k} and I_{0k}, are associated with the kth resistor. Suppose $U_0(I_0)$ to be the maximum of the $U_{0k}(I_{0k})$.

If the kth resistor is current controlled, we define U_{k+} as being the largest voltage that can be taken by the resistor within the interval $0 \leqslant i \leqslant I_0 b^2$, this being true for any λ with $0 \leqslant \lambda \leqslant 1$. Because the resistor voltage varies between the voltage of a linear resistor of value R and that of the original nonlinear resistor, U_{k+} is finite. Likewise, we define U_{k-} as being the smallest voltage that can be taken by the resistor within the interval $-I_0 b^2 \leqslant i \leqslant 0$. In figure 3.46, we have shown the constants U_{k+} and U_{k-}

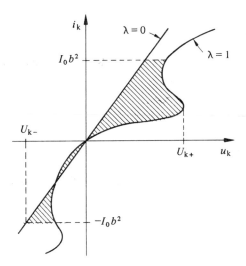

Fig. 3.46

182

for an example of a current-controlled resistor. The cross-hatched area represents the set of voltages taken for $-I_0b^2 \leqslant i \leqslant I_0b^2$ and $0 \leqslant \lambda \leqslant 1$.

If the kth resistor is voltage controlled, with a constitutive relation $i_k = g(u_k)$, and if it is strongly passive at $+\infty(-\infty)$, we define $U_{k+}(U_{k-})$ as being the smallest positive voltage (largest negative voltage) such that, for $u_k \geqslant U_{k+}(u_k \leqslant U_{k-})$, we have $i_k \geqslant I_0b^2 (i_k \leqslant -I_0b^2)$ for all the $0 \leqslant \lambda \leqslant 1$. In figure 3.47, U_{k+} and U_{k-} are marked for an example of a voltage-controlled resistor which is strongly passive at $+\infty$ and $-\infty$.

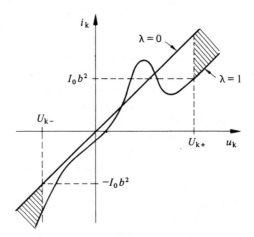

Fig. 3.47

The set of currents taken for $u_k \geqslant U_{k+}$ and for $u_k \leqslant U_{k-}$ is cross-hatched.

Let us define

$$U_+ = \max_k U_{k+}, \quad U_- = \min_k U_{k-} \tag{3.57}$$

where the maximum and the minimum extend to all the branches k where U_{k+} or U_{k-} is defined. For these branches, we thus have

$$u_k \geqslant U_+ \to i_k \geqslant I_0b^2 \tag{3.58}$$

and

$$u_k \leqslant U_- \to i_k \leqslant -I_0b^2 \tag{3.59}$$

Let us consider a solution (u,i) of the circuit with the modified resistors. Suppose that we have

$$u_m = \max_k |u_k| \tag{3.60}$$

By way of the colored-branch theorem, we show that there is a uniform cut set of branches with $|U_k| > U_m/b$. Indeed, if all the branches k with $|u_k| \geqslant U_m/b$ are colored green and oriented in the same way as their voltage, and if the other branches are colored red, then there is no uniform loop passing through the branch with the maximal voltage. If there were one, the Kirchhoff equation would be

$$\sum_{k \in A} u_k + \sum_{k \in B} u_k = 0 \tag{3.61}$$

where $A(B)$ includes all the green (red) branches of the loop. Because all the u_k with $k \epsilon A$ have the same sign, we have

$$\left| \sum_{k \in A} u_k \right| \geqslant U_m \tag{3.62}$$

and

$$\left| \sum_{k \in B} u_k \right| \leqslant \sum_{k \in B} |u_k| \tag{3.63}$$

$$\leqslant \sum_{k \in B} U_m/b \leqslant (b-1) U_m/b \tag{3.64}$$

We can see that (3.64) and (3.65) are in contradiction with (3.61), and we conclude that there is a uniform cut set of green branches passing through the maximal voltage branch.

According to the second hypothesis of theorem 3.4.14, there is, in such a cut set, either a current-controlled resistor, or a voltage-controlled resistor, which is strongly passive at ∞, and one which is strongly passive at $-\infty$. Because we have $|u_k| \geqslant U_m/b$ for any green branch, it follows from (3.58) and 3.59) that, if $U_m \geqslant b\max(U_+, |U_-|)$, there exists a branch which absorbs a power at least equal to $U_m I_0/b$. Conversely, no branch can produce more than the power $u_k I_0$, and the total power produced by the resistors located at an active operating point cannot exceed $U_m I_0(b-1)$. Therefore, the power balance (3.3) can only be verified if $U_m \geqslant b\max(U_+, |U_-|)$ and, consequently, we have

$$|u_k| < b \max (U_+, |U_-|) \tag{3.65}$$

for any k.

By analogous reasoning, we find that the currents are also bounded by a constant which does not depend on λ. Consequently, all the solutions of the circuit with modified resistors are located within a region $\|x\| < K$, where K does not depend on λ.

According to the reasoning of subsection 3.4.8, it follows that the original circuit has an odd number of solutions.

3.4.17 Examples

The exponential diode is a resistor that is both voltage controlled and current controlled, but for which not all currents are admissible. Furthermore, it is only strongly passive at $+\infty$ or $-\infty$, depending on its orientation. Therefore, in circuits composed of sources, resistors, and exponential diodes, attention should be given to cut sets composed of diodes and current sources.

In the circuit of figure 3.42, there is no cut set without resistor R. In the circuit of figure 3.44, there is one, but the diodes have opposite orientation within the cut set. Thus, theorem 3.4.14 guarantees the existence of a solution, as already verified in subsection 3.4.10. However, in the circuit of figure 3.43, the current source and the two diodes form a cut set, and the diodes have the same orientation within the cut set. If we orient the cut set in the same way as the current source, the two diodes are strongly passive at $-\infty$, but there is no resistor within the cut set that is strongly passive at $+\infty$. Thus, the second condition of theorem 3.4.14 is violated, and we cannot make any conclusion regarding the existence of a solution. Indeed, we observed in subsection 3.4.10 that this circuit has a solution for certain values I of the source and does not have a solution for others.

The circuit of figure 3.48 implements a combinatorial logic function (vol. VIII, subsec. 7.2.11). Each diode forms a loop with a resistor R_i, with the power supply source V and a signal source V_j. Due to corollary 3.1.15, there is no diode cut set. Consequently, the circuit has a solution for any combination of signal values. This remains true irrespective of the diode model, provided that it is a weakly active current-controlled resistor. This reasoning is valid for any circuit of that type.

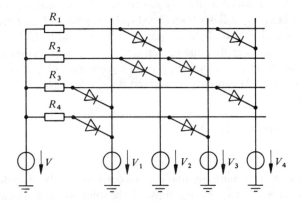

Fig. 3.48

3.4.18 Example

When nullators and norators are present, the existence of a solution is no longer guaranteed, even if we limit ourselves to locally passive resistors. This fact was previously observed in section 3.3 with regard to linear circuits.

Let us consider the circuit of figure 3.49, which constitutes a nonlinear version of the one in figure 3.33. The characteristic of the nonlinear resistor is shown in figure 3.50. It is a strictly locally passive resistor. The rest of the circuit is equivalent to a negative resistor of value $-R_2R_3/R_4$. Thus, the solutions of the circuit correspond to the intersections of the characteristic $i = g(u)$, with the straight line $i = R_4/R_2R_3 u$. If this straight line is like the continuous line in figure 3.50, there is no solution. Conversely, if it is like the dashed line, there are two solutions.

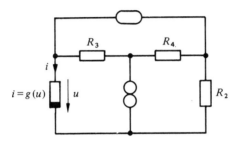

Fig. 3.49

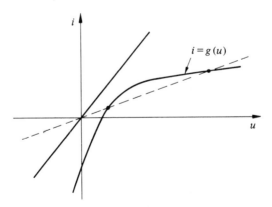

Fig. 3.50

3.4.19 Comment

It is interesting to compare the linear circuit of figure 3.33 with the nonlinear circuit of figure 3.49. The existence and uniqueness of the solution in both cases depend on the value of the elements. However, the linear circuit has either no solution, one solution, or an infinity of solutions. In the normal case, it has exactly one solution. The other two cases only occur when a certain relation between the elements holds. As opposed to this situation, the nonlinear circuit always has no solution or two solutions, except for the exceptional case in which the straight line touches the curve in figure 3.50.

Nevertheless, the following theorem shows that the behavior of the nonlinear circuit and the linear circuit of identical structure are closely related.

3.4.20 Theorem: Existence and Uniqueness of the Solution for Circuits with Operational Amplifiers

Let us suppose that we have a circuit composed of

- locally passive voltage-controlled resistors, with all the voltages admissible, including the current sources;
- locally passive current-controlled resistors, with all the currents admissible, including the voltage sources;
- operational amplifiers.

Let us consider the circuit obtained by replacing the nonlinear resistors with linear resistors having positive values, and replacing the operational amplifiers with pairs of nullators and norators. The sources are not replaced. Let us assume that this circuit:

- is always well posed for an arbitrary choice of positive values of the resistors;
- has a pair of conjugate trees such that the resistors of the tree are current controlled and those of the cotree are voltage controlled in the original circuit.

Then, the nonlinear circuit has exactly one solution.

3.4.21 Comments

This theorem reduces the nontrivial problem of the existence and uniqueness of the solution for a nonlinear circuit solution to the simpler problem of the existence and uniqueness of the solution for the linear circuit, or, more precisely, for a set of linear circuits obtained by varying the resistances. As a result of theorem 3.3.36, the existence and uniqueness of the solution of the linear circuit is further reduced to a purely topological criterion. Thus, we succeed in establishing the existence and uniqueness of the solution by simply verifying the elements' passivity and using a purely topological criterion concerning the elements' connection. This criterion amounts to checking to see whether there is a pair of conjugate trees such that the current-controlled (voltage-controlled) resistors are located on the tree (cotree), and making sure that there is no uniform partial orientation of the resistors.

3.4.22 Proof

The proof of theorem 3.4.20 goes by induction of the number n of nonlinear resistors, i.e., the number of resistors which are neither linear resistors nor sources.

If $n = 0$, the circuit is composed of resistors and sources only. The existence and uniqueness of the solution of this circuit is actually part of the hypotheses of the theorem. The solution is a continuous function of the source values. In fact, it is a linear combination of these values.

Let us now assume that the theorem is verified for the circuits with $n - 1$ nonlinear resistors. Furthermore suppose that the solution of the circuit with $n - 1$ nonlinear resistors is a continuous function of the source values. This is the induction hypothesis.

Let us consider a circuit C_n (fig. 3.51) with n nonlinear resistors. Let us assume that the circuit $C_{n,l}$ (fig. 3.52), obtained by replacing the nonlinear resistors with positive linear resistors, is always well posed. Thus, it has a pair of conjugate trees. Let us choose a branch k, including a nonlinear resistor in C_n, which becomes a tree resistor in $C_{n,l}$. By replacing it with a voltage source, we obtain a circuit C_{n-1} having $n - 1$ nonlinear resistors, which is used as an auxiliary circuit for the proof (fig. 3.53). We could also replace a nonlinear resistor of C_n, which becomes a cotree resistor in $C_{n,l}$, with a current source.

The circuit C_{n-1} satisfies the hypotheses of the theorem. Indeed, if we replace the nonlinear resistors of C_{n-1} with positive linear resistors, we obtain the circuit $C_{n-1,l}$ (fig. 3.54), which is identical to the circuit $C_{n,l}$, except for branch k, which

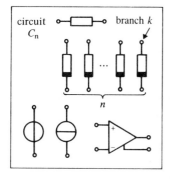

Fig. 3.51

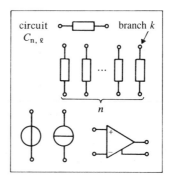

Fig. 3.52

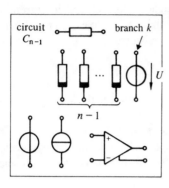

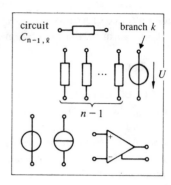

Fig. 3.53

Fig. 3.54

carries a voltage source instead of a resistor. Because k is a tree branch in $C_{n,l}$, the circuit $C_{n-1,l}$ is also always well posed, as a result of corollary 3.3.22. Based on the induction hypothesis, we conclude that the circuit C_{n-1} has exactly one solution, this being for any value U of the voltage source on branch k. We can write this solution as $i = \varphi(U), u = \psi(U)$, which defines the functions $\varphi:\mathbb{R} \to \mathbb{R}^b, \psi:\mathbb{R} \to \mathbb{R}^b$. According to the induction hypothesis, these functions are continuous.

Let us specifically consider $i_k = \varphi_k(U)$. The function φ_k is decreasing. Were this not the case, there would be two source values, $U < U'$, such that

$$\varphi_k(U) < \varphi_k(U') \tag{3.66}$$

Then, the two points (U, i_k) and (U', i'_k), where $i'_k = \varphi_k(U')$, would be located on the characteristic of a positive resistor:

$$R_k = (U' - U)/(i'_k - i_k) \tag{3.67}$$

which is in series with a voltage source having a value:

$$E = (i'_k U - i_k U')/(i'_k - i_k) \tag{3.68}$$

It is not difficult to see that the circuit C'_{n-1} obtained by replacing the nonlinear resistor on the branch k of C_n with this composite one-port (fig. 3.55), has exactly one solution, due to the induction hypothesis. Conversely, $i = \varphi(U), u = \psi(U)$ and $i' = \varphi(U'), u' = \psi(U')$ are two distinct solutions of this circuit, which is a contradiction.

If we find a voltage U such that (U, i_k) is situated on the characteristic curve of the nonlinear resistor located on the branch k of C_n, namely, if we find a pair (U, i_k), such that

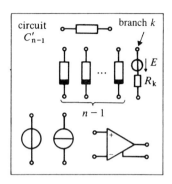

Fig. 3.55

$$U = h_k(i_k) \tag{3.69}$$

and

$$i_k = \varphi_k(U) \tag{3.70}$$

then $i = \varphi(U), u = \psi(U)$ is a solution of the original circuit C_n. In fact, the solutions of C_n bijectively correspond to the solutions of

$$F(U) = h_k(\varphi_k(U)) - U = 0 \tag{3.71}$$

Because h_k is an increasing function, insofar as we have assumed that the resistors are locally passive and we have just shown that φ_k is a decreasing function, the composition $h_k \circ \varphi_k$ is a decreasing function, as is F. We can even write, for $U < U'$, and, therefore, for any pair U, U':

$$F(U) - F(U') = h_k(\varphi_k(U)) - h_k(\varphi_k(U')) + U' - U \geqslant U' - U \tag{3.72}$$

$$|U' - U| \leqslant |F(U) - F(U')| \tag{3.73}$$

Furthermore, $F(U) \mapsto +\infty (-\infty)$ when $U \to -\infty (+\infty)$ and F is a continuous function. It follows that $F(U) = 0$ has exactly one solution. This proves that the circuit C_n has exactly one solution.

It remains to be proved that the solution of C_n continuously depends on the sources because this property is part of the induction hypothesis. Indeed, let us consider a source of value E in C_n. It is also part of C_{n-1}. Consequently, we can write

$$F(U, E) = h_k(\varphi_k(U, E)) - U = 0 \tag{3.74}$$

instead of (3.71). Due to the induction hypothesis, F is a continuous function of U and E. For each E, (3.72) has a solution $U(E)$. This solution leads to a solution $i(E) = \varphi(U(E)), u(E) = \psi(U(E))$ of C_n. If $U(E)$ is continuous, the solution of C_n is also continuous in E.

This last point is a consequence of (3.73), rewritten with parameter E as

$$|U' - U| \leqslant |F(U', E) - F(U, E)| \tag{3.75}$$

Indeed, suppose (U,E) and (U',E') to be two solutions of (3.74). Then, by using (3.75), we obtain

$$|U' - U| \leqslant |F(U', E) - F(U', E')| + |F(U', E') - F(U, E)| \tag{3.76}$$
$$= |F(U', E) - F(U', E')| \tag{3.77}$$

When $E' \to E, F(U', E') \to F(U', E)$, thanks to the continuity of F, and $U' \to U$, thanks to (3.77). This proves the continuity of $U(E)$.

3.4.23 Theorem: Regularity of the Solution

Let us assume that the hypotheses of theorem 3.4.20 are satisfied. Furthermore, let us suppose that the circuit depends on a parameter λ, i.e. that the constitutive relations of the voltage-controlled resistors are of the form:

$$i_k = g_k(u_k, \lambda) \tag{3.78}$$

and the constitutive relations of the current-controlled resistors are of the form:

$$u_k = h_k(i_k, \lambda) \tag{3.79}$$

If g_k and h_k are continuous (locally Lipschitz, smooth) functions **in both variables**, then the solution of the circuit is a continuous (locally Lipschitz, smooth) function of λ.

The proof of this theorem is obtained by examining the proof 3.4.22 in more detail. First, we note that the theorem is verified in the case $n = 0$, where the circuit is linear. Indeed, if we carry out the analysis of the circuit with respect to the pair of conjugate trees, according to subsection 3.3.19, we obtain the equation system (3.31):

$$y = (QM - 1) x \tag{3.80}$$

where y is a linear function of the source values and Q includes the resistances or their inverse values on the diagonal. According to Cramer's rule, we find

$$x_k = (\det(QM - 1)_k) / (\det(QM - 1)) \tag{3.81}$$

where $(QM - 1)_k$ is obtained by replacing the kth column of $QM - 1$ by y. Thus, the right-hand side is a rational function of the resistances and the source values. Such a function is smooth, except at the poles. However, because the circuit is never ill posed, $\det(QM - 1) \neq 0$, and the rational function does not present any poles in the range of positive resistance values. Consequently, x_k is a smooth function of the resistance and source values. This is also the case for the other voltages and currents of the circuit because they are linear functions of the x_k and the source values.

Still, for the case $n = 0$, if the constitutive relations of the elements depend on a parameter λ, such is only possible through the resistance and source values. Because the composition of a continuous (locally Lipschitz, smooth) function and a smooth function is continuous (locally Lipschitz, smooth), the theorem is verified for $n = 0$.

Hence, we make the hypothesis that the solution of C_{n-1} is continuous (locally Lipschitz, smooth) in λ. We prove that the solution of C_n is continuous (locally Lipschitz, smooth) in λ in the same way that we proved its continuity in E, in subsection 3.4.22. Indeed, instead of (3.77), we find

$$|U(\lambda') - U(\lambda)| \leqslant |F(U(\lambda'), \lambda) - F(U(\lambda'), \lambda')| \tag{3.82}$$

which shows that $U(\lambda)$ is continuous (locally Lipschitz), because, thanks to the hypothesis about C_{n-1}, $F(U, \lambda)$ is continuous (locally Lipschitz) in λ. Furthermore,

$$|U' - U| \leqslant |F(U', \lambda) - F(U, \lambda)| \tag{3.83}$$

implies

$$|\partial F / \partial U| \geqslant 1 \tag{3.84}$$

if F is smooth. By the implicit function theorem, we obtain that $U(\lambda)$ is smooth in this case.

3.2.24 Corollary: Uniqueness of the dc Operating Point

Suppose a circuit is composed of

- voltage-controlled capacitors;
- current-controlled inductors;
- time-dependent independent sources;
- voltage-controlled locally passive resistors, with all the voltages admissible;
- current-controlled locally passive resistors, with all the currents admissible;
- operational amplifiers.

If the autonomous resistive circuit obtained by replacing the capacitors (inductors) and the time-dependent current (voltage) sources by open circuits (short circuits) satisfies the hypotheses of theorem 3.4.20, then the following properties are verified:

- there is exactly one dc operating point;
- if the constitutive relations continuously depend on a parameter λ, then the dc operating point also is a continuous function of λ;
- if the circuit is smooth and the constitutive relations smoothly depend on λ, then the dc operating point also is a smooth function of λ.

This corollary results from theorem 3.4.20 because the dc operating points correspond one by one to the solutions of the resistive circuit.

3.4.25 Corollary: Existence of Global State Equations

Suppose a circuit composed of

- charge-controlled capacitors;
- flux-controlled inductors;
- time-dependent independent sources;
- voltage-controlled locally passive resistors, with all the voltages admissible;
- current-controlled locally passive resistors, with all the currents admissible;
- operational amplifiers.

If the autonomous resistive circuit obtained by replacing the capacitors (inductors) and the time-dependent voltage (current) by constant voltage (current) sources satisfies the hypotheses of theorem 3.4.20, then the following properties are verified:

- there exists a system of global state equations with the capacitor charges q_C and the inductor fluxes φ_L as state variables;
- starting from the arbitrary initial conditions $q_C(t_0), \varphi_L(t_0)$, there exists exactly one solution of the circuit, at least for a finite time interval around t_0;
- if the circuit is piecewise linear, the solution exists for all the times;
- if the circuit is smooth, the solution is a smooth function of t.

This theorem is a combination of theorems 3.4.20, 3.4.23, 2.4.15, and 2.4.16.

3.4.26 Circuits with Transistors

If we want to apply theorem 3.4.23 to circuits that include transistors, we must model the transistor by means of a circuit composed of locally passive resistors, nullators, and norators. For example, the three-pole of figure 3.56 complies with the Ebers-Moll equations (2.1) for a bipolar *npn* transistor if the inverse currents of the exponential diodes D_3 and D_4 are I_s, and if those of D_1 and D_2 are respectively I_s/β_F and I_s/β_R.

Thus, if a resistive circuit includes bipolar transistors, these should be replaced by three-poles of the type shown in figure 3.56. In order to check the hypotheses of theorem 3.4.20, we move on to the linear resistive circuit. As far as the transistor model is concerned, this amounts to replacing the diodes by linear resistors (fig. 3.57).

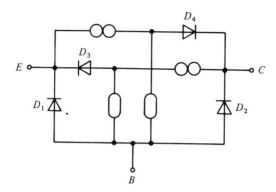

Fig. 3.56

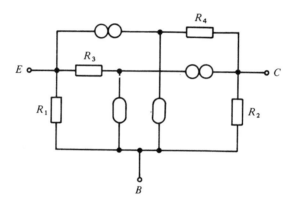

Fig. 3.57

Then, we must show the existence of a certain pair of conjugate trees. Because exponential diodes do not admit all currents, resistors R_1 to R_4 cannot be part of the tree.

Finally, we try to orient the resistors according to the requirements of theorem 3.3.36. If this is not possible, we are sure that the nonlinear circuit has exactly one solution. Because the circuit of figure 3.57 includes four resistors, there are $3^4 = 81$ possibilities to orient them, or not to orient them. However, without knowing the nature of the circuit surrounding the transistor, we can easily exclude a large number of combinations. Indeed, resistors R_1 and R_3 as well as resistors R_2 and R_4 form a loop with a nullator. If we want to avoid violating the first condition of property 3.3.33, the orientation of $R_1(R_2)$ sets the orientation of $R_3(R_4)$. Alternatively, if $R_1(R_2)$ is unoriented, neither can $R_3(R_4)$ be oriented. Therefore, we must provide an option for R_1 and R_2, and the rest will follow. This reduces the possibilities down to $3^2 = 9$.

The various combinations correspond to the operating regions of the transistor (vol. VIII). Indeed, the orientation of $R_1(R_2)$ corresponds to the orientation of the voltage at the terminals of diode $D_1(D_2)$. If it is oriented as $D_1(D_2)$, the base-emitter (base-collector) junction is conducting; in the opposite case, it is blocked. If $R_1(R_2)$ is unoriented, the corresponding voltage on $D_1(D_2)$ is zero, and it is reasonable to say that the base-emitter (base-collector) junction is also blocked. Thus, we assimilate the absence of orientation to the diode orientation in reverse direction.

Analogous reasoning could apply to other types of transistors. In particular, the model for a *pnp* bipolar transistor is that of figure 3.56, with its diodes reversed.

3.4.27 Property

Let us assume that a resistive circuit includes one or several bipolar transistors, modeled by the three-pole of figure 3.56, with its diodes reversed if it is a *pnp* transistor. Let us consider the circuit obtained by replacing the nonlinear resistors with linear resistors:

- If the circuit without transistors has a pair of conjugate trees complying with the condition of theorem 3.4.20, such is also the case for the circuit with transistors;
- If the resistors can be oriented in such a way as to comply with the conditions of theorem 3.3.36, then the subcircuits of figure 3.57 are oriented according to one of the four possibilities represented in figure 3.58. The zeros inside certain resistors indicate that these branches can also be without orientation. However, such a choice always concerns the pair R_1,R_3 or R_2,R_4. If we include these cases, there are nine possibilities.

Figure 3.58 refers to an *npn* transistor. For a *pnp* transistor, the nine possibilities remain, but their interpretation is different.

Indeed, the second property was previously proved in subsection 3.4.26. With regard to the first property, let us assume that the circuit without transistors has a pair of conjugate trees. If we add the two transistor norators (nullators) to the tree with norators (nullators), we obtain a pair of conjugate trees of the complete circuit such that resistors R_1 to R_4 of figure 3.57 are on the cotree.

3.4.28 Example

Let us return to the transistor amplifier of figure 2.1. First, we want to know whether it can have several dc operating points. Figure 3.60 represents the circuit in which the capacitors are replaced by open circuits and the transistor is replaced by the three-pole of figure 3.58.

We have also reported, in figure 3.59, an attempt to orient the resistors according to the requirements of theorem 3.3.36. To start with, we have chosen the forward active region for the transistor, according to figure 3.58. This sets the orientations of resistors R_6 to R_9. Then R_3 is oriented so as to avoid a uniform cut set R_9,N_2,R_8,R_6,R_3.

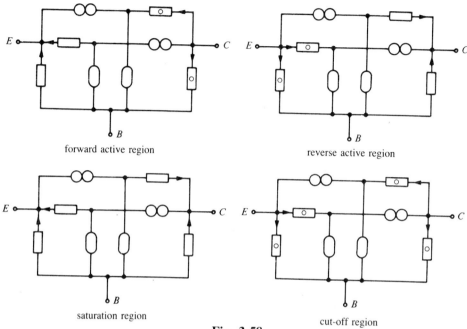

forward active region

reverse active region

saturation region

cut-off region

Fig. 3.58

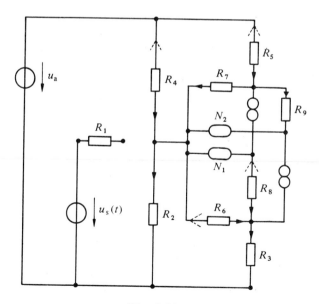

Fig. 3.59

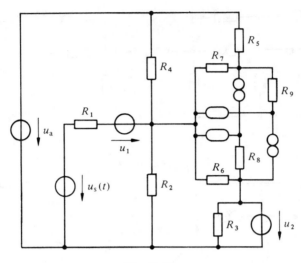

Fig. 3.60

Then, the orientation of R_2 must avoid the uniform loop R_2,R_6,R_3. In a symmetrical way, the orientation of R_5 avoids the uniform cut set R_5,R_7,R_9,N_1,R_8, and that of R_4 avoids the uniform loop R_4,R_7,R_5. However, we have created the uniform loop R_4,R_2,u_a. The alternative of not orienting R_7 and R_9 would not have improved the situation.

We start again with the transistor in the cut-off region. This attempt is marked by arrows in dotted lines wherever they differ from those of the preceding attempt. To avoid a uniform cut set R_2,R_4,R_7,N_2,N_1,R_6, at least one of the two resistors R_2 and R_4 must be oriented from the transistor base outward. In fact, both of them must be oriented this way, otherwise the loop R_2,R_4,u_a is uniform. Then, the possibility of uniform loops R_4,R_5,R_7 and R_2,R_3,R_6 imposes the orientation of R_5 and R_3. However, we have created the uniform cut set R_5,R_7,N_1,N_2,R_6,R_3. It is not difficult to see that the initial possibility for not orienting any of the two transistor junctions comes to a dead end. If we orient neither transistor junction, we necessarily come to the conclusion that no resistor can be oriented. This is also a failure. Finally, the other two operating modes of the transistor amount to an inversion of all the resistor orientations, which does not provide us with anything new. To conclude, corollary 3.4.24 guarantees the uniqueness of the dc operating point.

To apply corollary 3.4.25, we must consider the circuit of figure 3.60 rather than that of figure 3.59. In addition, it includes the sources u_1 and u_2. Resistor R_3 must remain without orientation, otherwise R_3 and u_2 would form a uniform loop. Except for this fact, the reasonings used for the circuit of figure 3.60 are similar to those preceding. To conclude, the circuit of figure 3.60 is always well posed. It follows from corollary 3.4.25 that there exists a system of global state equations for the circuit of figure 2.1, and that a unique solution of the circuit exists for arbitrary initial charges on the capacitors. This solution at least exists during a finite time interval, which excludes impasse points in particular.

3.4.29 Example

The Schmitt trigger (fig. 3.61) is a circuit with several dc operating points (vol. VIII, subsec. 6.2.7). Let us try to apply theorem 3.4.20. By replacing the two transistors with the three-pole of figure 3.57, we obtain the circuit of figure 3.62. In the same figure, the resistors are oriented according to the requirements of theorem 3.3.36. Consequently, several solutions can be expected for the circuit of figure 3.61.

In order to show that the orientations of figure 3.62 comply with the rules of theorem 3.3.36, we must find enough uniform norator-resistor-voltage source loops

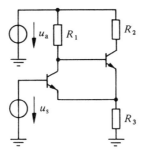

Fig. 3.61

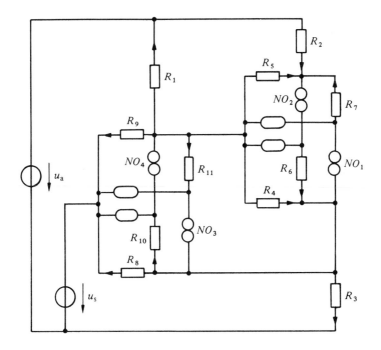

Fig. 3.62

and uniform norator-resistor-current source cut sets such that each resistor is included in at least one loop and one cut set. The following uniform loops are sufficient:

- R_6, NO_1, R_7, NO_2;
- $R_{10}, NO_4, R_{11}, NO_3$;
- $R_9, u_s, u_a, R_2, NO_2, R_6, R_{10}, NO_4$;
- $R_1, R_2, NO_2, R_6, R_{10}, NO_4$;
- R_3, u_a, R_2, NO_2, R_6;
- $R_8, u_s, u_a, R_2, NO_2, R_6$;
- R_4, R_{10}, NO_4;
- $R_5, NO_2, R_6, R_{10}, NO_4$.

The following uniform cut sets are sufficient:.

- $R_3, R_8, R_{10}, NO_3, R_{11}, NO_4, R_9, R_1, R_5, NO_2, R_7$;
- R_2, R_5, NO_2, R_7;
- $NO_1, R_6, R_4, R_{11}, NO_4, R_9, R_1, R_5, NO_2, R_7$.

3.5 PROPERTIES OF THE SOLUTIONS OF RESISTIVE CIRCUITS

3.5.1 Introduction

In this section, we shall study some general properties of the solutions for resistive circuits, the existence and uniqueness of which have been previously discussed. They are properties of any solution that exists.

What we are interested in is the exchange between the circuit environment, represented by the independent sources, and the inside of the circuit. More concretely, we ask whether the fact of knowing the voltages and currents of the sources is sufficient to specify bounds for the internal voltages and currents; or, also, whether a source voltage variation can generate larger voltage variations inside the circuit, which amounts to a *signal amplification*.

3.5.2 Examples

The circuit of figure 3.63 fulfills the function of an OR logic gate (vol. VIII, subsec. 7.2.3). Can voltage u_3 be higher than u_1 and u_2? We assume that the diodes are defined by a constitutive relation $i = g(u)$, with g being given by (1.15). Then,

$$u_3 = R(i_1 + i_2) \tag{3.85}$$
$$= R[g(u_1 - u_3) + g(u_2 - u_3)] \tag{3.86}$$

Let us distinguish several cases. If $u_1 > 0$ and $u_2 > 0$, then $u_3 > u_1$ and $u_3 > u_2$ cannot both be satisfied, because, in that case, we would have $g(u_1 - u_3) < 0$,

$g(u_2 - u_3) < 0$, and, by (3.86), $u_3 < 0$, which is a contradiction. Likewise, $u_3 < 0$ is not possible, because the right-hand member of (3.86) would be positive and, therefore, $u_3 > 0$. By analogous reasoning, for the case $u_1 < 0$, $u_2 < 0$, we exclude the possibilities $u_3 < \min\{u_1, u_2\}$ and $u_3 > 0$; and, for the case, $u_1 > 0$, $u_2 < 0$, we exclude the possibilities $u_3 > u_1$ and $u_3 < u_2$. In short, we have

$$\min\{u_1, u_2, 0\} \leqslant u_3 \leqslant \max\{u_1, u_2, 0\} \tag{3.87}$$

However, such an inequality depends on the circuit structure and the internal voltage considered. For instance, in the circuit of figure 3.64, voltage u_3 is close to $u_1 + u_2$ when the diode is conducting. It can thus exceed the highest voltage among u_1 and u_2. Nonetheless, we can easily see that

$$|u_3| \leqslant |u_1 + u_2| \tag{3.88}$$

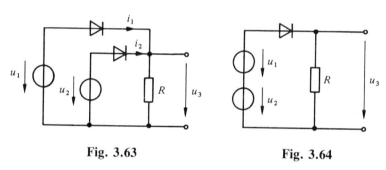

Fig. 3.63   **Fig. 3.64**

Note that the bounds (3.87) and (3.88) do not depend in any way on resistor R. However, this is not always so. The solutions for the circuit of figure 3.65 are given by the intersections of two curves in figure 3.66. We see that for $E < 0$, $u_1 \to \infty$, when $R \to \infty$, and thus $u_2 = -E - u_1 \to -\infty$, this being when the source voltage is held constant.

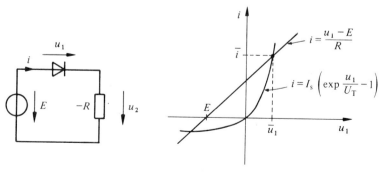

Fig. 3.65   **Fig. 3.66**

3.5.3 Theorem

Suppose a circuit is composed of

- strictly passive resistors;
- voltage and current sources.

Then, for any solution of the circuit, the voltage and the current of each branch that carries a resistor is bounded by

$$|u_k| \leq \sum_{\text{sources}} |u_j| \tag{3.89}$$

$$|i_k| \leq \sum_{\text{sources}} |i_j| \tag{3.90}$$

where the summation ranges over all the branches j that carry a source.

Furthermore, if the circuit is connected, includes at least one source, and all the sources are connected to ground, then the nodes with the maximum potential and the minimum potential are source terminals.

The proof of (3.89) is based on the colored-branch theorem. We color all the branches green and orient them in the same way as their current, which is part of the solution being considered. Any branch k is included in either a uniform cut set, or a uniform loop. For the uniform cut set, we have

$$\sum_{\text{cut set}} i_j = 0 \tag{3.91}$$

and $i_j \geq 0$, because all the branches are oriented as their currents, consequently $i_j = 0$ for all the branches of the cut set, in particular, for branch k. If that branch carries a resistor, we also have $u_k = 0$, due to the strict passivity. Equation (3.89) is thus verified in a trivial way.

If the resistor on branch k is part of a uniform loop, we have

$$0 = \sum_{\text{loop}} u_j \tag{3.92}$$

and, therefore,

$$\sum_{\substack{\text{loop,} \\ \text{resistors}}} u_j = -\sum_{\substack{\text{loop,} \\ \text{sources}}} u_j \tag{3.93}$$

Thanks to the passivity, the resistor voltages are nonnegative, and we can write

$$0 \leq u_k \leq \sum_{\substack{\text{loop,} \\ \text{resistors}}} u_j \leq \sum_{\text{sources}} |u_j| \tag{3.94}$$

The proof of (3.90) is obtained in a dual way by orienting the branches as their voltages.

For the proof of the last property, we assume that node P_0 carries the minimum potential and the branches $1, \ldots ,n$ link P_0 to the nodes $P_1, \ldots ,P_n$. The proof for the maximum potential is similar. In order to give a reference direction for the voltages and the currents, we orient the branches $1, \ldots ,n$ toward P_0 (fig. 3.67).

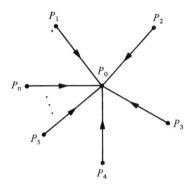

Fig. 3.67

With respect to that orientation, we have $u_k \geqslant 0$, $k = 1, \ldots ,n$. If a branch k carries a resistor, we also have $i_k \geqslant 0$, due to the passivity. Because

$$\sum_{k=1}^{n} i_k = 0 \tag{3.95}$$

there is either a source among the branches $1, \ldots ,n$, or $i_k = 0$ for $k = 1, \ldots ,n$. The latter case would imply, as a result of the strict passivity, that $u_k = 0$, $k = 1, \ldots ,n$, which means that the nodes $P_1, \ldots ,P_n$ are also at the minimum potential. We now repeat the reasoning for $P_1, \ldots ,P_n$, *et cetera*. Then, either we find a node with the minimum potential somewhere in the circuit to which a source is connected, or all the nodes are at the same potential. Because there is at least one source, the property holds, even in this case.

3.5.4 Comments

It should be noted that the sums in (3.89) and (3.90) range over *all* the sources, whether they are voltage or current sources. Bounds in terms of voltages of voltage sources and current of current sources necessarily involve the characteristics of the resistors. Bounds of this type will be given in theorem 3.5.6 and 3.5.11.

The examples of subsection 3.5.2 adequately illustrate theorem 3.5.3. In the circuit of figure 3.63, where both sources are connected to ground, the extreme potentials are indeed at the terminals of the sources, as shown by (3.87). In the circuit of figure 3.64, the two sources actually have a common terminal, but the voltage u_3 is the potential difference between two other nodes. Consequently, only the inequality (3.89) is applicable. Finally, the circuit of figure 3.65 includes an active resistor $-R$ and theorem 3.5.3 is no longer applicable.

When operational amplifiers and controlled sources are present, the theorem does not hold. For instance, the circuit of figure 3.68 transforms the input voltage into the output voltage:

$$u = -(R_2/R_1)E \qquad (3.96)$$

which tends toward infinity if $R_1 \to 0$.

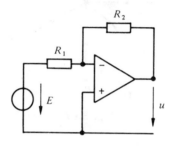

Fig. 3.68

Nevertheless, certain subcircuits, including operational amplifiers or controlled sources, for a given solution, behave as passive resistors. This is particularly the case for transistors. By using this approach, theorem 3.5.3 is applicable to a much wider class of circuits than that of its statement. We shall formulate the extension to the case of bipolar transistors in corollary 3.5.5. For other types of transistors, similar corollaries will be verified.

3.5.5 Corollary

Suppose a connected circuit is composed of

- strictly passive resistors;
- bipolar transistors modeled by Ebers-Moll equations;
- ground connected voltage and current sources.

There is at least one source.

Hence, the nodes with the minimum potential and the maximum potential are source terminals.

Indeed, the Ebers-Moll equations have the form (2.1), with function g being strictly increasing and passing through the origin, be it an *npn* or a *pnp* transistor. Let us assume that we have a solution of the circuit, and that the voltages and currents of a transistor are $\bar{u}_1$, $\bar{u}_2$, $\bar{i}_1$, $\bar{i}_2$. Let us consider the two-port of figure 3.69, composed of three linear resistors:

$$R_1 = \beta_F \bar{u}_1 / g(\bar{u}_1) \tag{3.97}$$

$$R_2 = \beta_R \bar{u}_2 / g(\bar{u}_2) \tag{3.98}$$

$$R_3 = (\bar{u}_1 - \bar{u}_2)/(\bar{i}_1 - (1/\beta_F) g(\bar{u}_1)) \tag{3.99}$$

Because the graph of $R_1 > 0$ does not pass across the first and the third quadrants, $R_2 > 0$. Relation (3.99) can be rewritten as

$$R_3 = (\bar{u}_1 - \bar{u}_2)/(g(\bar{u}_1) - g(\bar{u}_2)) \tag{3.100}$$

Because g is strictly increasing, $R_3 > 0$.

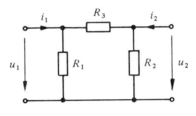

Fig. 3.69

We conclude that if we replace the transistor by the two-port of figure 3.69, having resistors given by (3.97) to (3.99), the solution of the original circuit under study remains a solution of the transformed circuit. By substituting all the transistors with the two-port of figure 3.69, each one having other resistors, we obtain a circuit composed exclusively of strictly passive resistors, and theorem 3.5.3 allows us to guarantee the inequalities of potentials.

3.5.6 Comment

Corollary 3.5.5 establishes in a rigorous way the well known fact that the potentials of the dc operating point of a circuit with transistors is located between the power supply potentials. As a result, neither an ideal operational amplifier nor a negative resistor can be synthesized by transistors, strictly passive resistors, and sources. Nevertheless, all the operational amplifiers we find on the market today are circuits of this type. Consequently, the only way they can fulfill the function of an ideal operational amplifier with good approximation is by having voltages much lower than the supply voltages. Outside this domain of almost ideal operation, a saturation effect occurs.

3.5.7   Theorem

Suppose a circuit is composed of

- strictly passive resistors such that any point $(u_k, i_k) \neq (0, 0)$ of their characteristics satisfies

$$R_{k-} \leqslant u_k/i_k \leqslant R_{k+} \tag{3.101}$$

where R_{k+} and R_{k-} are two finite resistors (fig. 3.70);
- voltage and current sources.

Then, for any solution of the circuit, the voltage and the current of each branch k that carries a resistor is bounded by

$$|u_k| \leqslant \max \left\{ \sum_{\text{voltage sources}} |u_j|, \ R_{k+} \sum_{\text{current sources}} |i_j| \right\} \tag{3.102}$$

$$|i_k| \leqslant \max \left\{ (1/R_{k-}) \sum_{\text{voltage sources}} |u_j|, \ \sum_{\text{current sources}} |i_j| \right\} \tag{3.103}$$

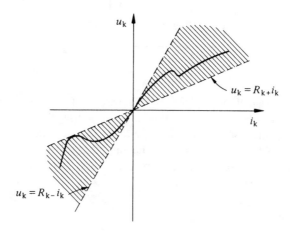

Fig. 3.70

The proof is slightly different from that of the inequalities (3.89) and (3.90). We only color the resistors in green and orient them as their voltage and current, while the voltage (current) sources are colored in red (blue). If the resistor on branch k is part of a uniform loop of green and red branches, we have

$$0 \leqslant u_k \leqslant \sum_{\text{loop, resistors}} u_j$$

$$= \sum_{\text{loop, sources}} u_j \leqslant \sum_{\text{voltage sources}} |u_j| \tag{3.104}$$

and, therefore,

$$0 \leqslant i_k \leqslant (1/R_{k-}) \sum_{\text{voltage sources}} |u_j| \tag{3.105}$$

On the other hand, if branch k is part of a cut set of green and blue branches, we have

$$0 \leqslant i_k \leqslant \sum_{\text{cut set, resistors}} i_j = \sum_{\text{cut set, sources}} i_j \leqslant \sum_{\text{current sources}} |i_j| \tag{3.106}$$

and, therefore,

$$0 \leqslant u_k \leqslant R_{k+}{}^* \sum_{\text{current sources}} |i_j| \tag{3.107}$$

The inequalities (3.104) to (3.107) imply (3.102) and (3.103).

3.5.8 Application

Let us consider the frequency multiplier of figure 3.30. If we replace the capacitors (inductors) by voltage (current) sources, we obtain the resistive circuit of figure 3.71. According to theorem 3.5.7, the power P_7 dissipated in the charge R_7 is bounded by

$$P_7 \leqslant \max \{(1/R_7)(|u_s|+|u_1|+|u_2|+|u_3|)^2, R_7(|i_4|+|i_5|)^2 \} \tag{3.108}$$

thanks to the inequality:

$$ab \leqslant (1/2)(a^2+b^2) \tag{3.109}$$

which is inferred from

$$(1/2)(a-b)^2 \geqslant 0 \tag{3.110}$$

we find

$$P_7 \leqslant \max \{(5/2R_7)(u_s^2+u_1^2+u_2^2+u_3^2), (3R_7/2)(i_4^2+i_5^2)\} \tag{3.111}$$

The energies stored in capacitor C_k and inductor L_j are, respectively,

$$W_k = C_k u_k^2/2 \tag{3.112}$$
$$W_j = L_j i_j^2/2 \tag{3.113}$$

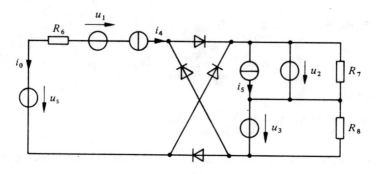

Fig. 3.71

If $C_- = \min\{C_1, C_2, C_3\}$, $L_- = \min\{L_4, L_5\}$, and $W_C = W_1 + W_2 + W_3$, $W_L = W_4 + W_5$, we find

$$P_7 \leqslant \max\{(5u_s^2/2R_7) + (5W_C/R_7C_-), 3R_7W_LL_-\} \tag{3.114}$$

Thus, at each instant, the power dissipated in the charge is bounded by an expression involving the source voltage and the energy stored in the capacitors and the inductors.

3.5.9 Theorem

Suppose a circuit is composed of

- resistors;
- voltage and current sources.

Let us consider a solution of the circuit such that the resistor voltages are bounded by U and the resistor currents by I. Then,

- if there is no loop of voltage sources nor any cut set of current sources, the voltages on the current sources are bounded by

$$|u_k| \leqslant N_R U + \sum_{\substack{\text{voltage sources}}} |u_j| \tag{3.115}$$

and the currents in the voltage sources are bounded by

$$|i_\varrho| \leqslant N_R I + \sum_{\substack{\text{current sources}}} |i_j| \tag{3.116}$$

- if there is no loop nor any cut set composed of voltage sources **and** current sources, the voltages on the current sources are bounded by

$$|u_k| \leqslant N_R U \qquad (3.117)$$

and the currents in the voltage sources are bounded by

$$|i_\varrho| \leqslant N_R I \qquad (3.118)$$

where N_R designates the number of resistors of the circuit.

To prove (3.115), we apply corollary 3.1.15 of the colored branch theorem. Let us assume that branch v is the current source located on branch k. The branches of B are the other current sources, and the branches of R are the resistors and the voltage sources. Because, by hypothesis, there is no cut set of current sources, there is a loop with v and branches of B. Kirchhoff's law for such a loop implies

$$|u_k| = \left| \sum_{\substack{\text{loop} \\ j \neq k}} u_j \right| \qquad (3.119)$$

$$\leqslant \sum_{\substack{\text{loop,} \\ \text{resistors}}} |u_j| + \sum_{\substack{\text{loop,} \\ \text{voltage sources}}} |u_j| \qquad (3.120)$$

$$\leqslant N_R U + \sum_{\text{voltage sources}} |u_j| \qquad (3.121)$$

The proof of (3.116) is dual.

The proof of (3.117) is similar to that of (3.115), except that the voltage sources are part of set B rather than set R. Finally, (3.118) is obtained by the dual argument.

3.5.10   Application

Let us return to the frequency multiplier of figure 3.30. Let us assume that the Graetz bridge is composed of piecewise-linear diodes defined by the constitutive relation (1.16). In addition, we suppose that the linear resistors of the circuit have values between the conduction resistance and the blocking resistance of the diodes:

$$g_0 \leqslant (1/R_k) \leqslant g_s, \quad k = 6, 7, 8 \qquad (3.122)$$

Then, according to theorem 3.5.7, the voltages on the resistors are bounded by

$$U = \max \{|u_s| + |u_1| + |u_2| + |u_3|, \ (|i_4| + |i_5|)/g_0\} \qquad (3.123)$$

and their currents are bounded by

$$I = \max\{g_s(|u_s|+|u_1|+|u_2|+|u_3|),\ |i_4|+|i_5|\} \tag{3.124}$$

Finally, we obtain the following bounds for the voltages on the inductors:

$$|u_k| \leqslant 8 \cdot \max\{|u_s|+|u_1|+|u_2|+|u_3|,\ (|i_4|+|i_5|)/g_0\},\quad k = 4, 5 \tag{3.125}$$

and for the currents in the voltage source and the capacitors

$$|i_k| \leqslant 8 \cdot \max\{g_s(|u_s|+|u_1|+|u_2|+|u_3|),\ |i_4|+|i_5|\},\quad k = 0, 1, 2, 3 \tag{3.126}$$

3.5.11 Comment

Unlike theorem 3.5.7, theorem 3.5.9 does not depend on the resistors' passivity. However, the bounds (3.115) to (3.118) are rather large, due to factor N_R. Nevertheless, they permit us to say that, for circuits such as the frequency multiplier of figure 3.30, **all** the voltages and **all** the currents remain bounded at any time, if the values of the sources, the voltages on the capacitors and the currents in the inductors remain bounded.

If a circuit composed of sources and resistors includes active resistors, some currents and voltages may become arbitrarily large, even if the sources remain bounded. The circuit of figure 3.65 illustrates this possibility. The phenomenon is, however, excluded if the active resistors are eventually passive. The reason for this is simple. As soon as a current or voltage becomes large, the corresponding point on the characteristic curve of the resistor concerned is located within the passive region.

3.5.12 Theorem

Suppose a circuit is composed of

- strictly eventually passive resistors, such that

$$u_k i_k > 0 \quad \text{for} \quad (u_k^2/R_0) + R_0 i_k^2 > K^2$$

with two finite positive constants R_0 and K, independent of k;

- voltage and current sources.

Hence, the following properties hold for any solution. If branch k carries a resistor, we have

$$|u_k| \leqslant \sum_{\text{sources}} |u_j| + N_{AP} K \sqrt{R_0} \tag{3.128}$$

$$|i_k| \leqslant \sum_{\text{sources}} |i_j| + N_{AP} K/\sqrt{R_0} \tag{3.129}$$

- If the resistor on branch k satisfies

$$R_k \leqslant u_k/i_k \leqslant R_{k+} \qquad \text{for} \qquad (u_k^2/R_0 + R_0 i_k^2) > K^2 \qquad (3.130)$$

then,

$$|u_k| \leqslant \max \left\{ \sum_{\text{voltage sources}} |u_j| + N_{AP} K \sqrt{R_0}, \right.$$

$$R_{k+} \cdot \left(\sum_{\text{current sources}} |i_j| + N_{AP} K/\sqrt{R_0} \right) \right\} \qquad (3.131)$$

$$|i_k| \leqslant \max \left\{ (1/R_{k-}) \left(\sum_{\text{voltage sources}} |u_j| + N_{AP} K \sqrt{R_0} \right) \right.$$

$$\left. \left(\sum_{\text{current sources}} |i_j| + N_{AP} K/\sqrt{R_0} \right) \right\} \qquad (3.132)$$

when N_{AP} stands for the number of strictly eventually passive resistors which are not strictly passive.

We prove (3.128) in almost the same way as (3.89). By coloring all the branches green and orienting them according to their current, we find either a uniform cut set, or a uniform loop through k. In the first case, $i_k = 0$. If the resistor is strictly passive, $u_k = 0$. Otherwise $|u_k| \leqslant K \sqrt{R_0}$. In any case, (3.128) is satisfied.

If there is a uniform loop, we part the loop resistors into two sets, A and B. Branch j belongs to A if $u_j i_j \leqslant 0$. This is only possible if $u_j = 0 = i_j$ or, because of (3.127), if $|u_j| \leqslant K \sqrt{R_0}$. If branch k belongs to A, $|u_k| \leqslant 0$, or $|u_k| \leqslant K \sqrt{R_0}$, and (3.128) is satisfied. If branch k belongs to B, we can write

$$0 < u_k \leqslant \sum_B u_j \qquad (3.133)$$

$$= -\sum_A u_j - \sum_{\text{sources}} |u_j| \qquad (3.134)$$

$$\leqslant N_{AP} K \sqrt{R_0} + \sum_{\text{sources}} |u_j| \qquad (3.135)$$

The proof of (3.129) is analogous. In a similar way, we transform the proofs of (3.102) and (3.103) into proofs of (3.131) and (3.132).

3.5.13 Example

In the rectifier of figure 3.72, we can combine the voltage source with the resistor R_1 into an eventually passive resistor. For different instants, its characteristic is different. The region of plane (u, i) covered by the characteristics by varying t, is cross-hatched in figure 3.73. Then, we have for any t:

$$u \cdot i > 0, \quad \text{if} \quad u^2/R_1 + R_1 i^2 > E^2/R_1 \tag{3.136}$$

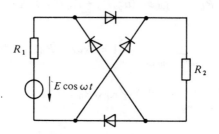

Fig. 3.72

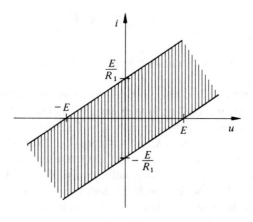

Fig. 3.73

Because $N_{AP} = 1$, $R_0 = R_1$, $K = E/\sqrt{R_1}$, we find, from (3.128) and (3.129), for all branches:

$$|u_k| \leqslant E \tag{3.137}$$

$$|i_k| \leqslant E/R_1 \tag{3.138}$$

This reasoning is generalized in the following corollary.

3.5.14 Corollary

Suppose we have a circuit composed of

- strictly passive resistors;
- voltage sources with the values $E_1, \ldots, E_n$ and the internal resistors $R_1, \ldots, R_n$.

Then, the voltage and the current of each branch are bounded by

$$|u_k| \leqslant \sum_{j=1}^{n} |E_j| \tag{3.139}$$

$$|i_k| \leqslant \sum_{j=1}^{n} |E_j|/R_j \tag{3.140}$$

The proof is that of (3.128) and (3.129), except that we use the bounds $|E_j|$ and $|E_j|/R_j$ directly for the voltage and current of source j. Also, note that (3.139) already results from theorem 3.5.3.

3.5.15 Signal Amplification

An amplifier such as the one in figure 2.1 includes reactances, and therefore it is not a circuit to be studied in this chapter. However, the reactive part normally fulfills a filtering function. The spectrum of the signal to be amplified is within the passband. As a good approximation, the operation of the amplifier in the passband, except possibly for a delay, is that of a resistive circuit. In the case of the amplifier in figure 2.1, the capacitors filter the dc components, and, as far as signal propagation is concerned, they can be replaced by short circuits.

If we consider signal amplification by a resistive circuit, the absolute values of the voltages are not what matters, but their variations as functions of the signal. Thus, the amplification possibilities of a resistive circuit are obtained by studying the internal voltage variations of the circuit as a function of a source variation.

3.5.16 Theorem

Suppose a circuit is composed of

- strictly locally passive resistors;
- voltage and current sources.

Let us consider a solution of this circuit and a solution obtained by changing the voltage of a voltage source by ΔE. If branch k carries a resistor, and Δu_k designates the difference between the voltages of the two solutions on the resistor, then

$$|\Delta u_k| \leqslant |\Delta E| \tag{3.141}$$

The proof is similar to that of theorem 3.5.7. We color the branches and orient them in the same way, but we replace u_j and i_j by Δu_j and Δi_j, respectively. Equation (3.104) becomes

$$0 \leqslant \Delta u_k \leqslant |\Delta E| \tag{3.142}$$

which proves the theorem for this case. In the other case, (3.106) becomes

$$0 \leqslant \Delta i_k \leqslant 0 \tag{3.143}$$

because the current sources are the same for the two solutions. Consequently, $\Delta i_k = 0$, and due to the strict local passivity, $\Delta u_k = 0$. Thus, in both cases, (3.141) is satisfied.

3.5.17 Comment

In view of the remarks of subsection 3.5.15, (3.141) implies that a circuit composed exclusively of strictly locally passive resistors (e.g., diodes) and sources cannot perform as an amplifier. Therefore, we must add to the list of elements either some locally active resistors, such as a tunnel diode, or some controlled sources, operational amplifiers, or transistors.

It is important to distinguish the incidence of passivity from the incidence of local passivity on the solution of a resistive circuit. If the resistors are passive, there will be bounds on voltages and currents. Such is the case for exponential diodes as well as tunnel diodes. Conversely, if the resistors are locally passive, there will be bounds on the voltage and current *increments*. If the *variations* of a signal source must be amplified, a locally active resistor, such as, for example, a tunnel diode, will be required.

With regard to the transistors, we have seen in subsection 3.5.5 that they behave like passive resistors, and therefore, their voltages and currents are bounded by the source voltages and currents, in particular, those of the power supply sources. Nonetheless, it is a well known fact—confirmed by the amplifier of figure 2.1—that transistors are capable of amplifying a signal. Thus, they behave like locally active resistors.

Chapter 4

Circuit Transformations

4.1 DUALITY

4.1.1 Introduction

The concept of duality is found in many scientific disciplines and circuit theory is no exception. In fact, there is a duality at the element level and a duality at the connection level. At the element level, it results from the exchange of the voltage and current variables. At the connection level, it stems from the correspondence between loops and cut sets.

We have already alluded to duality in some proofs. As a general rule, a theorem always has a dual theorem which is proved by the dual proof. One need only have an original-dual dictionary. Such a dictionary has been given in vol. IV, subsection 1.2.13; now, we shall extend it to nonlinear circuits.

4.1.2 Definitions

Let two N-ports Q_1 and Q_2 be defined by N constitutive relations involving the voltages at the ports $u_1(t), \ldots , u_n(t)$, and the currents to the ports $i_1(t), \ldots , i_N(t)$ as a function of time. Two *N-ports* Q_1 and Q_2 *are dual* if there is a resistance R_0, the *duality resistance*, such that the following property holds. If the column matrix composed of the functions $u_1(t), \ldots , u_N(t)$ and $i_1(t), \ldots , i_N(t)$ satisfies the constitutive relations of Q_1, then the column matrix composed of the functions $\tilde{u}_1(t), \ldots , \tilde{u}_N(t)$ and $\tilde{\imath}_1(t), \ldots , \tilde{\imath}_N(t)$ satisfies the constitutive relations of Q_2, and *vice versa*, where

$$\tilde{u}_k(t) = R_0\, i_k(t) \tag{4.1}$$
$$\tilde{\imath}_k(t) = u_k(t)/R_0 \tag{4.2}$$

4.1.3 Property

Consider a resistor defined by the constitutive relation:

$$f(u, i) = 0 \tag{4.3}$$

214

Then, any dual one-port is again a resistor. If the duality resistance is R_0, then the dual one-port complies with the constitutive relation:

$$0 = \tilde{f}(u, i) = f(R_0 i, u/R_0) \tag{4.4}$$

The dual resistor characteristic is constructed point by point from the original characteristic curve, according to figure 4.2.

For the proof, we apply definition 4.1.2. In the case of resistors, it is not necessary to consider the voltages and currents as functions of time, because (4.3) relates u and i separately at each instant t. By definition 4.1.2, the points of the plane u, i, which satisfy the constitutive relation of the dual one-port are $(\bar{u}, \bar{i})$, with $\bar{u} = R_0 i$ and $\bar{i} = u/R_0$, where (u, i) satisfies (4.3). This fact is illustrated by (4.4) and figure 4.1.

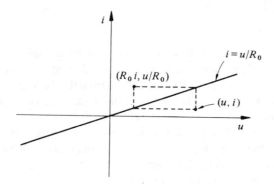

Fig. 4.1

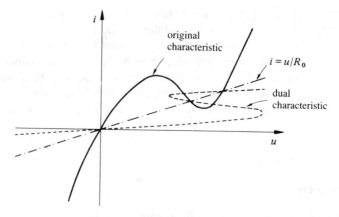

Fig. 4.2

4.1.4 Properties

If a resistor is voltage controlled, with the constitutive relation:

$$i = g(u) \tag{4.5}$$

then the dual resistors are current controlled, with the constitutive relations:

$$u = R_0 g(R_0 i) \tag{4.6}$$

(example of fig. 4.2). Conversely, if a resistor is voltage controlled, with the constitutive relation:

$$u = h(i) \tag{4.7}$$

then the dual resistors are voltage controlled, with the constitutive relations:

$$i = h(u/R_0)/R_0 \tag{4.8}$$

Indeed, the (4.6) and (4.8) are special cases of (4.4), by setting $f(u,i) = i - g(u)$ and $f(u,i) = u - h(i)$, respectively.

4.1.5 Property

The transformation of a resistor into its dual leaves invariant all of the concepts of passivity and local passivity introduced in section 3.2.

Indeed, we note that the dual resistor, at point $(\bar{u},\bar{\imath})$, absorbs the same power as the original resistor does at the corresponding point (u,i). Likewise, the incremental passivity absorbed by the dual resistor at the pair of points $(\bar{u}_1,\bar{\imath}_1)$, $(\bar{u}_2,\bar{\imath}_2)$ coincides with the incremental power absorbed by the original resistor at the corresponding pair of points (u_1,i_1) and (u_2,i_2). This proves the invariance of the local passivity concepts under duality transformation.

4.1.6 Properties

Consider a capacitor defined by the constitutive relation:

$$f(u, q) = 0 \tag{4.9}$$

Hence, any dual one-port is an inductor. If the duality resistance is R_0, then the dual one-port satisfies the constitutive relation:

$$0 = \tilde{f}(\varphi, i) = f(R_0 i, \varphi/R_0) \tag{4.10}$$

The construction of the inductance characteristic from the capacitance characteristic is analogous to that of figure 4.1. We need only simultaneously change the axis labels from u to φ, and from q to i.

If a capacitor is charge controlled, according to the constitutive relation:

$$u = h(q) \tag{4.11}$$

then the dual inductors are flux controlled, and they satisfy the constitutive relations:

$$i = h(\varphi/R_0)/R_0 \tag{4.12}$$

However, if the capacitor is voltage controlled, according to

$$q = g(u) \tag{4.13}$$

then the dual inductors are current controlled, according to the constitutive relations:

$$\varphi = R_0 g(R_0 i) \tag{4.14}$$

Conversely, the dual one-ports of an inductor are capacitors, and the corresponding formulas can be written.

For the proof, we note that two functions $u(t)$ and $i(t)$ satisfy the constitutive relation of the capacitor if and only if there exists a function $q(t)$, such that

$$f(u(t), q(t)) = 0 \tag{4.15}$$

and

$$i(t) = dq/dt \tag{4.16}$$

According to definition 4.1.2, it follows that the functions $\tilde{u}(t)$ and $\tilde{i}(t)$ satisfy the constitutive relation of the dual one-port, if and only if there exists a function $\tilde{q}(t)$, such that

$$f(R_0 \tilde{i}(t), \tilde{q}(t)) = 0 \tag{4.17}$$

and

$$\tilde{u}(t)/R_0 = d\tilde{q}/dt \tag{4.18}$$

By setting $\tilde{q}(t) = q(t)$ and $\tilde{\varphi}(t) = R_0 \tilde{q}(t)$, we obtain as a necessary and sufficient condition the existence of a function such that

$$f(R_0 \tilde{i}(t), \tilde{\varphi}(t)/R_0) = 0 \tag{4.19}$$

and

$$\tilde{u}(t) = d\tilde{\varphi}/dt \tag{4.20}$$

Thus, it is indeed an inductor with the constitutive relation (4.10). The rest of the proof is analogous to that of property 4.1.4.

4.1.7 Summary: Dual Basic Elements

The duality transformation applied to the linear elements has previously been covered in volume IV, and here we shall simply summarize the results together with those of subsections 4.1.3, 4.1.4, and 4.1.6 in figure 4.3. We observe that the set of basic elements introduced in section 1.3 remains invariant under duality.

4.1.8 Reminders: Dual Graph

The dual of an oriented graph has been introduced in volume IV (sec. 5.1). The branches of the dual graph correspond in a bijective manner to the branches of the original graph. Their orientation is fixed, except for one simultaneous inversion of all the branches.

If $u_1, \ldots, u_b$ is a set of voltages which satisfies the Kirchhoff voltage equations of the original graph, then the set of currents $\tilde{\imath}_1, \ldots, \tilde{\imath}_b$ with

$$\tilde{\imath}_k = u_k/R_0 \tag{4.21}$$

satisfies the Kirchhoff current equations of the dual graph for any duality resistance R_0. It is understood that $\tilde{\imath}_k$ and u_k are the current and the voltage of the corresponding branches, the signs of which are defined with respect to the corresponding orientations.

Likewise, if $i_1, \ldots, i_b$ is a set of currents which satisfies the Kirchhoff current equations of the original graph, then the set of voltages $\tilde{u}_1, \ldots, \tilde{u}_b$ with

$$\tilde{u}_k = R_0 i_k \tag{4.22}$$

satisfies the Kirchhoff voltage equations of the dual graph for any duality resistance R_0.

As stated in the form of a reminder in volume IV (sec. 5.1.4), the dual graph exists if and only if the original graph is planar.

4.1.9 Definition

Let us assume that the graph associated with a circuit is planar. Then, the *dual circuit* is obtained through replacing the graph by the dual graph, and placing on each branch (pair) the element that is the dual of the one located on the corresponding branch (pair) of the original circuit.

4.1.10 Comment

The transformation of a circuit by duality depends on an arbitrary duality resistance, as well as on the choice of an orientation in the transition from the graph to

Element		Dual element (duality res. R_0)	
Symbol	Constitutive relations	Symbol	Constitutive relations
	$f(u, q) = 0$ $u = h(q)$ $q = g(u)$		$f(R_0 i, \varphi/R_0) = 0$ $i = h(\varphi/R_0)/R_0$ $\varphi = R_0 g(R_0 i)$
	$f(\varphi, i) = 0$ $i = g(\varphi)$ $\varphi = h(i)$		$f(R_0 q, u/R_0) = 0$ $u = R_0 g(R_0 q)$ $q = h(u/R_0)/R_0$
	$u = e(t)$		$i = e(t)/R_0$
	$i = e(t)$		$u = R_0 e(t)$
	$f(u, i) = 0$ $i = g(u)$ $u = h(i)$		$f(R_0 i, u/R_0) = 0$ $u = R_0 g(R_0 i)$ $i = h(u/R_0)/R_0$
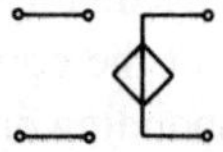	$i_1 = 0, \quad u_2 = \alpha u_1$	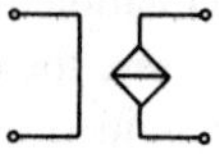	$u_1 = 0, \quad i_2 = \alpha i_1$
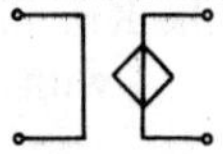	$u_1 = 0, \quad u_2 = r i_1$	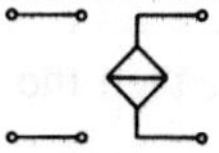	$i_1 = 0, \quad i_2 = r/R_0^2 u_1$
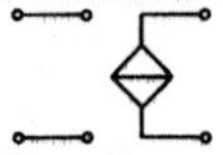	$i_1 = 0, \quad i_2 = g u_1$	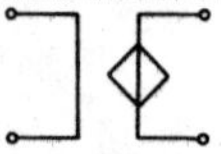	$u_1 = 0, \quad u_2 = R_0^2 g i_1$
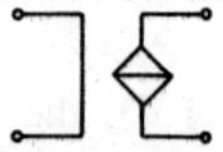	$u_1 = 0, \quad i_2 = \alpha i_1$	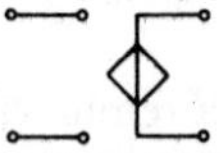	$i_1 = 0, \quad u_2 = \alpha u_1$
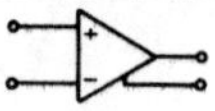	$u_1 = 0, \quad i_1 = 0$	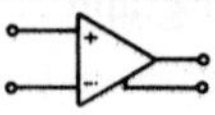	$u_1 = 0, \quad i_1 = 0$
	$u = 0, \quad i = 0$		$u = 0, \quad i = 0$
	–		–

Fig. 4.3

the dual graph. However, the various dual circuits differ only by their impedance level and a simultaneous inversion of all the branches, which is why they can all be designated by *the* dual circuit.

4.1.11 Theorem

If $\xi(t) = (u(t), i(t), q_C(t), \varphi_L(t))$ is a solution of a circuit with a planar graph, then $\tilde{\xi}(t) = (\tilde{u}(t), \tilde{i}(t), \tilde{q}_C(t), \tilde{\varphi}_L(t))$ is a solution of the dual circuit, and *vice versa*, where

$$\tilde{u}_k(t) = R_0 i_k(t) \tag{4.23}$$
$$\tilde{i}_k(t) = u_k(t)/R_0 \tag{4.24}$$
$$\tilde{q}_k(t) = \varphi_k(t)/R_0 \tag{4.25}$$
$$\tilde{\varphi}_k(t) = R_0 q_k(t) \tag{4.26}$$

Here, the numbering of the dual circuit branches is such that the corresponding branches of the original circuit carry the same index. Furthermore, the signs of the variables in both circuits are defined with respect to the corresponding orientations.

4.1.12 Comments

Theorem 4.1.11 means that two dual nonlinear circuits have the same qualitative behavior. For instance, if one has three dc operating points, then the other has three as well. Also, if one has impasse points, so too does the other. Except for the small changes in the transition from (4.23) to (4.26), the solutions are the same, even quantitatively. For instance, the output of the frequency multiplier of figure 4.4 is identical to that of figure 3.30 if the nonlinear resistors are the dual resistors of the diodes.

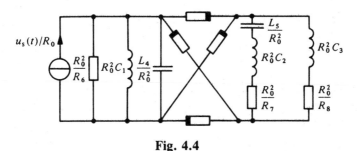

Fig. 4.4

If theorem 4.1.11 is only applicable to circuits with a planar graph, it remains that the duality concept has a much wider application field. As mentioned in section 4.1.1, a theorem and a proof, in principle, may always be transformed into a dual theorem and a dual proof, without necessarily requiring that the circuits under consideration be planar.

4.2 ADJOINT CIRCUIT, RECIPROCITY, AND CALCULATION OF THE SENSITIVITIES

4.2.1 Introduction

The concept of an *adjoint circuit* has been introduced to facilitate the calculation of the sensitivities. The *sensitivity* of a circuit voltage (current) with respect to a parameter is the voltage (current) variation rate per variation unit of the parameter. This quantity is precious information for the engineer who designs a circuit. For instance, if the output voltage of a filter is very sensitive to the value of one of its capacitors, a physical implementation of the filter might fail to satisfy the filtering requirements because a capacitor, as a physical element, does not exactly realize its nominal value.

Very roughly, the adjoint circuit is characterized by the following property. The influence of branch k on branch l in the original circuit is the same as the influence of branch l on branch k in the adjoint circuit. Just as roughly, a circuit which is equal to its adjoint is called *reciprocal*. Thus, this very important property guarantees a symmetry between the influence of branch k on branch l, and of branch l on branch k. This symmetry manifests itself, in the case of linear circuits, by the symmetry of the different impedance and admittance matrices.

We shall limit our study of the adjoint circuit to linear and nonlinear resistive circuits. For linear circuits, the definition of the adjoint circuit is easily generalized to circuits having capacitors and inductors in the domain of Laplace transforms. However, if a circuit with capacitors and inductors is nonlinear, we must stay within the time domain, and the definition of the adjoint circuit becomes complicated. We refer to [18].

In linear algebra, some properties of the symmetrical matrices are generalized by considering a nonsymmetrical matrix together with its transposed matrix. Analogously, we may simultaneously study a nonreciprocal circuit and its adjoint circuit. This shows that the concept of adjoint circuit is more than just a tool for the calculation of the sensitivities.

4.2.2 Definition

Let C and $\tilde{C}$ be two well-posed autonomous linear resistive circuits having the same graph. Circuit $\tilde{C}$ is an *adjoint circuit* of C if, for **any pair of branches k and l,** the following four properties are satisfied:

- If we add a voltage source E_k in series with branch k of C, and a voltage source E_l in series with branch l of $\tilde{C}$, then the current increments Δi_l and $\Delta \tilde{i}_k$, caused by E_k and $\tilde{E}_l$, respectively in branches l and k, satisfy

$$\Delta i_\varrho / E_k = \Delta \tilde{i}_k / \tilde{E}_\varrho ; \tag{4.27}$$

- If we add a current source $I_k(\tilde{I}_k)$ in parallel with branch $k(l)$ of $C(\tilde{C})$, we have

$$\Delta u_\varrho / I_k = \Delta \tilde{u}_k / \tilde{I}_\varrho; \tag{4.28}$$

- If we add a voltage source E_k (current source $\tilde{I}_l$) in series (in parallel) with branch $k(l)$ of $C(\bar{C})$, we have

$$\Delta u_\varrho / E_k = \Delta \tilde{i}_k / \tilde{I}_\varrho; \tag{4.29}$$

- If we add a current source I_k (current source $\tilde{E}_k$) in parallel (in series) with branch $k(l)$ of $C(\bar{C})$, we have

$$\Delta i_\varrho / I_k = -\Delta \tilde{u}_k / \tilde{E}_\varrho \tag{4 30}$$

These four properties are represented in figure 4.5. If circuit C is its own adjoint, it is a *reciprocal circuit*.

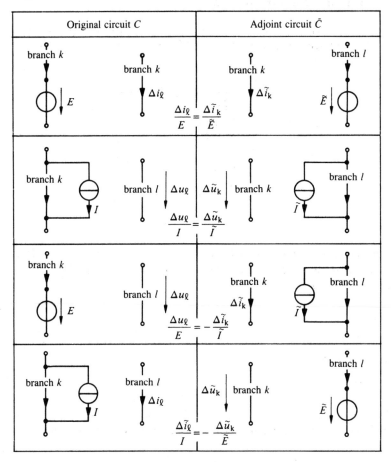

Fig. 4:5

222

4.2.3 Comments

Because the adjoint circuit is defined by four conditions, it is not guaranteed, *a priori*, that such a circuit exists. This fact shall be proved in subsection 4.2.5.

It is necessary, by definition 4.2.2, to impose the condition that C and $\bar{C}$ be well posed. In fact, if C is ill posed, it has either an infinity of solutions, or no solution. In both cases, Δi_l and Δu_l are not well defined. Similarly, if $\bar{C}$ is ill posed, then $\Delta \bar{u}_k$ and $\Delta \bar{\imath}_l$ are not well defined.

We can limit ourselves to **autonomous** resistive circuits, without restricting the generality. As mentioned in subsection 3.2.1, the study of a time-dependent resistive circuit amounts to studying a set of autonomous resistive circuits: one for each instant.

We can see that definition 4.2.2 is symmetrical in C and $\bar{C}$. If $\bar{C}$ is an adjoint circuit of C, C is also an adjoint circuit of $\bar{C}$.

4.2.4 Independence of the Conditions

We could imagine that the four conditions of subsection 4.2.2 are not independent. The following examples are intended to give evidence of their independence.

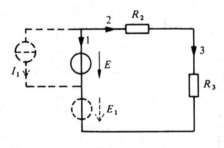

Fig. 4.6

Let us consider the circuit of figure 4.6. If we add in branch 1 a voltage source of value E_1, or a current source of value I_1, indicated by a dotted line in figure 4.6, we find

$$\Delta i_3/E_1 = 1/(R_2 + R_3) \tag{4.31}$$

$$\Delta u_3/I_1 = 0 \tag{4.32}$$

$$\Delta u_3/E_1 = R_3/(R_2 + R_3) \tag{4.33}$$

$$\Delta i_3/I_1 = 0 \tag{4.34}$$

Let the circuit of figure 4.7 be a candidate adjoint circuit. The sources added by dotted lines in figure 4.7 cause the following increments in branch 1

$$\Delta \tilde{i}_1/\tilde{E}_3 = 1/(R_2+R_3) \tag{4.35}$$
$$\Delta \tilde{u}_1/\tilde{I}_3 = 0 \tag{4.36}$$
$$\Delta \tilde{i}_1/\tilde{I}_3 = 0 \tag{4.37}$$
$$\Delta \tilde{u}_1/\tilde{E}_3 = 0 \tag{4.38}$$

By comparing (4.31) through (4.34) with (4.35) through (4.38), we can observe that the circuit in figure 4.7 is not an adjoint of the circuit in figure 4.6. However, only the third condition of subsection 4.2.2 is violated. Thus, this condition is independent of the other conditions.

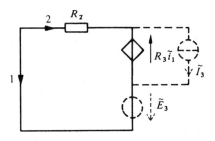

Fig. 4.7

By taking the dual circuit of figure 4.6 for C, and the dual circuit of figure 4.7 for $\tilde{C}$, conditions 1, 2, and 3 of subsection 4.2.2 are satisfied, whereas the last condition is not met.

Finally, if we replace resistor R_2 in the circuits of figure 4.6 and 4.7 by a current source, and if we consider the branches $k = 2$ and $l = 3$, it is only the second condition which is not satisfied, and, when passing to the dual circuits, it is the first condition which is not met.

In short, each condition in the definition 4.2.2 of the adjoint circuit is independent of the others.

4.2.5 Theorem

Let C be a well-posed autonomous linear resistive circuit. An autonomous linear resistive circuit $\tilde{C}$ having the same graph is an adjoint circuit of C if and only if it is obtained from C by the following element substitutions:

- the resistors are left unchanged;
- the independent voltage (current) sources remain independent voltage (current) sources, but they may change value;

- the voltage-controlled voltage sources are replaced by current-controlled current sources, and *vice versa*. The sign of the controlling coefficient is changed and the controlling branch and the controlled branch are interchanged;
- the voltage-controlled current sources and the current-controlled voltage sources change neither type nor value. However, the controlling branch and the controlled branch are interchanged;
- the operational amplifiers remain operational amplifiers, but the input and output branches are interchanged. In other words, the nullators are replaced by norators, and *vice versa*.

We decompose the proof into three parts. In the first part, we show that if the circuit $\tilde{C}$ is well posed, then it is an adjoint circuit of C. In the second part, we show that $\tilde{C}$ is well posed, and the third part provides the proof that any adjoint circuit of C originates from the substitutions listed above.

4.2.6 Property

Let C be a well-posed autonomous resistive circuit and let $\tilde{C}$ be a well-posed circuit obtained from C by the substitutions of subsection 4.2.5. Then, $\tilde{C}$ is an adjoint circuit of C.

For the proof, let us consider the following four circuits:

- the original circuit C;
- the circuit C_k obtained from C by inserting a voltage source of value E_k in series with branch k;
- the circuit $\tilde{C}$ obtained by the substitutions listed above;
- the circuit $\tilde{C}_l$ obtained from $\tilde{C}$ by inserting a voltage source of value $\tilde{E}_l$ in series with branch l.

The circuits C_k and $\tilde{C}_l$ have graphs that are different from C and $\tilde{C}$ because of the added source. By inserting short circuits, we can reduce the four circuits to the same graph, which includes two extra branches, branch $b + 1$ and branch $b + 2$, in contrast to C. The branches k, l, $b + 1$ and $b + 2$ of the four circuits are represented in figure 4.8.

Let $\Delta u_j(\Delta i_j)$ be the voltage (current) increment in branch j between the solution of circuit C and the solution of C_k. Similarly, let $\Delta \tilde{u}_j(\Delta \tilde{i}_j)$ be the voltage (current) increment in branch j between the solution of $\tilde{C}$ and the solution of $\tilde{C}_l$. Then,

$$\sum_{j=1}^{b+2} \Delta u_j \Delta \tilde{i}_j = 0 \tag{4.39}$$

and

$$\sum_{j=1}^{b+2} \Delta i_j \Delta \tilde{u}_j = 0 \tag{4.40}$$

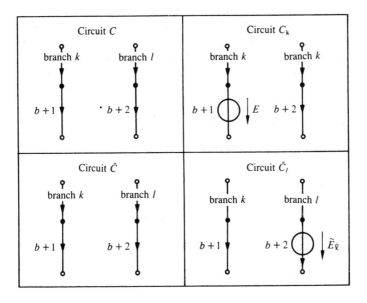

Fig. 4.8

This is, in fact, a generalization of the incremental power balance 3.1.8, and its proof results from Tellegen's theorem, according to comment 3.1.7. By forming the difference between (4.39) and (4.40), we obtain

$$\sum_{j=1}^{b+2} D_j = 0 \tag{4.41}$$

where

$$D_j = \Delta u_j \Delta \tilde{i}_j - \Delta i_j \Delta \tilde{u}_j \tag{4.42}$$

We first show that $D_j = 0$ for all the branches j which are not part of a controlled source. Indeed, if branch j of C carries a resistor R, branch j of $\tilde{C}$ also carries a resistor R, and we thus have

$$D_j = R \Delta i_j \Delta \tilde{i}_j - \Delta i_j R \Delta \tilde{i}_j = 0 \tag{4.43}$$

If branch j of C carries a voltage (current) source, such is also the case of branch j of C. Consequently, $\Delta u_j = \Delta \tilde{u}_j = 0$ ($\Delta i_j = \Delta \tilde{i}_j = 0$), which guarantees $D_j = 0$.

If branch j of C carries a nullator (norator), branch j of $\tilde{C}$ carries a norator (nullator), and we have $\Delta u_j = \Delta i_j = 0$ ($\Delta \tilde{u}_j = \Delta \tilde{i}_j = 0$), which again guarantees $D_j = 0$.

If branch j of C carries a source controlled by branch m, we show that $D_j + D_m = 0$. We limit the proof to a voltage-controlled voltage source, the controlling coefficient of which is α. Remember that, in this case, branch m is an open circuit. In $\bar{C}$, a current source is located on branch m, which is controlled by the current of branch j, which is a short circuit. The controlling coefficient is $-\alpha$. Then, $\Delta i_m = \Delta \bar{u}_j = 0$, and we have

$$D_j + D_m = \alpha \Delta u_m \Delta \tilde{i}_j - 0 + \Delta u_m (-\alpha) \Delta \tilde{i}_j - 0 = 0 \tag{4.44}$$

So far, we have shown that

$$\sum_{u=1}^{b} D_j = 0 \tag{4.45}$$

Finally, let us discuss D_{b+1} and D_{b+2}. From figure 4.8, we can see that

$$\Delta u_{b+1} = E_k, \ \Delta i_{b+1} = \Delta i_k, \ \Delta \tilde{u}_{b+1} = 0, \ \Delta \tilde{i}_{b+1} = \Delta \tilde{i}_k \tag{4.46}$$

and

$$\Delta u_{b+2} = 0, \ \Delta i_{b+2} = \Delta i_\varrho, \ \Delta \tilde{u}_{b+1} = \tilde{E}_\varrho, \ \Delta \tilde{i}_{b+1} = \Delta \tilde{i}_\varrho \tag{4.47}$$

thus,

$$D_{b+1} = E_k \Delta \tilde{i}_k \tag{4.48}$$

$$D_{b+2} = -\Delta i_\varrho \tilde{E}_\varrho \tag{4.49}$$

To conclude, (4.41), (4.45), (4.48), and (4.49) imply (4.27).

4.2.7 Property

Let C be a well-posed autonomous resistive circuit and let $\bar{C}$ be a circuit obtained by the substitutions of subsection 4.2.5. Then, $\bar{C}$ is well posed.

Let us assume the opposite. We set the sources of $\bar{C}$ to zero, this being compatible with the substitutions of the theorem. Thus, $\bar{C}$ has a nontrivial solution, the voltages and currents of which are $\Delta \bar{u}_j$ and $\Delta \bar{i}_j$. We chose the notation using Δ because we consider $\Delta \bar{u}_j$ and $\Delta \bar{i}_j$ to be increments with respect to the zero solution of $\bar{C}$. In fact, we can adopt the preceding context of the four circuits if we let $\bar{C}_l = \bar{C}$. Consequently, (4.41), (4.45), and even (4.48) hold. By contrast,

$$D_{b+2} = 0 \tag{4.50}$$

We conclude that

$$\Delta \tilde{i}_k = 0 \tag{4.51}$$

Because k is an arbitrary branch, (4.51) holds for all the branches. If the same argument were used with the proof of (4.28), we would also conclude that $\Delta \bar{u}_k = 0$ for all the k. This is in contradiction with the hypothesis that $\{\Delta u_j, \Delta i_j, j = 1, \ldots, b\}$ is a nontrivial solution of $\bar{C}$.

4.2.8 Property

Let C be a well-posed autonomous linear resistive circuit and let $\tilde{C}$ be an adjoint circuit of C. Then, $\tilde{C}$ is obtained starting from C by the substitutions of subsection 4.2.5.

We limit the proof to the case of a voltage source located on the branch k of C, which is controlled by the voltage of branch m, which is an open circuit. The other cases are proved in an analogous way.

We first show that the adjunction of a current source $\tilde{I}_l$ or a voltage source $\tilde{E}_l$ to an arbitrary branch l of $\tilde{C}$ causes voltage and current increments in branches k and m, satisfying the constitutive relations of a current source located on branch m, which is controlled by the current of branch k with a controlling coefficient $-\alpha$.

If we add a voltage source E_m in series with the branch m of C, the controlled source becomes the two-port of figure 4.9, which satisfies the constitutive relations:

$$i_1 = 0 \tag{4.52}$$
$$u_2 = \alpha(u_1 - E_m) \tag{4.53}$$

Thus, it is equivalent to the two-port of figure 4.10, which means that the adjunction of a source $E_k = -\alpha E_m$ in series with the branch k of C has the same effect, except on the branches k and m themselves. Thus, from (4.29), we find for any $l \neq k, m$:

$$\Delta \tilde{i}_k / \tilde{I}_\varrho = -\Delta u_\varrho / E_k = \Delta u_\varrho / \alpha E_m = -\Delta \tilde{i}_m / \alpha \tilde{I}_\varrho \tag{4.54}$$

and, therefore,

$$\Delta \tilde{i}_m = -\alpha \Delta \tilde{i}_k \tag{4.55}$$

The same relation holds if we excite $\tilde{C}$ by a source $\tilde{I}_k$ or $\tilde{I}_m$:

$$\Delta \tilde{i}_k / \tilde{I}_k = -\Delta u_k / E_k = \alpha(\Delta u_m / E_k) = -\alpha(\Delta \tilde{i}_m / \tilde{I}_k) \tag{4.56}$$

and an analogous proof is applied for $\tilde{I}_m$. Similarly, we show that (4.55) holds for any excitation by a source $\tilde{E}_l$.

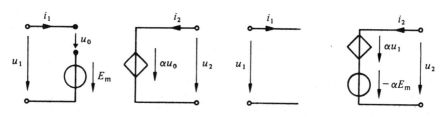

Fig. 4.9   Fig. 4.10

As a result of the equivalence of the one-ports in figure 4.11, we have

$$0 = \Delta u_\varrho / I_k = + \Delta \tilde{u}_k / \tilde{I}_\varrho \tag{4.57}$$

and, therefore,

$$\Delta \tilde{u}_k = 0 \tag{4.58}$$

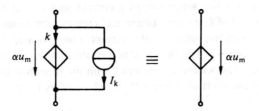

Fig. 4.11

thus, in an analogous way, we find (4.58) for any excitation by a source $\tilde{E}_l$.

Equations (4.55) and (4.58) are the constitutive relations of a current-controlled current source, exactly according to the substitution required by the theorem.

We now show that the solution of $\tilde{C}$ satisfies

$$\tilde{u}_k = 0 \tag{4.59}$$

and

$$\tilde{i}_m = -\alpha \tilde{i}_k \tag{4.60}$$

Indeed, we can cancel the independent sources of $\tilde{C}$ by connecting a voltage (current) source of opposite value in series (parallel) with each voltage (current) source of $\tilde{C}$. Because the solution of the resulting circuit is $u = i = 0$, (4.59) and (4.60) are obtained from (4.55) and (4.58).

Likewise, by connecting a voltage (current) source in parallel (series) with the arbitrary branch $\tilde{C}$, the solution of the resulting circuit always satisfies (4.59) and (4.60), due to (4.55) and (4.58).

4.2.9 Comment

We have not exactly proved property 4.2.8. In fact, we have shown the following property: if $\tilde{C}$ is an adjoint circuit of C, then its solution satisfies the system of circuit equations obtained by the substitutions of subsection 4.2.5. Furthermore, the same property holds if we connect a voltage (current) source in series (parallel) with an arbitrary branch of $\tilde{C}$. This is sufficient for the needs of this section.

In order to complete the proof, we should show that, for instance, if we connect the two-port constituted by the branches k and m of $\tilde{C}$, and insert it into another circuit, then (4.59) and (4.60) still hold.

4.2.10 Corollary

A well-posed autonomous linear resistive circuit is reciprocal if it contains only resistors and independent sources.

4.2.11 Comment

Because definition 4.2.2 for the adjoint circuit is expressed in terms of incremental variables, it is obvious that the value of the independent sources cannot play any role. Therefore, there is more than one adjoint circuit. Still, theorem 4.2.5 states that independent sources are the only indeterminacy of the adjoint circuit.

In view of theorem 4.2.5, we might be inclined to think that a reciprocal circuit can never contain a controlled source. A counterexample may be easily constructed. As given in volume IV (sec. 5.2), a circuit including resistors and ideal transformers is reciprocal, such as, for instance, the circuit of figure 4.12. However, an ideal transformer is equivalent to two controlled sources (fig. 4.13). If we apply the substitutions of theorem 4.2.5 to the circuit of figure 4.13, we observe that the two sources are interchanged; the controlling branches and the controlled branches are interchanged in such a way that the final result is the same circuit as we had at the start.

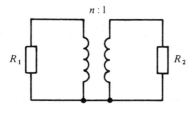

Fig. 4.12

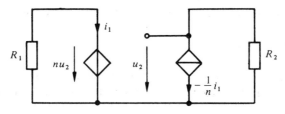

Fig. 4.13

4.2.12 Example

Let us consider the circuit C of figure 4.14. The adjoint circuit $\tilde{C}$ obtained by the substitutions of theorem 4.2.5 is represented in figure 4.15. Because the value of the voltage source on the branch 1 of $\tilde{C}$ is free, we have chosen a short circuit.

Let us verify property (4.27) for the pair of branches $k = 1$, $l = 4$. By direct calculation, we obtain

$$i_4 = -(R_3/R_2 R_4) E \qquad (4.61)$$

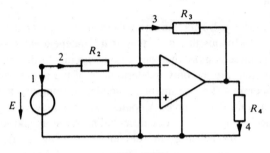

Fig. 4.14

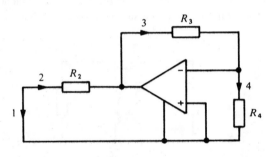

Fig. 4.15

The effect of a voltage source of value E_1 connected in series with branch 1 is the same as that resulting from the increase of E by E_1. Consequently,

$$\Delta i_4 = -(R_3/R_2 R_4) E_1 \qquad (4.62)$$

Conversely, if we insert a voltage source of value $\tilde{E}_4$ in series with branch 4 of $\tilde{C}$, we are back to the same circuit as C, except that R_2 and R_4 are interchanged. It follows that

$$\Delta i_4/E_1 = \Delta \tilde{i}_1/\tilde{E}_4 = -(R_3/R_2 R_4) \tag{4.63}$$

because only the product $R_2 R_4$ is involved in (4.63).

4.2.13 Definitions

Let C be a nonlinear autonomous resistive circuit, and let ξ be a dc operating point such that the circuit linearized around ξ, C_ξ, is well posed. A well-posed autonomous linear resistive circuit $\bar{C}_\xi$ is an *adjoint circuit of C in ξ* if it is an adjoint circuit of C_ξ, according to definition 4.2.2. If C_ξ is an adjoint circuit of itself at every operating point ξ, the nonlinear circuit C is *reciprocal*.

4.2.14 Theorem

Let C be an autonomous resistive circuit and let ξ be an operating point of C such that the circuit linearized around ξ, C_ξ, is well posed.

An autonomous linear resistive circuit $\bar{C}_\xi$ having the same graph is an adjoint circuit of C at ξ if and only if it is obtained from C by the following element substitutions:

- linear resistors, independent sources, controlled sources, and operational amplifiers are replaced according to the requirements of theorem 4.2.5;
- each nonlinear resistor is replaced by a linear resistor having a value which is the differential resistance at ξ.

This theorem is an immediate consequence of theorem 4.2.5 and definition 4.2.13.

4.2.15 Corollary

A circuit composed of resistors and independent sources is reciprocal.

4.3.16 Example

Let us consider the amplifier of figure 2.1. The nonlinear resistive circuit, the solution of which is the dc operating point of the amplifier, is represented in figure 4.16. We have replaced the *npn* bipolar transistor by the two-port of figure 4.17, which implements the Ebers-Moll model (2.1). After the substitutions of theorem 4.2.14, we obtain the two-port of figure 4.18, where r and r' are the differential resistances of the diodes D and D', respectively, at the operating point being considered. The complete adjoint circuit is represented in figure 4.19. As with example 4.2.12, we have replaced the source u_a by one of zero voltage, i.e., a short circuit.

We could have replaced the *npn* transistor by the two-port of figure 3.56, which is equivalent to that of figure 4.17, if we were to choose the inverse saturation currents

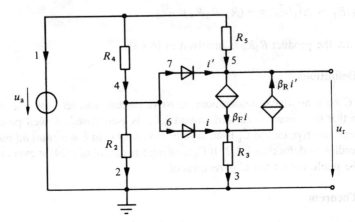

Fig. 4.16

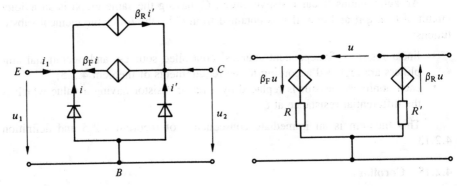

Fig. 4.17

Fig. 4.18

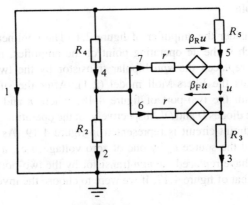

Fig. 4.19

of the diodes according to the requirements of subsection 3.4.26. The resulting circuit would be that of figure 3.60, with diodes instead of the resistors R_6 to R_9, but without the branches u_s,R_1,u_1 and u_2. The adjoint circuit would be obtained by simply interchanging the nullators and norators in figure 3.60.

4.2.17 Definition

Let $\xi(\alpha_1, \ldots ,\alpha_n)$ be a solution of a resistive circuit which depends in a differentiable way on $\alpha_1, \ldots ,\alpha_n$. Therefore, all the voltages u_k and all the currents i_k are differentiable functions of $\alpha_1, \ldots ,\alpha_n$. The *sensitivity* of u_k relative to α_1 at point $\bar{\alpha}_1, \ldots .\bar{\alpha}_n$ is defined by

$$S(u_k,\alpha_i) = (\bar{\alpha}_i/\bar{u}_k)(\partial u_k/\partial \alpha_i)(\bar{\alpha}_1, ..., \bar{\alpha}_n) \tag{4.64}$$

where

$$\bar{u}_k = u_k(\bar{\alpha}_1, ..., \bar{\alpha}_n) \tag{4.65}$$

Analogously, we define the sensitivity of i_k relative to α_i.
In the case where $\bar{\alpha}_i=0$ or $\bar{u}_k=0$, we use variants of (4.64) (vol. XIX, chap. 7).

4.2.18 Example

The circuit of figure 4.14 depends on the element values E,R_2,R_3,R_4. The sensitivities of the output current relative to these parameters are

$$S(i_4,E) = -S(i_4,R_2) = S(i_4,R_3) = -S(i_4,R_4) = 1 \tag{4.66}$$

We observe that, in this particular case, the sensitivities are independent of the parameter values.

4.2.19 Calculation of Sensitivities

Let us consider a nonlinear resistive circuit. Suppose that a resistor is located on branch k and depends on parameter α. Thus, in its most general form, the constitutive relation is

$$f_k(u_k,i_k,\alpha) = 0 \tag{4.67}$$

Let $\xi(\alpha)$ be a solution of the circuit which depends in a differentiable way on α. We wish to know the sensitivity $S(i_l,\alpha)$ at $\bar{\alpha}$.
Let α be close to $\bar{\alpha}$, and let Δu_j and Δi_j be the voltage and current increments between $\xi(\alpha)$ and $\xi(\bar{\alpha})$. By first approximation, (4.67) becomes

$$(\partial f_k/\partial u_k)\Delta u_k + (\partial f_k/\partial i_k)\Delta i_k + (\partial f_k/\partial \alpha)\Delta \alpha = 0 \tag{4.68}$$

where the partial derivatives are taken at the point:

$$(\bar{u}_k, \bar{i}_k, \bar{\alpha}) = (u_k(\bar{\alpha}), i_k(\bar{\alpha}), \bar{\alpha}) \tag{4.69}$$

Let us assume for now that $\partial f/\partial u_k \neq 0$ at $\bar{\xi}$ in (4.69). We shall return to case $\partial f/\partial u_k = 0$ later. Then, (4.68) can be rewritten as

$$\Delta u_k = \bar{r}_k \Delta i_k + \Delta E_k \tag{4.70}$$

where $\bar{r}_k$ is the differential resistance (2.180) at point $\bar{\xi} = \xi(\bar{\alpha})$ and

$$\Delta E_k = -\Delta\alpha \cdot [(\partial f_k/\partial\alpha)/(\partial f_k/\partial u_k)] \, (\bar{u}_k, \bar{i}_k, \alpha) \tag{4.71}$$

We note that (4.70) is the constitutive relation of the one-port represented in figure 4.20. It follows that *at first approximation, the alteration of α by $\Delta\alpha$ causes the same deviation of the solution as the insertion of a voltage source ΔE_k, given by (4.71), in series with branch k.*

If $\partial f_k/\partial i_k \neq 0$ at (4.69), we can choose an alternative method. Instead of (4.70), we write

$$\Delta i_k = \bar{g}_k \Delta u_k + \Delta I_k \tag{4.72}$$

where $\bar{g}_k$ is the differential conductance at $\bar{\xi}$, and

$$\Delta I_k = -\Delta\alpha \cdot [(\partial f_k/\partial\alpha)/(\partial f_k/\partial i_k)] \, (\bar{u}_k, \bar{i}_k, \bar{\alpha}) \tag{4.73}$$

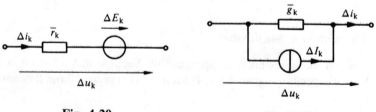

Fig. 4.20   **Fig. 4.21**

Equation (4.72) is the constitutive relation of the one-port of figure 4.21. Consequently, *at first approximation, the alteration of α by $\Delta\alpha$ has the same effect on the solution as the insertion of a current source ΔI_k, given by (4.73), in parallel with branch k.*

If $\partial f_k/\partial u_k = 0$ and $\partial f_k/\partial i_k = 0$ at $\bar{\xi}$, the linearized circuit at $\bar{\xi}$ is ill posed, and it cannot be used for calculating the sensitivities relative to α at $\bar{\xi}$.

As is the case of resistors, independent sources and controlled sources may also be dependent on parameters.

The discussion of the independent sources is trivial. If the value $E(I)$ of an independent voltage (current) source changes by $\Delta E(\Delta I)$, this is equivalent to adding a voltage source ΔE (current source ΔI) in series (in parallel) with the original source.

The discussion on controlled sources is only slightly more complicated. We limit ourselves to the voltage-controlled voltage source. If its controlling coefficient α is altered by $\Delta\alpha$, then the constitutive relation:

$$u_k = \alpha u_m \tag{4.74}$$

in incremental relations becomes

$$\Delta u_k = \bar{\alpha}\Delta u_m + \bar{u}_m \Delta\alpha \tag{4.75}$$

Thus, the alteration of α by $\Delta\alpha$ amounts to inserting a voltage source having a value of

$$\Delta E_k = \bar{u}_m \Delta\alpha \tag{4.76}$$

in series with the controlled source.

In short, *any parameter alteration in the circuit amounts, at first approximation, to inserting a voltage source or a current source into the circuit.*

4.2.20 Sensitivity Calculation Using the Linearized Circuit

Let α be a parameter which is involved in the constitutive relation of the branch k of an autonomous resistive circuit. Let us assume that a solution $\xi(\alpha)$ depends on α in a differentiable way, and that $\xi(\bar{\alpha})$ is an operating point with a well-posed linearized circuit. Then, we obtain the sensitivities $S(u_l,\alpha)$ and $S(i_l,\alpha)$ at $\bar{\alpha}$ by the following method:

- Determine the solution $\bar{\xi} = \xi(\bar{\alpha})$.
- Determine the linearized circuit around $\bar{\xi}$.
- Calculate the following value a_k or b_k:

If branch k carries a locally current-controlled resistor:

$$a_k = -[(\partial f_k/\partial\alpha)/(\partial f_k/\partial u_k)]\,(\bar{u}_k, \bar{i}_k, \alpha) \tag{4.77}$$

If branch k carries a locally voltage-controlled resistor:

$$b_k = -[(\partial f_k/\partial\alpha)/(\partial f_k/\partial i_k)]\,(\bar{u}_k, \bar{i}_k, \bar{\alpha}) \tag{4.78}$$

If branch k carries an independent voltage (current) source of value α, set $\alpha_k = 1(b_k = 1)$.

If branch k carries a voltage-controlled (current-controlled) voltage source of branch m, with a controlling coefficient α, set $a_k = \bar{u}_m (a_k = \bar{i}_m)$.

If branch k carries a voltage-controlled (current-controlled) current source of branch m, with a controlling coefficient ∂, set $b_k = \bar{u}_m (b_k = \bar{i}_m)$.

- Determine the solution $\Delta\xi$ of the linearized circuit with an arbitrary voltage source E_k (current source I_k) inserted in series (in parallel) with branch k;
- thus, the results are

$$S(u_\varrho, \alpha) \;=\; (\bar{\alpha}/\bar{u}_\varrho)(\Delta u_\varrho/E_k)\, a_k, \quad \text{if} \quad l \neq k \tag{4.79}$$

$$S(u_k, \alpha) \;=\; (\bar{\alpha}/\bar{u}_k)((\Delta u_k/E_k) + 1)\, a_k \tag{4.80}$$

$$S(i_\varrho, \alpha) \;=\; (\bar{\alpha}/\bar{i}_\varrho)(\Delta i_\varrho/E_k)\, a_k, \quad \text{all} \quad l \tag{4.81}$$

or

$$S(u_\varrho, \alpha) \;=\; (\bar{\alpha}/\bar{u}_\varrho)(\Delta u_\varrho/I_k)\, b_k, \quad \text{all} \quad l \tag{4.82}$$

$$S(i_\varrho, \alpha) \;=\; (\bar{\alpha}/\bar{i}_\varrho)(\Delta i_\varrho/I_k)\, b_k, \quad l \neq k \tag{4.83}$$

$$S(i_k, \alpha) \;=\; (\bar{\alpha}/\bar{i}_k)((\Delta i_k/I_k) + 1)\, b_k \tag{4.84}$$

This method is based on the observations of subsection 4.2.19 as well as on the following facts:

- the nonlinear circuit and the linearized circuit have the same behavior in first approximation;
- the solution of a linear circuit with a single source is proportional to the value of that source;
- the branch k with the parameter $\bar{\alpha} + \Delta\alpha$ corresponds to *the whole* equivalent branch, including the source.

4.2.21 Sensitivity Calculation Using the Adjoint Circuit

Let us use the same hypotheses as in subsection 4.2.20. The sensitivities $S(u_l, \alpha)$ and $S(i_l, \alpha)$ at α can be calculated by the following method:

- Determine the solution $\bar{\xi} = \xi(\bar{\alpha})$.
- Determine the adjoint circuit at $\bar{\xi}$.
- Calculate the value a_k or b_k as in subsection 4.2.20.
- Determine the solution $\Delta\tilde{\xi}$ of the adjoint circuit with, depending on the case, a voltage source $\bar{E}_l$ in series with branch l, or a current source I_l in parallel with branch l. The values $\bar{E}_l$ and $\bar{I}_l$ are arbitrary. Formulas (4.85) to (4.90) inform which one of the sources is to be inserted;
- thus, the results are

$$S(u_\varrho, \alpha) = -(\overline{\alpha}/\overline{u}_\varrho)(\Delta\widetilde{i}_k/\widetilde{I}_\varrho)a_k, \quad \text{if } l \neq k \tag{4.85}$$

$$S(u_k, \alpha) = -(\overline{\alpha}/\overline{u}_k)((\Delta\widetilde{i}_k/\widetilde{I}_k)-1)a_k \tag{4.86}$$

$$S(i_\varrho, \alpha) = (\overline{\alpha}/\overline{i}_\varrho)(\Delta\widetilde{i}_k/\widetilde{E}_\varrho)a_k, \quad \text{all } l \tag{4.87}$$

or

$$S(u_\varrho, \alpha) = (\overline{\alpha}/\overline{u}_\varrho)(\Delta\widetilde{u}_k/\widetilde{I}_\varrho)b_k, \quad \text{all } l \tag{4.88}$$

$$S(i_\varrho, \alpha) = -(\overline{\alpha}/\overline{i}_\varrho)(\Delta\widetilde{u}_k/\widetilde{E}_\varrho)b_k, \quad l \neq k \tag{4.89}$$

$$S(i_k, \alpha) = -(\overline{\alpha}/\overline{i}_k)((\Delta\widetilde{u}_k/\widetilde{E}_\varrho)-1)b_k \tag{4.90}$$

This method is directly derived from that of subsection 4.2.20 and definition 4.2.2 for the adjoint circuit.

4.2.22 Comment

Apparently, nothing is gained if we use the adjoint circuit for the calculation of the sensitivities rather than the linearized circuit itself. In both cases, the work required consists essentially of an analysis of the nonlinear circuit and an analysis of a linear circuit with the same complexity. These considerations certainly hold as long as no more than one sensitivity is calculated.

However, most of the time, the engineer is interested in the sensitivities of *one output quantity* with respect to *all parameters*. If we apply the method of subsection 4.2.20, we must carry out an analysis of the nonlinear circuit and as many analyses of the linear circuit as there are parameters. By contrast, the method of subsection 4.2.21, apart from the analysis of the nonlinear circuit, requires only a single analysis of the adjoint circuit because we insert the source at the output branch and calculate the increments on all the branches with an affected parameter to them. Because the number of parameters is usually of the same order of magnitude as the number of branches, the calculation time saved by using the adjoint circuit can be considerable.

4.2.23 Example

Let us again try to find the sensitivities (4.66) of the current i_4 in the circuit of figure 4.14 by using the adjoint circuit method. We follow the points of subsection 4.2.21 in the same order.

- The solution of the circuit of figure 4.14 is

$$\overline{u}_1 = \overline{u}_2 = E \tag{4.91}$$

$$-\overline{i}_1 = \overline{i}_2 = \overline{i}_3 = E/R_2 \tag{4.92}$$

$$\overline{u}_3 = -\overline{u}_4 = R_3 E/R_2 \tag{4.93}$$

$$\overline{i}_4 = -(R_3/R_2R_4)E \tag{4.94}$$

- The adjoint circuit is represented in figure 4.15.
- For each resistor, because parameter α is the resistance itself, we deduce from

$$u_k - R_k i_k = 0 \tag{4.95}$$

and by using (4.77), that

$$a_k = \overline{i_k} \ , \quad k = 2, 3, 4 \tag{4.96}$$

Furthermore, $\alpha_1 = 1$, because it is a voltage source.

- By inserting a voltage source of value $\tilde{E}$ in series with branch 4 in the circuit of figure 4.15, we obtain as a solution:

$$\Delta \tilde{u}_4 = \tilde{E} \tag{4.97}$$
$$\Delta \tilde{i}_4 = \Delta \tilde{i}_3 = \tilde{E}/R_4 \tag{4.98}$$
$$\Delta \tilde{u}_3 = -\Delta \tilde{u}_2 = R_3 \tilde{E}/R_4 \tag{4.99}$$
$$\Delta \tilde{i}_2 = -\Delta \tilde{i}_1 = (R_3/R_2 R_4) \tilde{E} \tag{4.100}$$

- The sensitivities (4.66) are derived from formula (4.87) with $l = 4, \alpha = E$ for $k = 1$, and $\alpha = R_k$ for $k = 2, 3, 4$.

4.2.24 Numerical Example

Let us return to the amplifier of figure 2.1, with the element values given in subsection 2.1.5. We intend to calculate the sensitivities of the output voltage relative to the resistors R_2 to R_5 and the supply voltage u_a. This example better illustrates the advantages of the adjoint circuit method as compared with the preceding example, the purpose of which was to clarify the different steps. Although SPICE allows direct calculation of the sensitivities of the dc operating point, we use the program for the calculation according to the method of subsection 4.2.21.

The steps, in the same order as subsection 4.2.21, are

- The branch currents and voltages at the dc operating point deduced from the information provided by SPICE are reported in columns 3 and 4 of table 4.1.
- The adjoint circuit at the dc operating point is represented in figure 4.19. Then, the differential resistance r is to be determined. The differential resistance r' is enormous, and we replace it with an open circuit. We have

$$1/r = (d/du_7) [(I_s/\beta_F(e^{u_7/U_T} - 1))] \tag{4.101}$$
$$1/r = (I_s/\beta_F U_T) e^{u_7/U_T} \tag{4.102}$$
$$1/r \cong \overline{i_7}/U_T = 2{,}673 \, \text{k}\Omega \tag{4.103}$$

- As in subsection 4.2.23, we have $\alpha_1 = 1$ and $\alpha_k = \bar{i}_k$ for $k = 2, \ldots, 5$. We do not need a_6 and a_7.
- The output voltage u_r is not a branch voltage. However, nothing forbids us to introduce a new branch, i.e., an open circuit between the collector and ground, the voltage of which is u_r. By connecting a current source of 1 mA between the same nodes of the adjoint circuit, we obtain the increments $\Delta\tilde{u}_k, \Delta\tilde{i}_k$, reported in columns 5 and 6 of table 4.1, by a SPICE analysis.
- The sensitivities are calculated by (4.85) and reported in column 7 of table 4.1. As a check on the results, we have calculated the output voltage shift by changing the element value by 1%. The results obtained from the SPICE analyses of the nonlinear circuit are reported in the last column of table 4.1. We note a good agreement with the sensitivities.

Table 4.1

Branch k	Parameter α	$\bar{u}_k$ [V]	$\tilde{i}_k$ [mA]	$\Delta\tilde{u}_k$ [V]	$\Delta\tilde{i}_k$ [mA]	$S(u_r, \alpha)$	$\dfrac{\bar{\alpha}}{\Delta\alpha}\dfrac{\Delta u_r}{\bar{u}_r}$
1	u_a	10.00	−1.472	0	−0.447	0.865	0.865
2	R_2	4.95	0.495	5.53	0.553	−0.530	−0.527
3	R_3	4.30	0.977	−4.87	−1.106	0.920	0.912
4	R_4	5.05	0.505	−5.53	−0.553	0.540	0.538
5	R_5	4.83	0.967	5.00	1.000	−0.936	−0.929

As mentioned in subsection 4.2.21, two SPICE analyses were needed to determine the sensitivities: one of the original circuit, and one of the adjoint circuit. Nonetheless, to obtain the check on the results of the last column of table 4.1, six analyses were required: one of the original circuit, and one for each modified element.

4.3 SUBSTITUTIONS

4.3.1 Introduction

One of the most common and most typical methods of circuit theory consists of substituting a one-port or a multiport by an equivalent one-port or multiport in a circuit. The goal aimed at by such an operation varies according to the application.

On the one hand, it is a tool for circuit analysis. For example, if we know that the one-port is composed of the resistors R_2, R_3, R_4, and the operational amplifier of figure 3.32 is equivalent to a negative resistor, it is easy to understand that the circuit of figure 3.32 is sometimes ill posed. Another example is the SPICE analysis of the circuit of figure 1.41. Because the memristor is not one of the basic SPICE elements, numerical calculation with that program is only possible if we substitute the memristor by an equivalent one-port, such as that of figure 1.59.

On the other hand, this method is almost indispensable for circuit synthesis. First, it permits us to extend an established synthesis method to a wider class of circuits. A typical case is the synthesis of piezoelectric resonator filters derived within the synthesis of LC filters (vol. XIX, sec. 8.1). We first carry out an LC synthesis, and then we substitute piezoelectric resonators to the resonant LC circuits, which are equivalent. Other applications aim at improving the dispersion of the element values or at lowering the sensitivity.

The operation of substituting one N-port in a circuit by another N-port is decomposed into three steps:

- we separate the circuit into two N-ports, P_1 and P_2;
- we establish the equivalence of P_2 with a third N-port, P_3;
- we connect P_1 and P_3.

Whereas the connection of N-ports was described in section 1.2, the reverse operation, the separation of a circuit into two N-ports, requires some explanations. Also, the concept of N-port equivalence must be rigorously defined. Finally, we shall establish precisely how the substitution affects the circuit.

After these general considerations, we shall discuss two classes of substitutions. The first is a generalization of the Thevenin-Norton equivalents to nonlinear resistive one-ports. The second is concerned with the removal of loops composed of capacitors and voltage sources, as well as cut sets composed of inductors and current sources. There would be, of course, many other applications that cannot be treated in this book.

4.3.2   Separation of a Circuit into Multipoles and Multiports

A circuit is *separated into two multipoles*, Q_1 *and* Q_2 in the following way. We separate the set of the graph branches into two subsets, B_1 and B_2, such that the two individual ports of the elementary two-ports of the circuit are always in the same subset. The graph of $Q_1(Q_2)$ is formed by branches of $B_1(B_2)$ and all the nodes to which the branches of $B_1(B_2)$ are incident. If a node simultaneously belongs to the graphs of Q_1 and Q_2, it is a *terminal* of Q_1 and Q_2. If a node belongs to $Q_1(Q_2)$, but not to $Q_2(Q_1)$, it is an *internal node* of $Q_1(Q_2)$. The construction of the graphs of Q_1 and Q_2, and the identification of the terminals are represented in figure 4.22.

The elements of the circuit are distributed between Q_1 and Q_2 in the same way as the branches of the graph. As a result of the hypothesis for the distribution of the branches, the elementary two-ports belong as well to either Q_1 or Q_2.

As mentioned in subsection 1.2.6, an M-pole can always be considered as an $M-1$-port, the ports of which all have a terminal in common. Therefore, by choosing from among the terminals of Q_1 and Q_2 the one which will be used as the common terminal, we associate Q_1 and Q_2 to two multiports P_1 and P_2. In such a way, we have *separated* the circuit *into two multiports*.

We observe that the decomposition of the set of branches into two subsets determines the multipoles Q_1 and Q_2 entirely, while there is some freedom for grouping

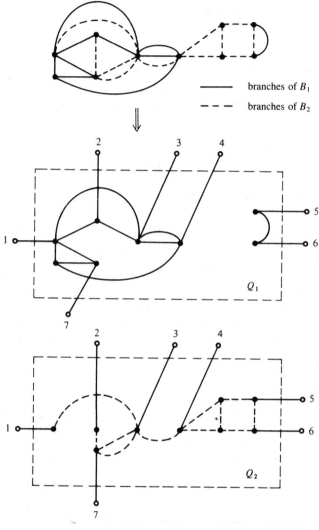

branches of B_1

branches of B_2

Q_1

Q_2

Fig. 4.22

the terminals by pairs into ports to reach the multiports P_1 and P_2. There would even be groupings of terminals into ports other than through a terminal common to all ports. The number of ports is generally equal to the number of terminals minus one. In the case where the graph of the multipole is not connected, we can reduce the number of ports by one unit per connected component, starting from the second component. For example, to the seven-pole Q_1 of figure 4.22, we can associate the five-port of figure 4.23.

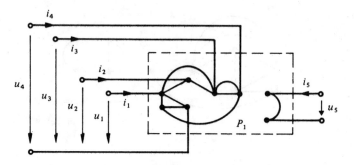

Fig. 4.23

4.3.3 System of Equations of Multiports

Once the circuit is separated into two N-ports, we need to establish their system of equations. This equation system involves the currents and voltages at the ports, as well as the currents, the voltages, and possibly the charges and fluxes, of the branches of their graph. These are called *internal variables*. The system of equations of each N-port is composed of

- the Kirchhoff equations of the N-port graph;
- the constitutive relations of the elements which are included in the N-port;
- the equations which express the voltages and currents at the ports as functions of the internal voltages and currents.

4.3.4 Example

We intend to separate the circuit of figure 4.24 into linear and nonlinear parts. Because there are three common nodes between the two parts, we separate the circuit into two two-ports (fig. 4.25 and 4.26). Let us establish the equations of the nonlinear two-port of figure 4.26 by following the points of subsection 4.3.3.

- Because there are no internal nodes, there are no Kirchhoff current equations. However, there is a loop:

$$u_3 - u_4 - u_5 = 0 \tag{4.104}$$

The constitutive relations of the elements are

$$q_3 = g(u_3) \tag{4.105}$$
$$dq_3/dt = i_3 \tag{4.106}$$
$$\varphi_4 = h(i_4) \tag{4.107}$$
$$d\varphi_4/dt = u_4 \tag{4.108}$$
$$i_5 = f(u_5) \tag{4.109}$$

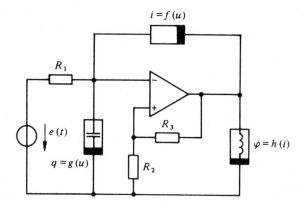

$i = f(u)$

R_1

$e(t)$

R_3

$q = g(u)$

R_2

$\varphi = h(i)$

Fig. 4.24

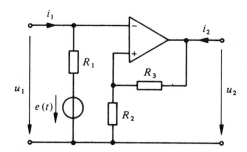

i_1

R_1

u_1

$e(t)$

R_3

R_2

i_2

u_2

Fig. 4.25

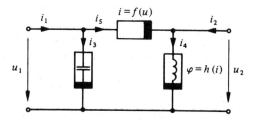

i_1

i_5

$i = f(u)$

i_2

i_3

i_4

u_1

$\varphi = h(i)$

u_2

Fig. 4.26

- The variables at the ports are expressed as functions of the internal variables in the following way:

$$u_1 = u_3 \tag{4.110}$$

$$i_1 = i_3 + i_5 \tag{4.111}$$

$$u_2 = u_4 \tag{4.112}$$
$$i_2 = i_4 - i_5 \tag{4.113}$$

4.3.5 Comment

According to definition 1.2.5, an N-port is described by N constitutive relations between the currents and the voltages at the ports. It is the "black box" N-port concept, where the internal structure is not known. However, by decomposing the circuit into two N-ports, we obtain N-ports which are themselves circuits, the equations of which involve internal currents and voltages. In principle, we obtain constitutive relations of these N-ports by eliminating the internal variables from their system of equations.

In the example of subsection 4.3.4, we obtain the constitutive relations of the two-port of figure 4.26 as follows. By combining (4.107), (4.113), (4.109), (4.104), (4.110), and (4.112), we find

$$\varphi_4 = h(i_2 + f(u_1 - u_2)) \tag{4.114}$$

By deriving (4.114) with respect to time, we obtain

$$u_2 = [(\mathrm{d}h/\mathrm{d}i)(i_2 + f(u_1 - u_2))] \\ [(\mathrm{d}i_2/\mathrm{d}t) + (\mathrm{d}f/\mathrm{d}u(u_1 - u_2))(\mathrm{d}u_1/\mathrm{d}t - \mathrm{d}u_2/\mathrm{d}t)] \tag{4.115}$$

Finally, we combine (4.111) with (4.106), (4.105), and (4.109):

$$i_1 = (\mathrm{d}g/\mathrm{d}u(u_1))\,\mathrm{d}u/\mathrm{d}t + f(u_1 - u_2) \tag{4.116}$$

Equations (4.115) and (4.116) only involve the voltages and currents at the ports. Thus, they are the constitutive relations of the two-port of figure 4.26.

If, for instance, the capacitor were not voltage controlled, this derivation of the constitutive relations would have failed. Thus, the derivation of the constitutive relations of an N-port from the system of equations which involves the internal variables is not an obvious operation. If we cannot express the constitutive relations of the N-port by starting from the constitutive relations of its components, then we must be satisfied with the system of equations of the N-port as it is. It is a *system of parametric equations* for the currents and voltages at the ports. The parameters are the internal currents and voltages.

We can go one step further and wonder whether nonparametric constitutive relations can exist at all. This amounts to asking whether the variables at the ports determine the internal variables entirely.

4.3.6 Definitions

An N-port is described by *parametric constitutive relations* if the voltages $u_1, \ldots u_N$ and the currents $i_1, \ldots i_N$ at the ports, as well as p parameters $\alpha_1, \ldots, \alpha_p$ must satisfy a system of $N+p$ equations as functions of time.

A *column matrix of 2N functions* of time $[u_1(t), \ldots, u_N(t), i_1(t), \ldots, i_N(t)]^T$ defined in a time interval $[t_0, t_1]$ is *admissible* for an N-port described by $N+p$ parametric constitutive relations if there are p functions of time $\alpha_1(t), \ldots \alpha_p(t)$ defined in $[t_0, t_1]$, such that the functions $u_1(t), \ldots, \alpha_p(t)$ satisfy the parametric constitutive relations. This definition also applies to the case $p=0$, where the constitutive relations are not parametric. If, for any admissible column matrix of voltages and currents, there is only one choice of parameters such that the constitutive relations are satisfied, the N-port is *transparent*.

4.3.7 Comment

The concept of a transparent N-port is related to the concepts of the controllable and observable system in system theory [19]. Roughly, a system is *controllable* if all the internal variables are determined by the signals. It is *observable* if it is possible to infer the internal variables from the responses. The signals, responses, and internal variables are understood to be functions of time.

As noted in section 1.5, we associate a circuit with a system by designating certain voltage-source voltages and current-source currents as signals, and certain voltages and currents of the circuit as responses. If we replace the signal sources by ports and set other ports in such a way that they capture the responses, we obtain an N-port to which a system is associated.

For example, we transform the amplifier of figure 2.1 into the two-port of figure 4.27. To that two-port, we associate the system of figure 4.28, the signals for which are u_s and i_r, and their responses are i_s and u_r. In fact we do not consider the entire system of figure 4.28, but only the response u_r to the signal u_s when $i_r = 0$.

The transition from figure 4.27 to figure 4.28 is typical. Half the variables at the ports of the multiport consist of the signals, the other half consist of the responses.

The system of figure 4.28 is controllable (observable) if the **two** functions $u_s(t), i_r(t)(i_s(t), u_r(t))$ determine all the voltages and all the currents of the internal branches at the two-port of figure 4.27. Conversely, the two-port is transparent if **four** functions $u_s(t), i_r(t), i_s(t), u_r(t)$ determine all the voltages and all the currents of the internal branches.

We see that the concept of transparency is much less restrictive than the concept of controllability or observability. Nevertheless, we shall see in subsection 4.3.9 that it is not difficult to give examples of two-poles or multipoles that are not transparent.

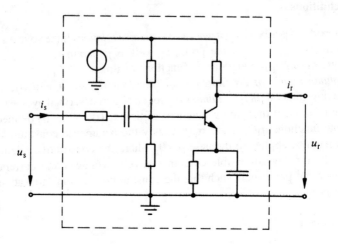

Fig. 4.27

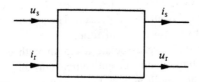

Fig. 4.28

Note, also, that in the particular case $p = 0$, there are no internal variables, and consequently an N-port described by nonparametric constitutive relations is always transparent.

4.3.8 Definitions

Two N-ports are *equivalent at the ports* if they have the same admissible column matrices of voltages and currents. If, in addition, both N-ports are transparent, then they are *equivalent*.

4.3.9 Example

The question at hand is whether connecting two resistors in parallel (fig. 4.29) constitutes a one-port that is equivalent to a single resistor.

Let us first assume that both resistors are voltage controlled, with the constitutive relations:

$$i_1 = g_1(u) \tag{4.117}$$

$$i_2 = g_2(u) \tag{4.118}$$

If we complete these two equations by using the relation:

$$i = i_1 + i_2 \tag{4.119}$$

we obtain a system of three parametric constitutive relations for the one-port of figure 4.29. Also, let us consider a voltage controlled resistor defined by the constitutive relation:

$$i = g(u) \tag{4.120}$$

with

$$g(u) = g_1(u) + g_2(u) \tag{4.121}$$

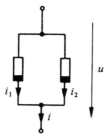

Fig. 4.29

If the functions $i_1(t), i_2(t), i(t), u(t)$ satisfy (4.117) to (4.119) for any $t \in [t_0, t_1]$, then naturally $i(t)$ and $u(t)$ satisfy (4.120) and (4.121). Conversely, if $i_1(t), i_2(t)$ satisfy (4.120) and (4.121), and if we define $i_1(t), i_2(t)$ by (4.117) and (4.118), then $i_1(t), i_2(t), i(t), u(t)$ satisfy (4.117) and (4.118). This shows that the one-port of figure 4.29 and the resistor are equivalent at that port. In fact, the choice of the parameters i_1 and i_2 from u is imposed by (4.117) and (4.118), which shows that the one-port of figure 4.29 is transparent and, therefore, equivalent to the resistor.

If one of the two resistors is current controlled rather than voltage controlled, with the constitutive relation:

$$u = h_2(i_2) \tag{4.122}$$

then the one-port of figure 4.29 is still transparent. Indeed, voltage u determines i_1 by (4.117); from i and i_1, we find i_2 by using (4.119). The one-port of figure 4.29 is

248

still equivalent to a single resistor, but this resistor is not usually voltage controlled any longer. Indeed, starting from (4.117), (4.119), and (4.122), we have the constitutive relation:

$$u - h_2\,(i - g_1\,(u)) = 0 \qquad (4.123)$$

of the equivalent resistor.

If both resistors of figure 4.29 are current controlled, the internal currents of the constitutive parametric relations cannot usually be removed any longer. In fact, the one-port transparency is no longer guaranteed. As an example, let us consider two resistors with the constitutive relations:

$$u = a_1 i_1^3 + b_1 i_1^2 + c_1 i_1 + d_1 \qquad (4.124)$$
$$u = a_2 i_2^3 + b_2 i_2^2 + c_2 i_2 + d_2 \qquad (4.125)$$

with

$$a_1 = a_2 = 1\,\text{V/A}^3 \qquad (4.126)$$
$$b_1 = -6\,\text{V/A}^2,\, b_2 = -8\,\text{V/A}^2 \qquad (4.127)$$
$$c_1 = 11\,\text{V/A},\, c_2 = 20\,\text{V/A} \qquad (4.128)$$
$$d_1 = -5\,\text{V},\, d_2 = -14\,\text{V} \qquad (4.129)$$

The characteristics of the two resistors are represented by the curves of figure 4.30. In figure 4.31, we have shown the set of the values of (u, i), such that there exist values of i_1 and i_2 which satisfy (4.124), (4.125), and (4.119). We note that said values consist of two curves intersecting at three points, P_1, P_2, and P_3. Any pair of admissible functions $u(t)$, $i(t)$ must possess values at any instant that correspond to a point on one of these curves.

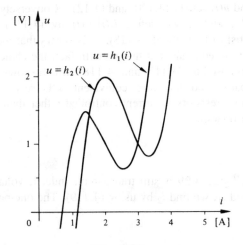

Fig. 4.30

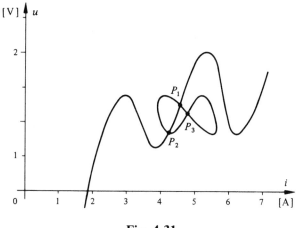

Fig. 4.31

The points P_j represent the pairs of values (u,i) for which there are two different pairs of values i_1,i_2. The respective numerical values are shown in table 4.2. Thus, the one-port of figure 4.29, with the resistors defined by (4.124) and (4.125) is not transparent; it is only equivalent at the terminals to a single resistor with the characteristic of figure 4.31.

However, this phenomenon does not necessarily occur in the presence of two resistors that are not voltage controlled. It is sufficient, for instance, to increase d_2 in (4.129) from -14 V to -13 V to obtain a transparent one-port, which is equivalent to a single current controlled resistor, the characteristic of which is represented in figure 4.32.

Table 4.2

Point	u [v]	i [A]	i_1 [A]	i_2 [A]
P_1	0.82	4.22	0.92	3.30
			2.89	1.33
P_2	1.23	4.55	1.75	2.80
			3.10	1.45
P_3	1.12	4.76	1.07	3.69
			1.87	2.89

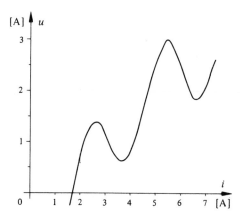

Fig. 4.32

250

4.3.10   Comments

We observe that, in the particular case of resistive multiports, the admissible voltages and currents are determined separately for each time. An analogous fact has already been mentioned in subsection 3.3.1.

4.3.11   Definition

A *column matrix* of 2N *numbers* $[u_1, \ldots ,u_N,i_1, \ldots i_N]^T$ is *admissible* for an autonomous resistive N-port, described by $N+p$ parametric constitutive relations, if there are p numbers $\alpha_1, \ldots ,\alpha_p$, such that the numbers $u_1, \ldots ,\alpha_p$ satisfy the constitutive relations.

4.3.12   Properties

A column matrix of 2N *functions* $[u_1(t), \ldots ,u_N(t),i_1(t), \ldots ,i_N(t)]^T$ defined in the interval $[t_0,t_1]$ is admissible for a resistive N-port if and only if, for any $t \in [t_0,t_1]$, the column matrix of the *values* $[u_1(t), \ldots ,u_N(t),i_1(t); \ldots ,i_N(t)]^T$ is admissible for the N-port. An autonomous resistive N-port is transparent if and only if, for any admissible column matrix of voltages and currents, there is only one choice of parameters such that the parametric constitutive relations are satisfied. A time-dependent resistive N-port is transparent if any autonomous resistive N-port, obtained by replacing the time-dependent sources by constant sources with arbitrary values, is transparent.

Indeed, the parametric constitutive relations of a resistive N-port only relate voltages and currents of the same instant.

4.3.13   Theorem: Thevenin-Norton Equivalent

An autonomous resistive one-port is always equivalent at the port to a single resistor. If it is transparent, it is equivalent to the resistor. The resistor characteristic is composed of the pairs (u,i) that are admissible for the one-port.

This theorem is an immediate consequence of definition 4.3.8 and property 4.3.12.

4.3.14   Comments

We could extend theorem 4.3.13 to time-dependent resistive one-ports. The equivalent resistor would also be time dependent. Nonetheless, there is nothing analogous to the Thevenin or Norton equivalent for nonlinear circuits that include capacitors and inductors, unless we introduce one-ports having a constitutive relation which is an arbitrarily complicated function of derivatives and integrals.

The Thevenin and Norton equivalents for linear circuits (vol. IV, sec. 5.4) are distinguished by the type of independent source. In the context of nonlinear circuits, an independent source is simply a particular resistor. For this reason, we absorb the source into the resistor and end up with only one type of equivalent.

4.3.15 Examples

Figure 4.31 shows the characteristic of the Thevenin-Norton equivalent of the one-port, obtained by connecting the two resistors (4.124) and (4.125) in parallel. In this case, it is only an equivalence at the port.

The characteristic of the Thevenin-Norton equivalent of the one-port of figure 4.33 has been calculated by SPICE. The result is represented in figure 4.34. Note the similarity to the characteristic of a tunnel diode.

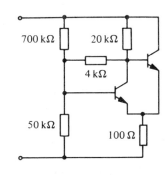

Fig. 4.33

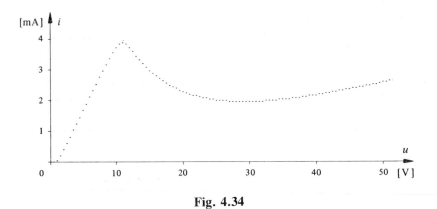

Fig. 4.34

4.3.16 Substitution Theorem

Let us assume that a circuit C is separated into two N-ports Q_1 and Q_2. If we connect Q_1 with an N-port $\tilde{Q}_2$, which is equivalent to Q_2 at the ports, then the resulting circuit $\tilde{C}$ has the same solutions on Q_1. This means that the same currents and voltages of Q_1, which are part of a solution of C, are also part of a solution of $\tilde{C}$, and *vice versa*. Furthermore, if Q_2 and $\tilde{Q}_2$ are transparent, this correspondence of the solutions of C and $\tilde{C}$ is bijective.

For the proof, let us consider a solution of C in a time interval $[t_0,t_1]$. Let $u_1(t), \ldots, u_N(t), i_1(t), \ldots, i_N(t)$ be the voltages and currents at the ports of Q_1 that are part of that solution. Then, $u_1(t), \ldots, u_N(t), -i_1(t), \ldots, -i_N(t)$ are the voltages and the currents at the ports of Q_2. Hence, they form a column matrix of admissible functions for Q_2, and also for $\tilde{Q}_2$, if $\tilde{Q}_2$ is equivalent to Q_2 at the ports. Thus, there are internal voltages and currents of $\tilde{Q}_2$ such that, together with $u_1(t), \ldots, -i_N(t)$, they satisfy the equations for $\tilde{Q}_2$. These internal voltages and currents complement the voltages and currents of Q_1 into a solution of $\tilde{C}$. In the same way, we pass from a solution of $\tilde{C}$ to a solution of C. If $\tilde{Q}_2$ is transparent, there is only one solution of $\tilde{C}$ that can be obtained from a solution of C. Conversely, if Q_2 is transparent, only one solution of C corresponds to a solution of $\tilde{C}$.

4.3.17 Comment

The advantage derived from the Thevenin-Norton equivalent is based on the substitution theorem. If a transmission chain is composed of cascade-connected two-ports, closed by one-ports at the end, then the dc operating point of a two-port in the chain can be determined by replacing the whole circuit on the source side and the whole circuit on the charge side with their respective equivalent resistors. It is in exactly the same way that the Thevenin-Norton equivalents are used in linear circuit theory.

It should be emphasized that the substitution theorem is very general. It is not limited to one-ports, nor to resistive multiports.

For instance, let us consider a circuit composed of a linear two-port and a nonlinear two-port (fig. 4.35). Neither one is supposed to be resistive. If the linear two-port has a Thevenin equivalent and we replace it by that equivalent, we obtain the circuit of figure 4.36. As far as the nonlinear part is concerned, the circuits of figures 4.35 and 4.36 have the same solutions.

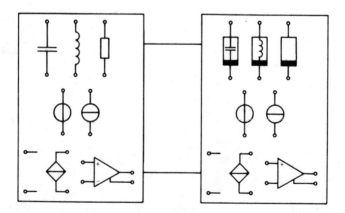

Fig. 4.35

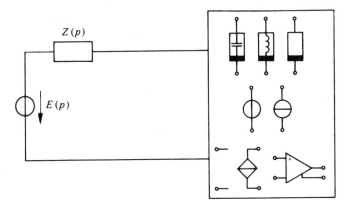

Fig. 4.36

Let us note, also, that the substitution of the linear part in a nonlinear circuit amounts to a question of equivalence between linear multiports. This is a problem which fully belongs to linear circuit theory.

4.3.18 Removal of Capacitor Loops and Inductor Cut Sets

In section 2.5, we have seen that the circuits which include loops composed of capacitors and voltage sources, or cut sets composed of inductors and current sources cause problems in establishing the state equations. More precisely, the solutions of the circuit evolve in a subset of lower dimension of the configuration space. If we wish to establish state equations, we must characterize the parameters of this subset. This operation was carried out in subsection 2.5.28 for the circuit of figure 2.65.

In the following subsections, we shall address the same problem in a different way. Instead of taking into account a space of initial conditions that is too small, we take the point of view that the configuration space is too large. Accordingly, we attempt to replace the whole subcircuit of the capacitors (inductors) and voltage (current) sources by an equivalent N-port in such a way that the resulting circuit has a configuration space wherein the initial conditions are not subject to any constraints. An additional advantage of this approach is that impasse points can be detected by means of the associated linear resistive circuit. Remember that in the original circuit the associated linear resistive circuit is always ill posed. No distinction can be made between the points where the solution exists and the impasse points.

The price to be paid for these advantages is the introduction of multiport capacitors and inductors. In other words, the N-port which replaces the subcircuit of the capacitors (inductors) and the voltage (current) sources is no longer a connection of elements of the same type, but rather it is a capacitor (inductor) with N ports. We shall briefly explain hereafter how the expansions of sections 2.4 and 2.5 can be generalized to circuits that include multiport capacitors and inductors.

4.3.19 Definitions

An *N-port capacitor* is an *N*-port having constitutive relations of the form:

$$f_1(u_1, ..., u_N, q_1, ..., q_N, t) = 0$$
$$\vdots \tag{4.130}$$
$$f_N(u_1, ..., u_N, q_1, ..., q_N, t) = 0$$

$$dq_1/dt = i_1$$
$$\vdots \tag{4.131}$$
$$dq_N/dt = i_N$$

In compact notation, they are written as

$$f(u, q, t) = 0 \tag{4.132}$$
$$dq/dt = i \tag{4.133}$$

If time t does not explicitly appear in (4.130), it is an *autonomous capacitor*. The concepts of charge-controlled or voltage-controlled *N*-port capacitor are defined by generalizing (1.21) and (1.22) to N dimensions.

An *N-port inductor* is an N-port having constitutive relations of the form:

$$f(\varphi, i, t) = 0 \tag{4.134}$$
$$d\varphi/dt = u \tag{4.135}$$

where φ, i, and u are the column matrices of N fluxes, currents, and voltages. The definitions of a flux-controlled and current-controlled autonomous N-port inductor are analogous to those of an N-port capacitor.

4.3.20 Example

Let us add parasitic capacitors to the Schmitt trigger of figure 3.61. They are linear capacitors C_k in parallel with the resistors R_k, for $k = 1,2,3$, and nonlinear capacitors with the constitutive relation:

$$q = g(u) \tag{4.136}$$

in parallel with the base-emitter and base-collector junctions of the transistors (fig. 4.37). The resulting circuit is composed of resistive and capacitive parts. Because the two parts are connected to each one of the six nodes, the circuit is separated into a resistive five-port and a capacitive five-port. The latter is represented in figure 4.38. We have associated the voltage sources to the capacitive five-port. Several capacitor-voltage source loops are inside the five-port.

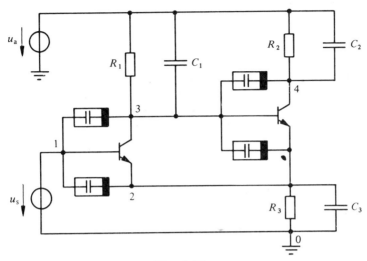

Fig. 4.37

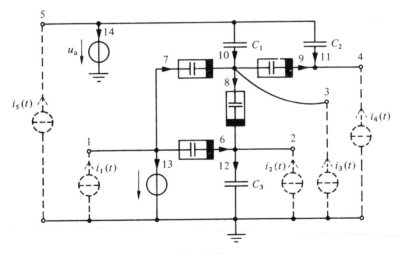

Fig. 4.38

Is this five-port equivalent to a five-port capacitor? In order to be able to answer this question, we will try to remove the internal variables of the five-port's parametric system. To that effect, we close the five ports on the current sources represented by a dotted line in figure 4.38.

The linear capacitors and the voltages sources form a tree of the resulting circuit. The currents of the linear capacitors are expressed by the currents of the cotree in the following way:

256

$$i_{10} = -i_3 - i_7 + i_8 + i_9 \tag{4.137}$$

$$i_{11} = -i_4 - i_9 \tag{4.138}$$

$$i_{12} = i_2 + i_6 + i_8 \tag{4.139}$$

By integrating (4.137) to (4.139) from t_0 to t, we obtain the same relations, but in terms of $q_k(t)$ instead of $i_k(t)$, and up to a constant. Since the charges associated with the current sources are an artificial construction because they do not appear in any constitutive relation, we can arbitrarily choose the initial charges $q_2(t_0), q_3(t_0)$, $q_4(t_0)$. It is convenient to choose them such that (4.137) to (4.139) also hold for the charges:

$$q_2(t_0) = -q_6(t_0) - q_8(t_0) + q_{12}(t_0) \tag{4.140}$$

$$q_3(t_0) = q_7(t_0) + q_8(t_0) + q_9(t_0) - q_{10}(t_0) \tag{4.141}$$

$$q_4(t_0) = -q_9(t_0) - q_{11}(t_0) \tag{4.142}$$

By using the constitutive relations of the capacitors, we can rewrite (4.137) to (4.139), with i_k being replaced by q_k, in terms of voltages and charges at the ports:

$$C_1(u_5 - u_3) = -q_3 - g(u_1 - u_3) + g(u_3 - u_2) + g(u_3 - u_4) \tag{4.143}$$

$$C_2(u_5 - u_4) = -q_4 - g(u_3 - u_4) \tag{4.144}$$

$$C_3 u_2 = q_2 + g(u_1 - u_2) + g(u_3 - u_2) \tag{4.145}$$

If we take into account the constraints between the port voltages imposed by voltage sources, we have

$$u_1 = u_s \tag{4.146}$$

$$u_5 = u_a \tag{4.147}$$

hence,

$$q_2 = C_3 u_2 - g(u_s - u_2) - g(u_3 - u_2) \tag{4.148}$$

$$q_3 = C_1(u_3 - u_a) - g(u_s - u_3) + g(u_3 - u_2) + g(u_3 - u_4) \tag{4.149}$$

$$q_4 = C_2(u_4 - u_a) - g(u_3 - u_4) \tag{4.150}$$

If complemented with

$$dq_k/dt = i_k , \quad k = 2, 3, 4 \tag{4.151}$$

then these equations are the constitutive relations of a voltage-controlled three-port capacitor. It is time dependent because of the presence of u_s in the constitutive relations.

Thus, we have shown that the admissible voltages and currents for the five-port of figure 4.38 are also admissible for the five-port of figure 4.39.

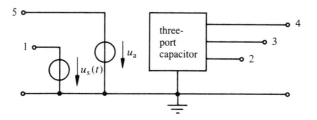

Fig. 4.39

Conversely, if we have functions $u_k(t),i_k(t),q_k(t),k=1,\ldots,5$, which satisfy (4.146) to (4.151), we infer from u_k the voltages of the internal branches of the five-port of figure 4.38, and then the capacitor charges by the constitutive relations. The internal currents are obtained by derivation. They satisfy (4.137) as a result of (4.147) to (4.149). This shows that the functions $u_k(t),i_k(t),q_k(t)$ are also admissible for the five-port of figure 4.38.

Actually, the five-port of figure 4.38 is transparent. Indeed, we have just observed that the port voltages determine all the internal variables. This completes the proof that the five-ports of figures 4.38 and 4.39 are equivalent.

According to the substitution theorem, we obtain the same solutions in the resistive part of the circuit of figure 4.37 if we replace the capacitive part by the five-port of figure 4.39. The solutions of the original circuit and the modified circuit correspond in a bijective way. In fact, the correspondence is explicated by (4.140) to (4.142), which relate the initial conditions on the capacitors in figure 4.37 with the initial conditions on the three-port capacitor of figure 4.39, which substitutes them.

The graph of the five-port of figure 4.39 is represented in figure 4.40. We observe that we have actually removed the loops of capacitors and voltage sources. In fact, the graph is the same as that of the chosen tree.

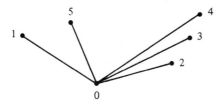

Fig. 4.40

4.3.21 Example

The three-port of figure 4.41 includes two capacitors having the constitutive relation of $i=g(u)$. Let us try to remove the loop of capacitors and voltage sources by the method of subsection 4.3.20. We must, therefore, choose a tree and then express the currents of the tree branches by those of the cotree branches. Let us choose the tree composed of branches 4, 5, and 7. Because this tree only includes one capacitor, we should obtain a three-pole composed of a capacitor with two terminals and two voltage sources. Then,

$$i_7 = i_2 + i_3 + i_6 \qquad (4.152)$$

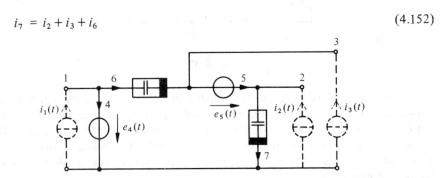

Fig. 4.41

By integrating this equation, we would find the charges of ports 2 and 3, which indicates the presence of a two-port capacitor. However, only the sum of the two currents is involved in (4.152). For this reason, we introduce a new current:

$$i = i_2 + i_3 \qquad (4.153)$$

which, by integrating from t_0 to t, leads to

$$q = q_7 - q_6 \qquad (4.154)$$

if we choose

$$q(t_0) = q_7(t_0) - q_6(t_0) \qquad (4.155)$$

By using the constitutive relations, we obtain

$$q = g(u_1) - g(e_4(t) - e_5(t) - u_2) \qquad (4.156)$$
$$dq/dt = i \qquad (4.157)$$
$$u_1 = e_4(t) \qquad (4.158)$$
$$u_3 = u_2 + e_5(t) \qquad (4.159)$$

This system of equations, complemented by (4.153), describes the three-port of figure 4.42. By the same reasoning as used in subsection 4.3.20, we establish that the three-poles of figures 4.41 and 4.42 are equivalent. In this particular case, we could remove the loop of capacitors and voltage sources without introducing any multiport capacitor. Again, the graph of the resulting three-port is the same as that of the chosen tree in the original three-port.

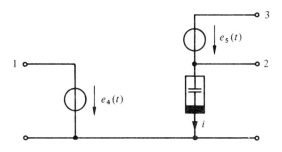

Fig. 4.42

4.3.22 Theorem

Consider an N-port without internal nodes, composed of voltage-controlled capacitors and S voltage sources. Let us assume that there is no source loop. Then, for each tree which includes all the sources, there is an equivalent N-port composed of the same voltage sources, as well as an $N\text{-}S$ port voltage-controlled capacitor, having the following properties:

- its graph is identical to the tree;
- the location of the voltage sources is the same as in the original circuit;
- the branches of the tree which carry a capacitor in the original circuit become the ports of the multiport capacitor.

The dual theorem holds for the N-ports composed of current-controlled inductors and current sources.

For the proof, we first assume that we have a column matrix u_p, i_p of admissible functions for the given N-port. Consequently, there is a set of functions for the internal variables such that the constitutive relations are satisfied. We adopt the following notation:

- u_T, i_T, q_T: voltages, currents, and charges of the tree capacitors;
- u_C, i_C, q_C: voltages, currents, and charges of the cotree capacitors;
- u_s, i_s: voltages and currents of the voltage sources;
- u_p, i_p: voltages and currents of the ports.

By decomposing the fundamental loop matrix with respect to the tree, we can write

$$\begin{pmatrix} i_T \\ i_s \end{pmatrix} = \begin{pmatrix} C & D \\ E & F \end{pmatrix} \begin{pmatrix} i_C \\ i_p \end{pmatrix} \tag{4.160}$$

$$\begin{pmatrix} u_c \\ u_p \end{pmatrix} = -\begin{pmatrix} C^T & E^T \\ D^T & F^T \end{pmatrix} \begin{pmatrix} u_T \\ u_s \end{pmatrix} \tag{4.161}$$

In particular, we have

$$i_T = Ci_C + Di_p \tag{4.162}$$

Let us set

$$i = Di_p \tag{4.163}$$

and, therefore,

$$i = i_T - Ci_C \tag{4.164}$$

By integrating (4.164), starting from an arbitrary initial time t_0, and by defining

$$q(t) = q(t_0) + \int_{t_0}^{t} i(t)\, dt \tag{4.165}$$

with

$$q(t_0) = q_T(t_0) - Cq_C(t_0) \tag{4.166}$$

we find

$$q = q_T - Cq_C \tag{4.167}$$

Consider the constitutive relations of the capacitors

$$q_k = g_k(u_k) \tag{4.168}$$

We group them in the following way:

$$q_T = g_T(u_T) \tag{4.169}$$
$$q_C = g_C(u_C) \tag{4.170}$$

Combined with (4.167), we obtain

$$q = g_T(u_T) - Cg_C(u_C) \tag{4.171}$$

According to (4.161), we find

$$q = g_T(u_T) - Cg_C(-C^T u_T - E^T u_s) \tag{4.172}$$

This is the constitutive relation of a voltage controlled capacitor with as many ports as there are components of u_T.

If we add (4.163) and (4.164) to (4.172), as well as the expression u_p as a function of u_T and u_s of (4.161), and $i = dq/dt$, we thus obtain the required parametric constitutive relations of the N-port.

Conversely, if $u_p(t),i_p(t)$ are admissible for this transformed N-port, we have functions of time $q(t),u_T(t),i(t)=dq/dt$ such that (4.163), (4.164), and the second line of (4.161) are satisfied. For the parametric constitutive relations of the original N-port to be satisfied, we simply define u_C by the first line of (4.161), q_C by (4.170), $i_C=dq_C/dt$, q_T by (4.169) and $i_T=dq_T/dt$.

The two N-ports are transparent because the port voltages determine the tree voltages, and, from u_T,u_s, all the internal variables are determined. With this, the proof of the equivalence of the two N-ports is completed.

4.3.23 Comment

The constitutive relation of the N-S port capacitor is (4.172). The correspondence between the solutions of the two circuits, which have a different capacitive part, according to theorem 4.3.22, is given by the correspondence of the initial conditions on the capacitors (4.166).

If the capacitive part of a circuit has internal nodes, we still separate it into an N-port composed of capacitors and voltage sources without internal node, and an N-port which contains the rest of the circuit. The latter will, however, include isolated terminals. An example is represented in figure 4.43.

The removal of the loops of capacitors and voltage sources as represented here has been published in [20].

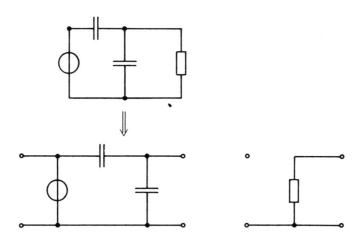

Fig. 4.43

4.3.24 State Equations in the Presence of Multiport Capacitors and Inductors

What happens to the developments of sections 2.4 and 2.5 if a circuit includes multiport capacitors or inductors? Nothing is changed! Let us take, for example, theorem 2.4.15. The functions $h_C(q_C)$ and $g_L(\varphi_L)$ in (2.143) and (2.144) simply cover the constitutive relations of all the capacitors and all the inductors, whether they have one port or several ports.

Similarly, the concept of associated resistive circuit is easily adapted. Instead of replacing each capacitor (inductor) by a voltage (current) source, we must replace *each port* of any capacitor (inductor) by a voltage (current) source.

When applying the capacitor-loop and inductor-cut-set removal method, the only difficulty is that we obtain capacitors (inductors) with several ports that are voltage (current) controlled, whereas theorem 2.4.15 requires them to be charge (flux) controlled. Thus, to realize the state equations (2.148) and (2.149), we must invert the constitutive relations of the multiport capacitors and inductors obtained by theorem 4.3.22. If, at a generalized operating point, the inverse function does not even exist locally, we should expect an impasse point for the solutions.

Chapter 5

Asymptotic Behavior and Stability of the Solutions

5.1 STATEMENT OF THE PROBLEM

5.1.1 Introduction

In chapter 2, we discussed the conditions that ensure the existence of the solution of a circuit, and we indicated the parameters, i.e., the initial conditions, which determine the solution. As a general rule, the theorems of chapter 2 only guarantee the existence of the solution for a limited period, which starts from the initial time. However, theorem 2.4.16 shows that under conditions that are not too restrictive, we can be assured that the solutions exist up to $t \to +\infty$.

The existence and uniqueness of the solutions having been established, we can now proceed with the study of the actual behavior of the circuit. We will limit this study to qualitative properties, since the quantitative properties are within the area of numerical computer calculation. One of the most important characteristics of the behavior of a circuit is its *steady-state solution.*

What is a steady-state solution? It is the solution that is reached after a certain time has elapsed, during which the *transient phenomena* occurred. If formulated this way, however, the concept of the steady-state solution is too vague to suit rigorous analysis. Indeed, we should at least give the order of magnitude for the transient time if we wish to characterize the steady-state solution in such a way. Further, transient time is only known from the study of transient phenomena, which is more detailed than that of steady-state phenomena.

We can distinguish transient from steady-state phenomena by defining the steady-state solution as being the asymptotic solution, when $t \to +\infty$. Adopting this viewpoint does not imply that an infinitely long period of time must elapse before a steady state is reached. On the contrary, a solution cannot be distinguished from the steady state after a time t, which may occasionally seem very short to the observer. Nevertheless, we set $t \to \infty$ because said time is not known at this stage of circuit analysis.

The steady-state solution is primarily dictated by the asymptotic behavior of the time-dependent sources. Among the infinity of possible time dependencies, autonomous circuits and circuits having sinusoidal sources play a principal role, as a result of the large number of circuits used in practice, the model of which falls within these categories.

Autonomous circuits are primarily used to model the behavior of electronic circuits in the absence of signals. Therein we study the equilibrium state of a circuit before it is excited by a signal, and the return to equilibrium after injection of a signal of finite duration. Circuits with sinusoidal sources are primarily used to model electric power distribution systems and signal modulated transmission systems. In the field of linear circuits, the role played by sinusoidal sources is even more important. Thanks to the superposition principle, we obtain the circuit response to an arbitrary signal, starting from the responses to all sinusoidal signals, by using Fourier analysis. On the contrary, in the context of nonlinear circuits, the superposition principle does not hold, and consequently the *study of the steady-state of nonlinear circuits cannot be limited to circuits with sinusoidal sources*.

The precise definition of the steady-state solution is given in section 5.3. In this section, we state a conjecture concerning the asymptotic behavior of the solutions. This is a natural generalization of some well known properties of linear circuits. We shall then show, by a series of counterexamples, that the conjecture is wrong.

5.1.2 Reminder: Asymptotic Behavior of Linear Circuits

All the solutions of a strictly stable linear circuit that includes only a single source, the time dependence of which is sinusoidal, converge toward the same sinusoidal solution. This sinusoidal steady state is unique, and it has the same period as the source. However, the voltages and currents of the branches have phases which differ from that of the source (vol. IV, sec. 2.3).

This property also holds in the particular case of zero frequency, i.e., autonomous circuits, where the steady-state solution is constant.

If several sources are present in a circuit, the steady-state solution is the sum of the steady-state solutions of each source considered separately, while the other sources are set to zero. All the solutions converge toward this steady-state solution.

5.1.3 Comment

What can be expected from the steady state of a nonlinear circuit? Because the superposition principle does not hold, from the start we must give up the idea that the steady state of a circuit with several sources is the sum of the steady states of different sources taken one by one, nor do we expect a purely sinusoidal steady state if a nonlinear circuit is excited by a sinusoidal source. However, the steady state should reflect the time symmetries of the sources.

We refer to a *symmetry of a circuit* as any transformation which leaves the equations of the circuit invariant. We allow for current and voltage transformations, time transformations, and combined transformations. A precise definition is given in subsection 5.3.13. For example, any time translation is a symmetry of an autonomous circuit. Similarly, the time translation by T is a symmetry of a circuit having sources that are periodic of period T. These are two examples of *time symmetry*.

We refer to a *symmetry of a solution* as any transformation which leaves the solution invariant. A precise definition is given in subsection 5.3.13.

If the equations are invariant under a transformation, such is generally not the case for the solutions. For example, most of the solutions of an autonomous circuit are not constant. Conversely, we would expect the **steady-state solution** to have the same symmetries as the circuit. In particular, the steady-state solution of an autonomous circuit should be constant and the steady-state solution of a circuit having periodic sources should be periodic.

These considerations lead to the following conjecture.

5.1.4 Conjecture

If the solutions of a nonlinear circuit excited by a sinusoidal source remain bounded, they all converge toward the same periodic steady-state solution, the fundamental period of which is identical to that of the source.

In the case of zero frequency, i.e., if the circuit is autonomous, the steady-state solution is constant and, thus, is located at a dc operating point.

5.1.5 Comment

The purpose of this section is to show by a series of counterexamples that conjecture 5.1.4 is wrong. Even very simple circuits can prove it wrong. All the results to be presented here have been obtained by numerical calculation.

In conjecture 5.1.4, as opposed to conjecture 5.1.2 for linear circuits (property 5.1.2), we do not use the term "stable." In fact, it is no longer possible to define stability by the position of the poles of the Laplace transform of the zero-input response, since these concepts become meaningless in the context of nonlinear circuits. The concept of stability becomes much more subtle, and later we shall devote a good part of section 5.2 to its definition.

In fact, the concepts of stability and instability of a linear circuit have both global and local aspects. The most spectacular property of an unstable linear circuit is that its solutions tend toward infinity when $t \to \infty$. Conversely, a stable linear circuit has bounded solutions. We thus introduce this property instead of stability as a condition for conjecture 5.1.4.

The local aspect is that two solutions of an unstable linear circuit are almost always diverging from one another, although they start from very close initial conditions. Without defining said property precisely, we shall say in this section that it is unstable if no other solution of the circuit can remain indefinitely in its neighborhood, even if the selected initial conditions are very close. Otherwise, we shall say that it is stable. The precise definition will be given in subsection 5.2.2.

5.1.6 Example: Amplifier

The transistor amplifier of figure 2.1, with a signal $u_s(t) \equiv 0$, is an autonomous circuit. We thus expect a constant steady state. A constant solution of an autonomous circuit is frozen at a dc operating point. We have already shown in subsection 3.4.28 that this amplifier has only one dc operating point. The numerical analysis confirms conjecture 5.1.4, according to which any solution, starting from any initial condition, tends toward the dc operating point when $t \to \infty$.

If we apply a signal of finite duration to the amplifier that returns to zero at the instant T, then the solution also tends toward the dc operating point when $t \to \infty$. Indeed, starting from T, we return to the autonomous circuit, and the effect of the signal is the same as that of a certain initial condition at T.

The return to the dc operating point after the injection of a signal is illustrated in figure 5.1, which represents a solution calculated by SPICE. The signal $u_s(t)$ is represented in figure 5.2.

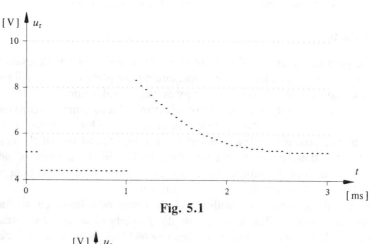

Fig. 5.1

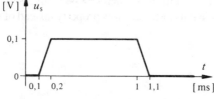

Fig. 5.2

5.1.7 Counterexample: Circuit with a Tunnel Diode

We observed in section 2.1 that the circuit of figure 2.10 can have three dc operating points, two of which are stable and one is unstable. We have proved, and the numerical analysis has confirmed, that in the absence of a signal $u_s(t)$ any solution converges toward one of the stable dc operating points, with the choice depending on the initial condition. The only exception is the solution that starts exactly at the unstable dc operating point and which consequently remains there indefinitely.

Thus, there are three steady-state solutions, one of which is unstable. This is a first counterexample for conjecture 5.1.4, since the steady-state dc operating point is not unique.

5.1.8 Counterexample: Oscillator

The Wien-bridge oscillator of figure 2.56 is an autonomous circuit having only one dc operating point. It is the operating point at which all the voltages and all the currents vanish. This fact may be easily inferred from the equation system (2.152) to (2.155) and the constitutive relations of the elements by setting $i_1 = i_2 = 0$, which characterizes the dc operating points. According to conjecture 5.1.4, we expect all solutions to tend toward zero.

Conjecture 5.1.4 is verified by numerical calculation if we choose the values $R_1 = R_2 = R_3 = 10$ kΩ, $C_1 = C_2 = 10$ nF, and a nonlinear resistor, which has the characteristic:

$$i_4 = g(u_4) = au_4 + bu_4^3 \tag{5.1}$$

with

$$a = 1/10\,\text{k}\Omega \ , \quad b = 1\,\text{V}^2/10\,\text{k}\Omega \tag{5.2}$$

The solution of this circuit, with the initial conditions $u_1 = 0$, $u_2 = 10$V, was calculated by SPICE, and the voltage of the operational amplifier's output node is represented in figure 5.3.

However, if we reduce R_3 to 1 kΩ, without changing the other elements, nor the initial conditions, we obtain the solution shown in figure 5.4.

Rather than tending toward zero, the solution converges toward a periodic solution. By calculating many solutions starting from different initial conditions, we can be convinced that in fact all the solutions converge toward a periodic solution, which is the same as that of figure 5.4, except for a time shift. The only exception is the solution that remains frozen at the dc operating point $u = i = 0$. This dc operating point is always unique, but in the present case it is unstable. A physical circuit could not remain indefinitely at an unstable dc operating point because of the inaccuracies of the model and the initial conditions. Nonetheless, this type of solution can generally be reproduced by numerical calculation.

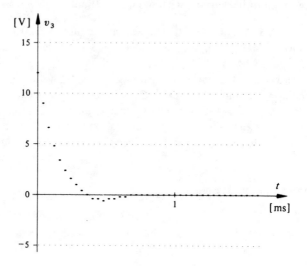

Fig. 5.3

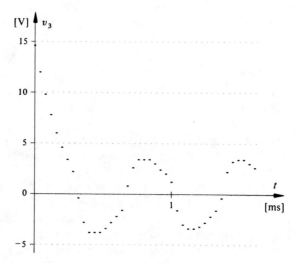

Fig. 5.4

This circuit constitutes a second counterexample for conjecture 5.1.4 in that the steady-state solution is not unique. Strictly speaking, there is an infinity of periodic steady states, distinguished by a translation in time. However, their orbits in the state space (u_1, u_2) are all identical (fig. 5.5). Were we to count all periodic steady states as one, the steady state solution would still not be unique, since there remains the unstable dc operating point.

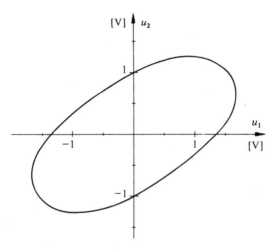

Fig. 5.5

5.1.9 Symmetry Breaking

What distinguishes counterexample 5.1.8 from counterexample 5.1.7 is the occurrence of a new phenomenon, *symmetry breaking*. The steady-state solution does not exhibit the same symmetries as the circuit.

Because the circuit is autonomous, it is invariant under an ***arbitrary translation*** in time. Were conjecture 5.1.4 true, the steady-state solution would be constant and thus invariant under an arbitrary translation in time. Conversely, the steady-state solutions of subsection 5.1.8 are only invariant under ***certain translations***, more precisely this is under translations of a multiple of the period. These steady-state solutions thus exhibit less time symmetries than the circuit. We thus say that the circuit symmetry has been *broken*.

A steady-state solution which exhibits less symmetries than the circuit cannot be unique. This general fact is easily explained by way of an autonomous circuit such as the oscillator of subsection 5.1.8. Let us assume that $\xi(t)$ is a solution of an autonomous circuit. Also, let us introduce a new time scale:

$$t' = t + \tau \tag{5.3}$$

With this time scale, the solution $\xi(t)$ becomes

$$\xi_1(t') = \xi(t' - \tau) \tag{5.4}$$

Additionally, because the circuit is autonomous, changing from t to t' does not modify it. Its equations remain the same, except that d/dt is replaced by d/dt' in (2.84) and (2.86). Consequently, if $\xi(t)$ is a solution of the circuit, so too is the following:

$$\xi_2(t') = \xi(t') \tag{5.5}$$

By comparing (5.4) with (5.5), we infer that if $\xi(t)$ is a solution of an autonomous circuit, so too is $\xi(t - \tau)$ for an arbitrary τ.

Two cases may then occur. Either $\xi(t - \tau)$ is identical to $\xi(t)$, or the two solutions are different. If the circuit symmetry is not broken, $\xi(t - \tau)$ is always identical to $\xi(t)$. Conversely, if the symmetry is broken, some of the functions $\xi(t - \tau)$ constitute new steady-state solutions. This is the case of periodic steady states wherein τ is not a multiple of the period.

5.1.10   Counterexample: Ferroresonant Circuit

Let us return to the circuit of figure 2.52, the inductor of which is assumed to have the piecewise linear characteristic of figure 1.37. This inductor constitutes a simplified model of a coil with a ferromagnetic core. Therefore, the nonlinear phenomena observed in circuits having inductors with a saturation characteristic as the only nonlinear elements, such as those of figures 1.36 and 1.37, are described by the term *ferroresonance*. However, the same phenomena occur in other nonlinear circuits.

The fact that the circuit of figure 2.52 is piecewise linear and of order two makes numerical calculation much easier. The results of this subsection were obtained from a special program, in which the explicit solution was programmed in each linear domain. This process is convenient from the viewpoint of numerical precision. Hereafter, we shall describe the asymptotic behavior of the solutions for the following choice of parameters: $C_1 = 1.69 \ \mu F$, $R_4 = 50 \ \Omega$, $R_5 = 10 \ k\Omega$; the inductor characteristic is determined by the parameters $L_0 = 33.3 \ H$, $L_1 = 1.28 \ H$, $\varphi_0 = 0.92$ Wb; and the time-dependent source is sinusoidal, according to

$$e_3(t) = E \cos \omega t \tag{5.6}$$

with $\omega/2\pi = 50$ Hz. The amplitude varies from 0 to 10 kV. We have calculated a large number of solutions for a large number of values of E.

As long as E remains lower than 91 V, conjecture 5.1.4 holds. All the solutions converge toward a single steady-state solution, which is entirely confined in the linear domain $|\varphi_2| < \varphi_0$. Thus, it is the same steady-state solution as that of the linear circuit obtained from the circuit in figure 2.52 by replacing the nonlinear inductor with a linear inductor of value L_0. This is the very special case of a nonlinear circuit excited by a sinusoidal source, the steady-state solution of which is purely sinusoidal. The orbit of the steady-state solution for $E = 80$ V is represented in figure 5.6.

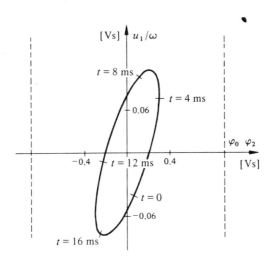

Fig. 5.6

If the amplitude E is between 91 V and 246 V, we observe that there are three steady-state solutions, two of which are stable and one is unstable. Their orbits are represented in figure 5.7 for $E = 160$ V. They can be distinguished by the size of their orbits. The smallest one is stable. This coincides with the steady-state solution of the linear circuit with the inductor L_0, such as that in figure 5.6. The largest one is stable as well. It crosses the three linear domains, and therefore it is not purely sinusoidal, but it is periodic. The third steady-state solution, the orbit of which is located between the other two, is unstable. It also crosses the three linear domains. Its instability is shown in figures 5.8 and 5.9, where we have chosen two initial conditions at time $t = 0$, which are very close to that of the unstable steady-state solution. We observe that one of the solutions tends toward the small steady-state solution, and the other toward the large one.

The behavior of this circuit is similar to that of subsection 5.1.7. Rather than three constant steady-state solutions, it presents three that are periodic. In both cases, the steady-state solutions comply with the time symmetry of the circuit.

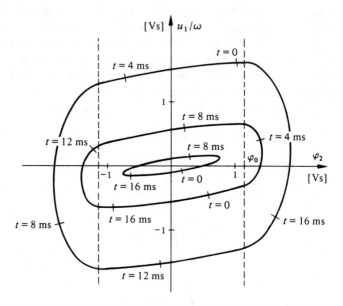

Fig. 5.7

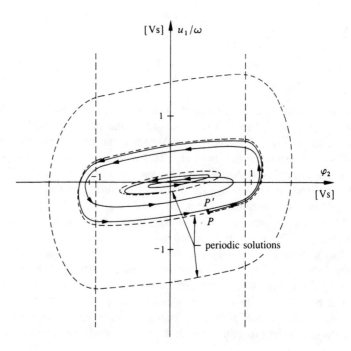

Fig. 5.8

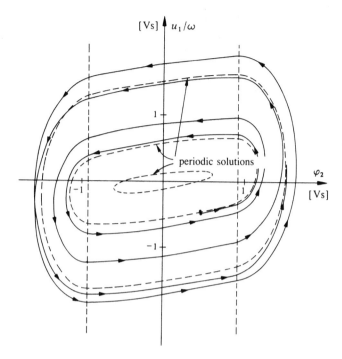

Fig. 5.9

Among the initial conditions at time $t = 0$, we may wonder which of these converge toward the small steady-state solution, and which toward the large one. By grouping the initial conditions at $t = 0$ according to their asymptotic behavior, we obtain the *basins of attraction* of the two stable steady-state solutions. By looking at figure 5.7, we might think that the unstable steady-state solution separates the two basins of attraction. According to this conjecture, any initial condition inside (outside) the unstable orbit would converge toward the small (large) solution. This is not so. The basins of attraction are represented in figure 5.10. Any initial condition located inside (outside) the cross-hatched area produces a solution which converges toward the small (large) steady-state solution. The initial conditions at $t = 0$ of the three steady-state solutions are indicated by a cross in figure 5.10. As might be expected, the cross corresponding to the unstable steady-state solution is located on the boundary between the two basins of attraction. In fact, all the points on this boundary are exceptions in that the corresponding solutions do not converge toward a stable steady-state solution, but rather toward the unstable steady-state solution. What is surprising in figure 5.10 is that there are points very far from the origin having solutions which tend toward the small steady-state solution. We must note that the scale of the axes in figure 5.10 is smaller than that in figure 5.7.

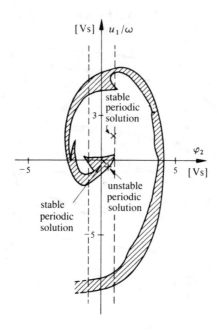

Fig. 5.10

Again, the circuit with E between 246 V and 1694 V has only one steady-state solution toward which all the solutions converge. It is represented in figure 5.11, for $E = 300$ V.

If the amplitude E is between 1694 V and 4389 V, once again we have three steady-state solutions, two of which are stable and one is unstable. These three solutions are represented in figure 5.12, for $E = 2300$ V. The unstable orbit is indicated by a dotted line in figure 5.12. Their nature is different from the periodic solutions of figure 5.7. We can understand this if we consider the transformation that advances the time by one-half period of the source, $t \rightarrow t + \pi/\omega$, and which reverses the sign of all the electrical variables. This transformation leaves the equation system of the circuit invariant. All the solutions encountered until now are also invariant under this symmetry transformation. However, the symmetry is broken in the case of two steady-state solutions, one of which is necessarily the image of the other under the symmetry transformation, according to the remarks of subsection 5.1.9.

If we start from $E = 4389$ V, we can again observe a single periodic steady state toward which all the solutions converge. It is represented in figure 5.13, for $E = 5000$ V.

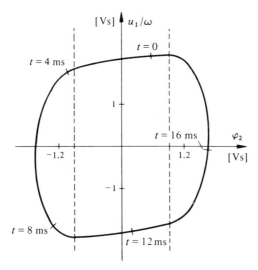

Fig. 5.11

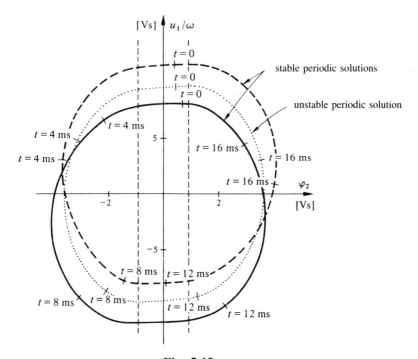

Fig. 5.12

276

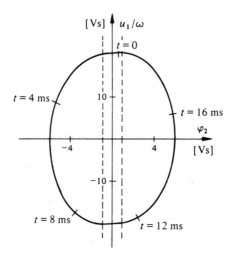

Fig. 5.13

5.1.11 Subharmonic Periodic Solutions

All the periodic solutions encountered in subsection 5.1.10 have the same period as the source, which means: $T = 2\pi/\omega$. This is not the case if we choose the following parameters for the circuit: $C_1 = 5\ \mu F$, $R_4 = 5\ \Omega$, $R_5 = 10\ k\Omega$, $L_0 = 100\ H$, $L_1 = 5\ H$, $\varphi_0 = 0.92\ Wb$, $\omega/2\pi = 50\ Hz$, $E = 300\ V$. We observe the presence of periodic steady-state solutions the fundamental period of which is not T, but $3T$. Because their frequency is a fraction of the source frequency, they are called *subharmonic periodic solutions*. Remember that the time translation by T is a symmetry of the circuit. In the case of subharmonic steady-state solutions, this symmetry is broken. Only if we repeat the translation three times can such a steady-state solution be reproduced. According to the reasoning of subsection 5.1.9, translation by T, as well as translation by $2T$, must produce new subharmonic steady-state solutions. Indeed, there are three stable subharmonic solutions having an identical orbit, and which are delayed relative to one another by T.

Furthermore, there are three unstable subharmonic solutions with period $3T$, as well as one stable periodic solution with period T. The two different orbits of the four stable periodic solutions are represented in figure 5.14. Their basins of attraction are shown in figure 5.15. We note that their form is rather complicated.

The same circuit with different parameters can include an even larger number of periodic steady states. In one case, we have obtained, by numerical calculation, 41 periodic solutions, of which 17 are stable [21]. Other numerical results are reported in [21,22].

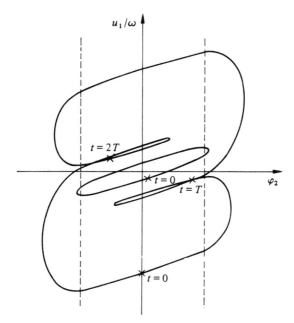

Fig. 5.14

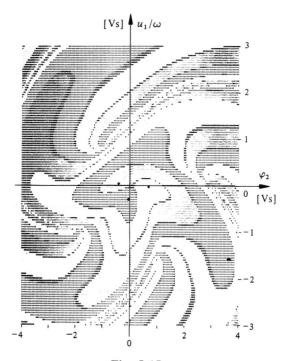

Fig. 5.15

5.1.12 Bifurcation

If we follow the evolution of the steady-state solutions as a function of amplitude E in the case of the ferroresonant circuit of subsection 5.1.10, we observe the following phenomena. In general, the steady-state solutions, indeed as all solutions, are continuously deformed as a function of E, although their nature remains unchanged. Conversely, at very specific values of E, the steady-state solutions change in a *qualitative* way. These values are called *bifurcation points*. The concept of the bifurcation point is used for any physical or mathematical system having solutions which change in nature as a function of a parameter. Such a parameter is called a *bifurcation parameter*.

A graphic representation of the steady-state solutions as functions of a parameter, which shows the bifurcation points, is called a *bifurcation diagram*. If we study the dc operating points of a circuit as functions of a parameter λ, we can represent a current or a voltage of the dc operating point as a function of λ. In the case of a steady-state solution of a circuit excited by periodic sources, we represent a current or a voltage at the instants $t_0 + nT$ as a function of λ, where t_0 is an arbitrary time and T is the period of the sources. An example of such a bifurcation diagram will be given for the next counterexample. At the same time, we shall see why the qualitative change of the solutions is called a bifurcation.

5.1.13 Counterexample: Circuit with a Junction Diode

The circuit of figure 5.16 includes a junction diode, as the only nonlinear element. If the diode is modeled by a strictly locally passive resistor, such as the exponential diode (1.15) or the piecewise linear diode (1.16), then conjecture 5.1.4 holds. This fact shall be proved in subsection 5.7.5. However, if we take into account the nonlinear junction capacity, we cannot guarantee a unique steady-state any longer. Indeed, if we choose certain element values, we may observe the same type of phenomenon as in the case of the ferroresonant circuit.

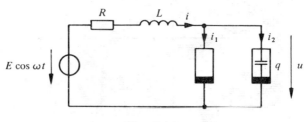

Fig. 5.16

We shall describe the asymptotic behavior of the solutions for the following choice of elements: $R = 25\ \Omega$, $L = 10$ mH, $\omega/2\pi = 92$ kHz. The resistor and the nonlinear capacitor have the piecewise linear characteristics of figures 5.17 and 5.18, respectively, with the parameters $G_1 = 0$, $G_2 = 1/2\ \Omega$, $U_j = 0.75$ V, $C_1 = 300$

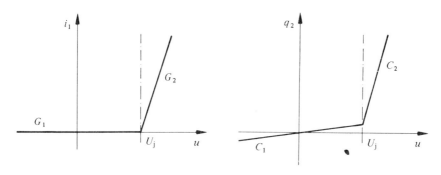

Fig. 5.17   **Fig. 5.18**

pF, $C_2 = 2\ \mu\text{F}$, and amplitude E varying between 0 V and 6 V. With the parameters G_1, G_2, C_1, C_2, U_j, we have roughly approximated the model of a junction diode as incorporated in SPICE. Indeed, the numerical calculation of the same circuit by SPICE has given asymptotic behaviors similar to those we are to be describing, which have been obtained by an *ad hoc* computer program [23].

A bifurcation diagram permits us to visualize simultaneously the asymptotic behavior of all the values of E (fig. 5.19). For each value of the bifurcation parameter, i.e., the amplitude E, we represent the values of a variable, the current i in our case, at discrete instants that are separated from one another by a period $T = 2\pi/\omega$ of the source. In figure 5.19, we have marked on the ordinate, for each abscissa value of E, the 20 values $i(t_0 + nT)$, $t_0 = T/5$, $n = 80, \ldots, 99$. The initial time t_0 has been chosen in such a way as to obtain large amplitudes for i. We have not marked the first 80 values so that the solution can reach a steady state. The solutions calculated from different initial conditions have confirmed that the values of the current represented in figure 5.19 can be considered as belonging to a steady-state solution. If there are several steady-state solutions, we superpose the corresponding series of points $i(t_0 + nT)$.

If the steady-state solution is unique and periodic by period T, the values $i(t_0 + nT)$ are identical and, for the amplitude E concerned, only one point appears as an ordinate. This is the case for $E < 1.02$ V. In figure 5.20, we have represented the orbit of the steady-state solution for $E = 0.8$ V in the state space (q,i).

Conversely, if a steady-state solution is subharmonic, with a period $2T$, the currents $i(t_0 + nT)$ alternately take two values, and two points appear as ordinates for the corresponding abscissa E. According to the reasoning of subsection 5.1.9, there is a second steady-state solution of the same type. However, its currents $i(t_0 + nT)$ are identical to those of the first subharmonic solution, up to a translation by T. The pair of subharmonic solutions of period $2T$ is present for E between 1.02 V and 2.12 V, except for the intervals (1.29 V, 1.41 V) and (1.41 V, 1.52 V), which are to be discussed later. The orbit of the steady-state solution for $E = 1.8$ V is represented in figure 5.21.

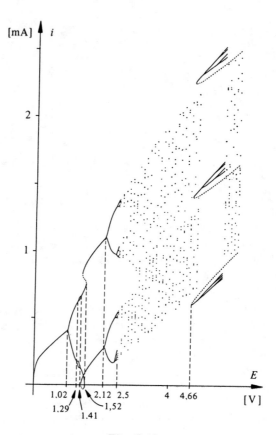

Fig. 5.19

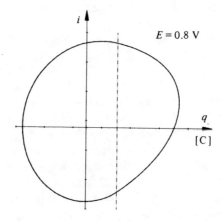

Fig. 5.20

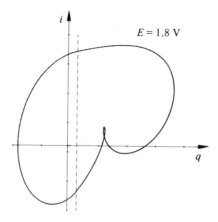

Fig. 5.21

In the case of a steady-state solution of period $4T$, four different points appear in figure 5.19 for the same E. Such a solution exists for E between 2.12 V and 2.40 V. The other three solutions of the same period, which necessarily go with it, because of the breaking of symmetry, yield the same four values $i(t_0 + nT)$. The orbit of the steady-state solution for $E = 2.2$ V is represented in figure 5.22.

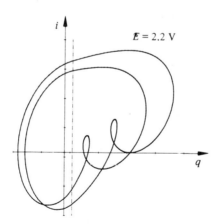

Fig. 5.22

Bifurcations occur at the amplitudes $E = 1.02$ V, $E = 2.12$ V, and $E = 2.40$ V. Figure 5.19 well justifies this terminology. In other instances, the term bifurcation is not as descriptive. For example, between 1.41 V and 1.52 V, there is a reversal of the bifurcation curve. In fact, here we have marked the points $i(t_0 + nT), n = 80, \ldots,$ 99 several times, starting from different initial conditions. While there normally are two periodic solutions of period $2T$, the values of which taken by the currents $i(t_0 + nT)$ are identical, we find between 1.41 V and 1.52 V four stable periodic solutions of period $2T$, grouped by two with the same values $i(t_0 + nT)$. The dotted line, which joins the two branches, belongs to an unstable periodic solution of period $2T$. To conclude, at $E = 1.41$ V and $E = 1.52$ V, there are indeed bifurcations, but the curves $i(t_0 + nT)$ as functions of E do not split into two branches at these points. Thus, they do not display any bifurcation in the sense of everyday language.

We also notice a bifurcation of the periodic solutions of period $2T$ into periodic solutions of $4T$ at $E = 1.29$ V, the effect of which is cancelled by the second bifurcation at $E = 1.41$ V.

Period doublings follow one another at points that become ever closer on the abscissa. Obviously, with 20 points $i(t_0 + nT)$, it becomes impossible to represent a periodic solution of a period exceeding $20T$ in the bifurcation diagram. The scale of figure 5.19 would also be too small. The limitation at 20 points primarily arises from problems of numerical accuracy. The more E is increased, the less accurate is the numerical calculation of the solutions. This fact is explained in subsection 5.1.14.

Even if we cannot distinguish the steady-state solutions of periods over $20T$ from an even more complicated asymptotic movement, we have good reason to believe that, beyond about $E = 2.5$ V, a new phenomenon occurs. One reason is that the period doublings that can still be discriminated indicate a convergence of these bifurcation points on the axis of the E. Beyond the limiting point, we would thus have steady-states having a period which would be $> \infty$! In subsection 5.1.14, we shall try to describe the corresponding asymptotic behavior.

So as not to overload the bifurcation diagram, we have generally not represented the points $i(t_0 + nT)$ of the unstable solutions. In fact, at each period doubling, there is an unstable periodic solution that extends the stable periodic solution which existed before the bifurcation point. Therefore, the correct description of the period doubling is as follows: a stable period mT solution becomes unstable and, at the same time, two stable $2mT$ solutions are created (fig. 5.23).

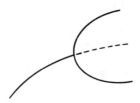

Fig. 5.23

5.1.14 Chaotic Solutions

What happens beyond $E = 2.5$ V? We observe that the values $i(t_0 + nT)$ are all different, and if we start with different initial conditions, each time we obtain another series of points $i(t_0 + nT)$. However, all the series are located within the same current interval. In figure 5.19, we have only represented one series.

The orbit in the plane (q,i) of a solution for $E = 4$ V is represented in figure 5.24. The solution follows an erratic movement, the orbit of which never closes on itself. Such solutions are called *chaotic*. A characteristic property of chaotic solutions is that they are unstable. This fact is shown in figure 5.25. We have represented the current $i(t)$ of the solution, the orbit of which is drawn in figure 5.24, for $t \in [t_0 + 80T, t_0 + 99T]$. The current of another solution is represented by the fine line, the charge $q(t_0 + 80T)$ of which is the same and its current $i(t_0 + 80T)$ differs by only 0.1%. At the start, both solutions coincide in the scale of figure 5.25, but after a time $15T$, they are already 50% apart. This instability of the solutions actually disturbs numerical calculation. Although we were taking advantage of the fact that the circuit is piecewise linear by using an appropriate program, we had to limit ourselves to 20 points $i(t_0 + nT)$ in figure 5.19.

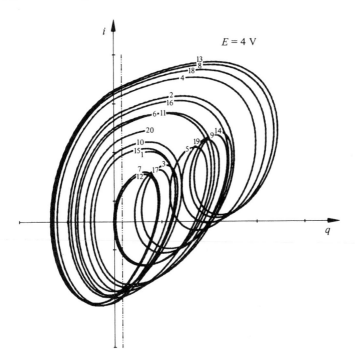

Fig. 5.24

284

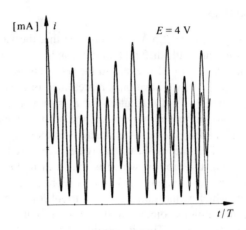

Fig. 5.25

Is there still some reason to speak of an asymptotic behavior in the presence of such erratic solutions? Surprisingly, the answer is yes. Indeed, if we mark the points $(q(t_0 + nT), i(t_0 + nT)$ in the plane (q,i), we observe that they accumulated on a curve with a strange shape, which we call an *attractor*. In figure 5.26, we have reported a series of 420 points without regard to numerical inaccuracies. If such inaccuracies were distorting that which is essential in this figure, there would just be a bunch of points instead of a curve.

A global description of the asymptotic behavior of all the solutions is as follows. There is a contractive movement in a direction perpendicular to the curve, and an expanding movement on the curve. Because the curve is bounded, the expanding movement on the curve cannot continue indefinitely. Therefore, in a sufficiently large time scale, the movement on the curve appears to be erratic.

At $E = 4.66$ V, a stable periodic solution of period $3T$ appears. Later, this solutions undergoes bifurcations which are doublings of the period. It appears that these bifurcation points again converge onto the axis of the E, and we have returned to the chaotic solutions. In figure 5.19, we have marked with a broken line the unstable solution of period $3T$, which accompanies the stable periodic solution. There is a zone where the chaotic solutions have disappeared. We call it a *window in the chaos*. We also note that for a small amplitude interval E, there is coexistence between the chaotic solutions and a stable periodic solution of period $3T$. In figure 5.27, we have represented the points $(q(t_0 + nT), i(t_0 + nT))$ in the plane (q,i), for $n = 80, \ldots, 500$, belonging to a chaotic solution of the circuit, by $E = 4.70$ V. Furthermore, we have marked with a cross the three points corresponding to the stable subharmonic solution. We note that, around the crosses, there are no points of the chaotic solution. Indeed, around the crosses, there is a small attraction zone of the stable subharmonic solution.

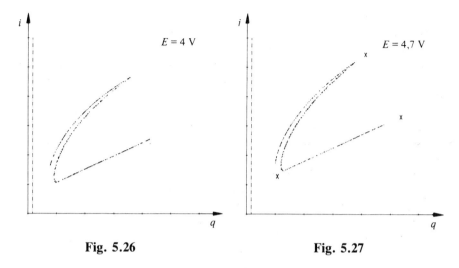

Fig. 5.26   Fig. 5.27

5.1.15 Comments

The ferroresonant circuit also exhibits chaotic solutions for certain choices of parameters [22].

Chaotic solutions have been discovered in many different fields of study since the 1970s. Most of them are the results of computer calculations. It is not surprising that before the arrival of computers such phenomena were not suspected, because it is contrary to intuition that a very simple equation system has such complicated solutions.

One remarkable exception is the work of Levinson [24] on the forced van der Pol oscillator. This circuit is the series connection of a linear inductor, a linear capacitor, a sinusoidal source, and a nonlinear resistor. The nonlinear resistor is asymptotically passive, but locally active close to the origin. Levinson showed the existence of solutions with an erratic appearance. Such solutions exist in combination with stable periodic solutions of different periods. We can imagine that these erratic solutions cannot determine toward which subharmonic steady-state solution they are to converge. They sometimes adopt one of the periods, sometimes the other, which actually gives them a random aspect. Thus, they are indeed solutions that are exactly on the boundary between the basins of attraction of subharmonic solution of different periods. The initial conditions which generate them form a curve in the plane of the two state variables. Because we cannot supply a computer, which has a finite word length, with initial conditions that are located *exactly* on this curve, hence it is not possible to find these chaotic solutions by numerical calculation [25].

It is not surprising that chaotic solutions have been found in meteorology and hydrodynamics, since these are systems with very high degrees of freedom. Each air or water molecule can move with a certain independence relative to other molecules. Because we are studying objects with an enormous number of molecules, we would expect very complicated solutions, unless the initial conditions and the boundary conditions impose a regular movement. Such a naive idea that chaotic solutions are due to the high degrees of freedom is wrong. Indeed, the equation systems of degree 2 or 3 frequently used as highly simplified models in these disciplines already exhibit chaotic solutions [26]. In such cases, even very simple initial and boundary conditions cannot force the solution to follow a regular movement.

We must emphasize that there is not, at present, any precise definition of a chaotic solution. Often, this term is associated with any movement appearing to be erratic and unstable.

In the study of electrical circuits, the chaotic solutions are undesirable, and methods must be found to exclude them. However, subharmonic steady-state solutions can be useful, depending on the application. While they are troublesome in high-voltage installations, they are evidently indispensable for a frequency divider.

One of the simplest dynamic systems that can be imagined is the iteration of a function which transforms the interval $[-1, 1]$ into itself; for instance,

$$f(x) = 1 - \lambda x^2 , \qquad 0 < \lambda < \sqrt{2} \tag{5.7}$$

If we consider, as the trajectory, the point sequence x, $f(x)$, $f(f(x))$, *et cetera*, we observe that, according to the value of λ, it tends toward a limit point, two limit points between which it oscillates, that it asymptotically oscillates between four points, *et cetera*, or that it also follows a chaotic movement. The corresponding bifurcation diagram, with the bifurcation parameter λ, shows a striking resemblance to that of figure 5.19. For this type of example, not only numerical results, but also rigorous results are available [27].

The dynamic systems which we master in theory show that the attractors of chaotic solutions have a very complicated geometry. It would be incorrect to refer to the object on which they are located in figure 5.26 as a curve, which was done in subsection 5.1.14. In fact, it is a curve full of gaps. Such an object is known as a *strange attractor* [26, 28].

5.2 CONCEPTS OF STABILITY AND THE LIAPUNOV FUNCTIONS

5.2.1 Introduction

A linear circuit is stable (strictly stable) if all its eigenfrequencies are within the left half-plane (without the imaginary axis). If one eigenfrequency is in the right half-plane, it is unstable. If it is unstable, the solutions tend toward infinity when $t \rightarrow \infty$ and solutions with close initial conditions are diverging more and more from one another. Only in the exceptional case where no eigenfrequency is excited in the right half-plane does the solution remain bounded.

If we compare the stability-instability alternative for linear circuits with the different behaviors of the examples in section 5.1, we notice that in the domain of nonlinear circuits, the range of possible behaviors is much wider.

For example, the ferroresonant circuit of subsection 5.1.10, with $E = 160$ V has an unstable periodic solution. At $t = 0$, it is located at point P in figure 5.8. If we choose an initial condition at $t = 0$, at a point P' as close as we with to P, the solution always eventually diverges from the unstable periodic solution. A linear circuit can also exhibit this phenomenon. However, the solution that starts from P' remains bounded, which is peculiar to nonlinear circuits. In fact, it tends toward the small periodic solution.

Stability is both a local and global concept. Local instability means a divergence of solutions with close initial conditions, and global instability means an unlimited increase of the solutions. In order to avoid confusion with other concepts, we shall not use the terms of global stability and global instability. When speaking of global stability, we shall refer to bounded solutions; for local stability, we shall simply refer to stability.

Let us return to figure 5.10. Which initial conditions at $t = 0$ give way to unstable solutions? They are the points located on the boundary of the cross-hatched area. The other initial conditions generate stable solutions. Indeed, if a point P is located inside (outside) the cross-hatched area, so is a whole neighborhood. Consequently, if we start from a sufficiently close initial condition P', the solutions starting from P and P' diverge only slightly from one another. Here, the instability is an exception. Conversely, the circuit of figure 5.16, with the parameters given in subsection 5.1.13 and $E = 4$ V, has only unstable solutions. Nevertheless, none of them tends toward infinity. If $E = 4.7$ V, there are simultaneous stable and unstable solutions. However, neither of these cases constitute exceptions. The plane of the initial conditions at t_0 is divided into two parts, neither area of which is zero.

These examples clearly show that the concept of a stable or unstable *circuit* is no longer adequate. We should speak instead of a stable or unstable *solution*.

The jump of a solution starting from an impasse point, as we have noted in section 2.1, could also be considered instability. However, in this chapter, we study the asymptotic properties when $t \rightarrow \infty$, while jumps occur at finite times. Because these jumps are not to be studied, we shall initially adopt the following hypotheses:

- The hypotheses of theorem 2.4.15 hold. Thus, there exists a system of global state equations, the solution of the circuit starting from an arbitrary operating point of the circuit exists, and it is unique.
- All the solutions exist indefinitely in the future.

These properties are proved by using the methods of chapter 2, combined with the results of chapter 3.

If we assume continuous dependence on the initial conditions, does this not mean that we *a priori* exclude unstable solutions? This is not the case. The solution at *a fixed finite time* depends continuously on the initial conditions. If we require that, at time t_1, two solutions be separated from one another only by a distance of $\epsilon > 0$, it is sufficient to choose their initial conditions at t_0 at a sufficiently small distance $\delta > 0$. However, when one of the two solutions is unstable, δ must be chosen smaller and smaller as t_1 increases. There will be no distance $\delta > 0$ that guarantees a maximum distance of ϵ for *all the times* $t_1 > t_0$.

These considerations suggest, in fact, the good definition of a stable solution. It must be required that **with sufficiently close initial conditions, the solutions remain close throughout the future.**

5.2.2 Definition

A solution $\xi(t)$ of a circuit, defined in the time interval $[t_0, \infty)$ is *stable*, if, for every $\epsilon > 0$, there exists a $\delta > 0$ such that any solution $\tilde{\xi}(t)$ with

$$\| \tilde{\xi}(t_0) - \xi(t_0) \| < \delta \tag{5.8}$$

satisfies

$$\| \tilde{\xi}(t) - \xi(t) \| < \epsilon \tag{5.9}$$

for any $t > t_0$. If this is not the case, the solution $\xi(t)$ is *unstable*.

5.2.3 Comments

This concept of stability is also known as *Liapunov stability*. It is illustrated in figure 5.28 for the trajectory of a single time function. By fixing $\epsilon > 0$, we provide a stripe with a vertical width 2ϵ around the given solution $\xi(t)$. In the case of a stable solution, it is sufficient that, at the stripe's entry, on $t = t_0$, the solution $\tilde{\xi}(t)$ is located in a vertical interval of length 2δ centered on $\xi(t_0)$, and it will never leave the stripe. On the contrary, if the solution $\xi(t)$ is unstable, we can choose $\delta > 0$ as small as we want, and the solution will always eventually leave the stripe.

It is not difficult to show that this concept of stability does not depend on the initial time t_0 [29].

5.2.4 Example

Let us consider a resonant LC circuit that is lossless (fig. 5.29). The parameters of its configuration space Λ are the state variables q_1 and φ_2:

$$u_1 = u_2 = q_1/C \tag{5.10}$$
$$i_2 = -i_1 = \varphi_2/L \tag{5.11}$$

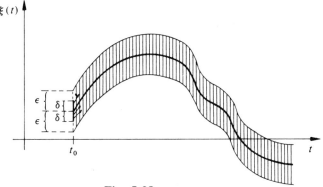

Fig. 5.28

The resulting state equations are

$$dq_1/dt = -\varphi_2/L \tag{5.12}$$
$$d\varphi_2/dt = q_1/C \tag{5.13}$$

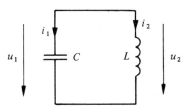

Fig. 5.29

The solution of (5.12) and (5.13), with the initial conditions $q_1(0) = q_0$, $\varphi_2(0) = \varphi_0$, is

$$q_1(t) = q_0 \cos \omega t - \varphi_0 \sin \omega t / L\omega \tag{5.14}$$
$$\varphi_2(t) = \varphi_0 \cos \omega t + q_0 \sin \omega t / C\omega \tag{5.15}$$

where $\omega = 1/\sqrt{LC}$. The energy stored in the circuit is

$$W(q_1, \varphi_2) = q_1^2/2C + \varphi_2^2/2L \tag{5.16}$$

It follows from (5.12) and (5.13) that

290

$$\frac{d}{dt} W(q_1(t), \varphi_2(t)) = 0 \tag{5.17}$$

Consequently, if we introduce the state variables:

$$x = q_1/\sqrt{2C} \tag{5.18}$$
$$y = \varphi_2/\sqrt{2L} \tag{5.19}$$

then the orbits in Λ, projected on the plane (x,y), are concentric circles (fig. 5.30):

$$x^2 + y^2 = E \tag{5.20}$$

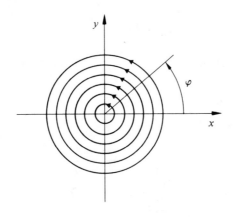

Fig. 5.30

The movement on the orbit as a function of time is obtained by introducing the angle:

$$\psi = \arcsin(y/\sqrt{x^2 + y^2}) \tag{5.21}$$

By way of an elementary calculation, we find that

$$d\psi/dt = \omega \tag{5.22}$$

This is a movement with constant angular speed on the circle.

The energy difference ΔE, and the angle difference, $\Delta \psi$, of two solutions remain constant as functions of time. If these two differences are small, the two solutions are close to one another. Because the differences ΔE and $\Delta \psi$ do not change with time, both solutions will remain close. We thus expect all solutions to be stable.

Now, we must establish the stability of the solutions given by (5.14) and (5.15) in addition to (5.10) and (5.11), according to definition 5.2.2. To this end, we must explicate the distance $\| \tilde{\xi}(t) - \xi(t) \|$ between two solutions $\xi(t)$ and $\tilde{\xi}(t)$, with respect to the differences ΔE and $\Delta \psi$. If we choose as normalization constants C, L, ω, and

$$R = \sqrt{L/C} \tag{5.23}$$

for the norm (2.117), we find for $\xi \in \Lambda$:

$$\| \xi \|^2 = \|(u_1, u_2, i_1, i_2, q_1, \varphi_2)\|^2 \tag{5.24}$$

$$\| \xi \|^2 = 2u_1^2 \sqrt{C/L} + 2i_2^2 \sqrt{L/C} + q_1^2 \omega/C + \varphi_2^2 \omega/L \tag{5.25}$$

$$\| \xi \|^2 = 3q_1^2 \omega/C + 3\varphi_2^2 \omega/L \tag{5.26}$$

$$\| \xi \|^2 = 6\omega (x^2 + y^2) \tag{5.27}$$

We observe that the norm (2.117) on Λ, with our choice of normalization constants, is equal to the Euclidean norm in the plane (x,y), up to a factor of 6ω. Thus, we need only establish the stability of the projections of the solutions in the plane (x,y) by using the Euclidean norm.

If $(x(t), y(t))$ and $(\tilde{x}(t), \tilde{y}(t))$ are the projections of the solutions $\xi(t)$ and $\tilde{\xi}(t)$, we need to find a bound:

$$(\tilde{x}(t) - x(t))^2 + (\tilde{y}(t) - y(t))^2 < \epsilon^2 \tag{5.28}$$

under the condition that

$$(\tilde{x}(0) - x(0))^2 + (\tilde{y}(0) - y(0))^2 < \delta^2 \tag{5.29}$$

Let the circle of radius $\sqrt{E}$ be the projection of the orbit of $\xi(t)$, and let $\psi(0)$ be the angle at the instant $t = 0$ (fig. 5.31). The points $(\tilde{x}(0), \tilde{y}(0))$ that satisfy (5.19) lie on the disk of radius δ, which is cross-hatched in figure 5.31. At the instant t, the corresponding points $(\tilde{x}(t), \tilde{y}(t))$ are located on the disk of the same radius, whose center is $(x(t), y(t))$. This last point is located on the circle of radius $\sqrt{E}$ and its angle is $\psi(t) = \psi(0) + \omega t$. Consequently, (5.28) is satisfied, with $\epsilon = \delta$.

If we had chosen other normalization constants for the norm (2.117), we would have found ellipses instead of circles, and thus found $\epsilon > \delta$. However, nothing would have modified the conclusion that all the solutions are stable.

5.2.5 Example

Let us modify example 5.2.4 by introducing a nonlinear inductor with saturation, according to the constitutive relation:

292

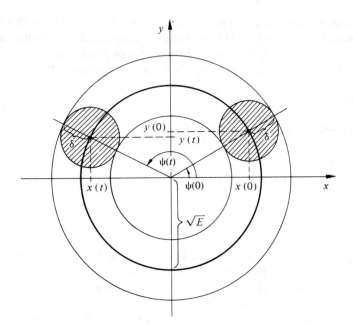

Fig. 5.31

$$i_2 = (\varphi_2 + \alpha\varphi_2^3)/L \tag{5.30}$$

The configuration space of this circuit (fig. 5.32) is still characterized by the parameters q_1 and φ_2:

$$u_1 = u_2 = q_1/C \tag{5.31}$$
$$i_2 = -i_1 = (\varphi_2 + \alpha\varphi_2^3)/L \tag{5.32}$$

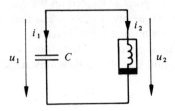

Fig. 5.32

The state equations become

$$dq_1/dt = -(\varphi_2 + \alpha\varphi_2^3)/L \qquad (5.33)$$
$$d\varphi_2/dt = q_1/C \qquad (5.34)$$

As we shall see in section 5.3, the energy stored in the circuit is

$$W(q_1, \varphi_2) = q_1^2/2C + \varphi_2^2/2L + \alpha\varphi_2^4/4L \qquad (5.35)$$

We verify, with the help of (5.33) and (5.34), that

$$\frac{d}{dt} W(q_1(t), \varphi_2(t)) = 0 \qquad (5.36)$$

Again by using the normalized state variables x, y, according to (5.18) and (5.19), we can write

$$W(x, y) = x^2 + y^2 + \alpha L y^4 \qquad (5.37)$$

The orbits in Λ, projected on the plane (x, y), are closed curves defined by

$$x^2 + y^2 + \alpha L y^4 = E \qquad (5.38)$$

A set of such curves is represented in figure 5.33. All the solutions of the circuit are periodic functions.

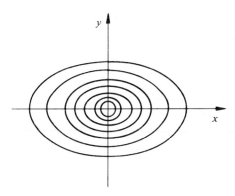

Fig. 5.33

Let us introduce the angle ψ by way of (5.21). By using (5.31) and (5.32), we find, after some calculations, that

$$\frac{d\psi}{dt} = \frac{1}{\sqrt{LC}} \cdot \frac{x^2 + y^2 + 2\alpha L y^4}{x^2 + y^2} \tag{5.39}$$

If we set $r^2 = x^2 + y^2$, we have $y = r \sin \psi$ and

$$d\psi/dt = (1 + 2\alpha L r^2 \sin^4 \psi)/\sqrt{LC} \tag{5.40}$$

This implies that the angular speed on the curves (5.38), at a given angle ψ, increases as a function of r, and therefore as a function of E. Consequently, the solution periods decrease as a function of the stored energy, while in example 5.2.4 it was constant. If we choose initial solutions of two solutions that are very close, but have different stored energies, they eventually diverge from one another, if we wait long enough, because of their different periods.

We have just shown that all solutions of the circuit of figure 5.32 are unstable, with one exception. The solution that is tied to the dc operating point $u = i = q = \varphi = 0$ is stable because the orbits of solutions with initial conditions close to this point remain in its neighborhood.

5.2.6 Comments

The example 5.2.5 shows that the solutions of the circuit of figure 5.29 are at the stability limit. The nonlinearity (5.30) destabilizes them, even if it is very small. Furthermore, their asymptotic behavior depends on the initial conditions. Normally, we expect a convergence of the solutions toward one another when $t \to \infty$, at least if the initial conditions are close. This is a stronger stability concept.

5.2.7 Definition

A solution $\xi(t)$ of a circuit, defined in the time interval $[t_0, \infty)$ is *asymptotically stable*, if it is stable and if there exists a $\delta > 0$ such that any solution $\tilde{\xi}(t)$ with

$$\| \tilde{\xi}(t_0) - \xi(t_0)\| < \delta \tag{5.41}$$

satisfies

$$\lim_{t \to \infty} \| \tilde{\xi}(t) - \xi(t)\| = 0 \tag{5.42}$$

5.2.8 Examples

The examples of section 5.1 are full of asymptotically stable solutions. To clarify this idea, let us take the ferroresonant circuit of subsection 5.1.10 with $E = 160$ V.

Apart from the solutions having initial conditions at $t = 0$ on the boundary of the cross-hatched area of figure 5.10, all the solutions are asymptotically stable. In fact, at $t = 0$ they are situated in either the interior or exterior of the cross-hatched area. With these solutions, an entire neighborhood is on the same side, and hence all the corresponding solutions converge toward one another.

The resonant circuit of figure 5.29 has solutions which are stable without being asymptotically stable. The solutions for the oscillator of subsection 5.1.8 with $R_3 = 1 \text{ k}\Omega$ have the same property. In fact, all the solutions $\xi(t) \not\equiv 0$ converge toward a periodic solution. This solution depends on $\xi(t_0)$. However, the orbit of the solution is always the same, which is what makes the nonlinear oscillator distinct from the linear resonant circuit.

5.2.9 Comments

In figure 5.34, the concept of asymptotic stability is illustrated for a single function of time $\xi(t)$. If another solution, $\tilde{\xi}(t)$, is located at a sufficiently short distance δ at the instant t_0, it will remain within a stripe of width 2ϵ around $\xi(t)$ as a result of stability. Furthermore, it converges toward $\xi(t)$ when $t \to \infty$.

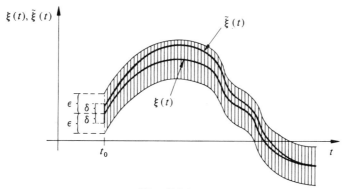

Fig. 5.34

Is it not superfluous to require, in definition 5.2.7, that the solution be stable in addition to having asymptotic convergence (5.42)? This is not so. A counterexample of an autonomous system of order two is mentioned in [30].

Only those solutions that are initially located in a small neighborhood are guaranteed to converge toward an asymptotically stable solution, according to definition 5.2.7. Given a circuit and a solution that are asymptotically stable, we may wonder how far it is possible to extend this neighborhood without affecting the asymptotic behavior of the solutions. This brings us to the following definitions.

296

5.2.10 Definitions

Let $\xi(t)$ be a solution of a circuit defined in the time interval $[t_0,\infty]$ and asymptotically stable. The *basin of attraction of ξ_0 at t_0* is the set of the operating points $\tilde\xi_0$ at t_0 such that the solution $\tilde\xi(t)$ with $\tilde\xi(t_0) = \tilde\xi_0$ converges toward $\xi(t)$ when $t \to \infty$. If the basin of attraction of $\xi(t)$ at t_0 includes the whole configuration space at t_0, $\xi(t)$ is *globally asymptotically stable*.

5.2.11 Examples

The basins of attraction at $t = 0$ for the two asymptotically stable periodic solutions of the ferroresonant circuit of subsection 5.1.10, for $E = 160$ V, are represented in figure 5.10. That of the small periodic solution is cross-hatched, and the one of the large periodic solution is left unlined.

If for the same circuit we choose the amplitude E in one of the intervals (0 V, 91 V), (246 V, 1694 V) and (4389 V, ∞), there is only one periodic solution, which is globally asymptotically stable.

The variant of the ferroresonant circuit of subsection 5.1.11 has four stable periodic solutions with their basins of attraction at $t = 0$ is represented in figure 5.15.

5.2.12 Definitions

Two operating points $\tilde\xi_0$ and ξ_0 at t_0 are *asymptotically equivalent* if the solutions $\tilde\xi(t)$ and $\xi(t)$ with the initial conditions $\tilde\xi(t_0)$ and $\xi(t_0)$ satisfy

$$\lim_{t \to \infty} \| \tilde\xi(t) - \xi(t) \| = 0 \tag{5.43}$$

In the above case, we write $\tilde\xi_0 \sim \xi_0$. Likewise, we write for the *solution* $\tilde\xi(t) \sim \xi(t)$, and we say that they have the *same asymptotic behavior*. If all the solutions of a circuit have the same asymptotic behavior, the *circuit exhibits a unique asymptotic behavior*.

5.2.13 Property

The relation between two operating points of a circuit at t_0, which was introduced in subsection 5.2.12, is an *equivalence relation* [31], which means that

- it is reflective:

$$\xi_0 \sim \xi_0 \tag{5.44}$$

- it is symmetrical:

$$\tilde\xi \sim \xi_0 \Rightarrow \xi_0 \sim \tilde\xi_0 \tag{5.45}$$

- it is transitive:

$$\tilde{\xi}_0 \sim \xi_0 \text{ and } \xi_0 \sim \hat{\xi}_0 \Rightarrow \tilde{\xi}_0 \sim \hat{\xi}_0 \tag{5.46}$$

The relation between two solutions of a circuit, which is essentially the same thing, therefore is an equivalence relation as well.

This property results directly from definition 5.2.12.

5.2.14 Comments

What makes equivalence relations interesting is that they decompose the sets in which they are defined into disjointed subsets. These subsets are called *equivalence classes* [31]. Two elements belong to the same equivalence class if and only if there is a relation between them. In our present case, two operating points at t_0 belong to the same equivalence class if and only if they are asymptotically equivalent. In addition, two solutions belong to the same equivalence class if and only if they have the same asymptotic behavior.

The configuration space at t_0 and the set of solutions are thus decomposed into equivalence classes. The equivalence classes that include the operating points which generate asymptotically stable solutions are the basins of attractions for such solutions. From property 5.2.13, it follows that the basins of attraction for the two asymptotically stable solutions are either identical or disjoint. They are open subsets within the configuration space because, at each point ξ_0 of the basin of attraction for a solution $\xi(t)$, there corresponds one solution $\tilde{\xi}(t)$, which is itself asymptotically stable. Thus, according to definition 5.2.7, $\tilde{\xi}_0$ is surrounded by an entire neighborhood of equivalent operating points.

Figure 5.10 gives evidence to this fact. The global parameters of the configuration space are characterized by the state variables u_1 and φ_2, as we observed in subsection 2.4.4. The two basins of attraction at $t = 0$ are the cross-hatched part and the unlined part **without the boundary between them.** The boundary is the third equivalence class to which the operating point of the unstable solution at $t = 0$ belongs.

If there is only one equivalence class, the circuit has a unique asymptotic behavior. This is the case of a linear circuit having its eigenfrequencies located within the left half-plane.

One additional point still deserves further attention. Neither definition 5.2.2 of stability, nor definition 5.2.7 of asymptotic stability, prohibits the solution from tending toward infinity when $t \to \infty$.

5.2.15 Definition

A solution $\xi(t)$ of a circuit defined in the time interval $[t_0, \infty)$ is *bounded* if there exists a constant K, such that

$$\|\xi(t)\| \leqslant K \text{ for } t \geqslant t_0 \tag{5.47}$$

298

5.2.16 Comment

The concept of a bounded solution is illustrated in figure 5.35 for the case of a single time function $\xi(t)$. If inequality (5.47) is satisfied, the solution must remain within the cross-hatched area of figure 5.35.

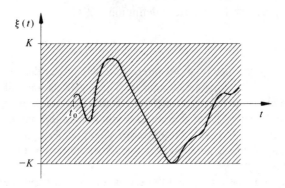

Fig. 5.35

5.2.17 Liapunov Functions

A very simple method is available to prove that a solution is bounded, or that a solution is asymptotically stable; it is the *Liapunov function* method. In some texts, it is called *the second method of Liapunov*.

The idea is as follows. We introduce an auxiliary function W with real values in the space of the circuit variables. Then, we prove that function W decreases along any solution of the circuit. If the level surfaces $W = E$ are concentric having a size which increases with E, then the solutions of the circuit cross these surfaces from the outside to the inside as a function of time. It follows that a solution located inside a level surface at a given instant is no longer able to escape to the outside. Consequently, the solution is bounded. In two-dimensional space, the level surfaces are curves (fig. 5.36).

If the level surfaces $W = E$ enclose a finite connected domain having a size which increases with E, then W necessarily possesses a minimum E_0, and the level surface $W = E_0$ consists of a single point ξ_0. If the decrease of W along the solutions is strict apart from ξ_0, then W tends toward its minimum value and, consequently, the solutions tend toward ξ_0 (fig. 5.36). In fact, this case is only possible if ξ_0 is a dc operating point. It follows that this is a globally asymptotically stable dc operating point.

If the strict decrease of W along the solutions is only guaranteed for $E > E_1 > E_0$, then we are no longer sure whether the solutions enter the area encompassed by the level surface $W = E_1$. We can imagine solutions tending toward this area without entering it (fig. 5.37). Then, $W \leq E_1$ is said to be a *globally attractive set*. This set is cross-hatched in figure 5.37.

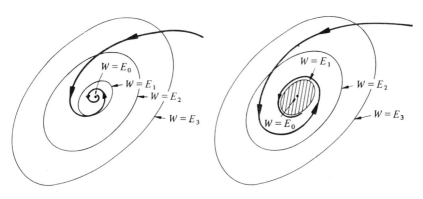

Fig. 5.36   **Fig. 5.37**

If the decrease, whether strict or not, is only guaranteed for $W < E_2$, then the conclusions that can be drawn from the existence of function W only apply to the solutions having initial points which are located inside the level surface $W = E_2$. In that case, we can only guarantee that the dc operating point ξ_0 is asymptotically stable, without being ***globally*** asymptotically stable (fig. 5.38). Similarly, we only know that the set $W = E_1$ is attractive, without being sure whether it is ***globally*** attractive.

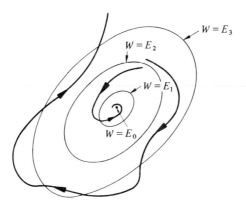

Fig. 5.38

To accommodate the inference of these properties from the asymptotic behavior of the solutions by introducing a Liapunov function, we must introduce some technical hypotheses.

5.2.18   Definitions

Let $\Lambda \subset \mathbb{R}^{2b+N_C+N_L}$ be the generalized configuration space of a circuit. We designate s as the voltage (current) column matrix of the time-dependent voltage (current) sources. A function $W: \mathbb{R}^{2b+N_C+N_L} \to \mathbb{R}$ is a *Liapunov function* of the circuit if it satisfies the following conditions:

- W is continuous and differentiable.
- W is *bounded from below on* Λ, this means that there is a constant E_0 such that $W(\xi) \geqslant E_0$ for all the $\xi \in \Lambda$.
- For any pair of constants E and A, the set of the $\xi \in \Lambda$ such that $W(\xi) \leqslant E$ and $\|s\| \leqslant A$ is bounded. The bound depends in general on E and A. We designate it by $K(A,E)$.
- There exists a continuous function $V: \Lambda \times \mathbb{R} \to \mathbb{R}$ such that, for any solution $\xi(t)$ of the circuit, we have

$$dW(\xi(t))/dt = V(\xi(t), t) \tag{5.48}$$

- For any $\xi \in \Lambda$ and any t, we have

$$V(\xi, t) \leqslant 0 \tag{5.49}$$

If, instead of inequality (5.49), we have

$$V(\xi, t) \leqslant -P(W(\xi)) \tag{5.50}$$

where $P: \mathbb{R} \to \mathbb{R}$ is a continuous function, such that

$$P(E) > 0 \tag{5.51}$$

for $E > E_0$, then W is a *strict Liapunov function* of the circuit.
If inequality (5.49) is only valid if

$$E_1 \leqslant W(\xi) < E_2 \tag{5.52}$$

where the two constants E_1 and E_2 satisfy

$$E_0 \leqslant E_1 < E_2 \leqslant \infty \tag{5.53}$$

we say that W is a *Liapunov function in the interval* $[E_1,E_2)$. If the more restrictive inequality (5.50) is satisfied by a function P, which is positive for $E_1 < E < E_2$, then W is a *strict Liapunov function in* $[E_1,E_2)$.

5.2.19 Comments

The energy stored in the capacitors and inductors is the prototype for a Liapunov function of a circuit. It has already been used in subsection 2.2.13 to prove that the solutions of a circuit are bounded. Its systematic use as a Liapunov function is to be covered in sections 5.5 and 5.6.

The Liapunov functions, such as defined in subsection 5.2.18, are not adequate to prove the uniqueness of the asymptotic behavior of the solutions in the case of circuits with time-dependent sources. In section 5.7, we shall adapt the concept of Liapunov function to this problem.

The third condition of definition 5.2.18 does not mean that the set of the points $\xi \in \mathbb{R}^{2b + N_C + N_L}$ with $W(\xi) \leqslant E$ and $\|s\| \leqslant A$ must be bounded. **This property is only required from the subset composed of the generalized operating points** $\xi \in \Lambda$. For instance, the sets of the points $\xi \in \mathbb{R}^{2b + N_C + N_L}$, such that the amplitudes of the time-dependent sources are bounded by E, in general, is not bounded. It is only by using the Kirchhoff equations and the constitutive relations of the elements that we can derive a bound for the currents and the voltages of the resistors.

5.2.20 Calculation of the Derivative

At first sight, the fourth condition of definition 5.2.18 seems to be difficult to establish because it involves the solutions of the circuit. The solutions are explicitly known in very rare cases only. However, if function W depends solely on the capacitor charges q_C and the inductor fluxes φ_L, as is the case of the stored energy, we can write

$$dW(q_C(t), \varphi_L(t))/dt = (\partial W/\partial q_C)(dq_C/dt) + (\partial W/\partial \varphi_L)(d\varphi_L/dt) \quad (5.54)$$

$$= (\partial W/\partial q_C) i_C(t) + (\partial W/\partial \varphi_L) u_L(t) \quad (5.55)$$

Consequently, if we define the function $V: \mathbb{R}^{N_C + N_L} \to \mathbb{R}$ by

$$V(q_C, \varphi_L) = \frac{\partial W}{\partial q_C}(q_C, \varphi_L) i_C + \frac{\partial W}{\partial \varphi_L}(q_C, \varphi_L) u_L \quad (5.56)$$

then

$$dW(\xi(t))/dt = V(\xi(t)) \quad (5.57)$$

for any solution.

To check the last condition, we need to study the restriction of the function V on the generalized configuration space.

If W also depends on other variables in addition to q_C and φ_L, we can also establish (5.49) without knowing the solutions. We use the global parameterization for the generalized configuration space by the state variables q_C and φ_L, and by the amplitudes s of the time-dependent sources. Such a parameterization exists as a result of the first hypothesis in subsection 5.2.1. It is given by $\psi = (\pi/\Lambda)^{-1}$, where π is the projection (2.138). Thus, the restriction to Λ of W is a function of q_C, φ_L, and s:

$$\tilde{W}(q_C, \varphi_L, s) = W(\psi(q_C, \varphi_L, s)) \tag{5.58}$$

Consequently, we can write

$$dW(\xi(t))/dt = d\tilde{W}(q_C(t), \varphi_L(t), e(t))/dt \tag{5.59}$$

$$= (\partial \tilde{W}/\partial q_C)(dq_C/dt) + (\partial \tilde{W}/\partial \varphi_L)(d\varphi_L/dt)$$
$$+ (\partial \tilde{W}/\partial s)(de/dt) \tag{5.60}$$

$$= (\partial \tilde{W}/\partial q_C) i_C(t) + (\partial \tilde{W}/\partial \varphi_L) u_L(t) + (\partial \tilde{W}/\partial s)(de/dt) \tag{5.61}$$

If we define the function $V: \mathbb{R}^{2b + N_C + N_L} \to \mathbb{R}$ by

$$V(q_C, \varphi_L, t) = \frac{\partial \tilde{W}}{\partial q_C}(q_C, \varphi_L, e(t)) i_C + \frac{\partial \tilde{W}}{\partial \varphi_L}(q_C, \varphi_L, e(t)) u_C$$
$$+ \frac{\partial \tilde{W}}{\partial s}(q_C, \varphi_L, e(t)) \frac{de}{dt} \tag{5.62}$$

where i_C and u_L are to be expressed by q_C, φ_L, and $e(t)$ by using function ψ, then we can write

$$d\tilde{W}(q_C(t), \varphi_L(t), e(t))/dt = V(q_C(t), \varphi_L(t), t) \tag{5.63}$$

for any solution of the circuit. The fourth condition is satisfied, and establishing (5.49) amounts to studying function V, which is explicitly known.

5.2.21 Property

Let us assume that function W satisfies the first four conditions of definition 5.2.18, and that function V *does not explicitly depend on time or that* $V(\xi, t)$ *is periodic in* t. If V satisfies the *strict* inequality (5.49) for all the ξ such as $E_1 < W(\xi) < E_2$, and for all the times, then W is a strict Liapunov function in $[E_1, E_2)$.

Indeed, we need only define $P(E)$ as the minimum of $-V(\xi, t)$ on the set $\{(\xi, t) \,|\, W(\xi) = E \text{ and } 0 \leq t \leq T\}$ for $E_1 < E < E_2$. Because this set is compact, this minimum exists [6] and it is positive.

5.2.22 Example

The energy stored in the two capacitors of the Wien bridge oscillator (fig. 2.56) is

$$W(q_1, q_2) = (q_1^2/2C_1) + (q_2^2/2C_2) \tag{5.64}$$

It is a continuous function that is differentiable and bounded below by 0. This bound not only is valid on Λ, but in the entire space of the circuit variables.

The set of the q_1, q_2 such that $W(q_1, q_2) \leq E$ is bounded in $\mathbb{R}^2$. It is the inside of an ellipse. Because q_1 and q_2 are global parameters for Λ, the other variables of the circuit are bounded as well.

According to (5.55), we have

$$\mathrm{d}W(q_1(t), q_2(t))/\mathrm{d}t = q_1(t)\, i_1(t)/C_1 + q_2(t)\, i_2(t)/C_2 \tag{5.65}$$

We look for a function V, such that

$$\mathrm{d}W(q_1(t), q_2(t))/\mathrm{d}t = V(q_1(t), q_2(t)) \tag{5.66}$$

Thus i_1 and i_2 must be expressed by q_1 and q_2. In the configuration space, (2.158) and (2.159) are satisfied, with u_1 and u_2 instead of E_1 and E_2.

Consequently, we find

$$
\begin{aligned}
V(q_1, q_2) = {} & \frac{q_1}{C_1}\left(-\frac{G_1}{C_1}q_1 + G_1 h\!\left(\frac{G_3}{C_2}q_2\right)\right) \\
& + \frac{q_2}{C_2}\left(-\frac{G_1}{C_1}q_1 - \frac{G_2}{C_2}q_2 + G_1 h\!\left(\frac{G_3}{C_2}q_2\right)\right)
\end{aligned}
\tag{5.67}
$$

$$R(i_4) = h(i_4)/i_4 \tag{5.68}$$

which is associated with the nonlinear resistor at each operating point. Then, we can write

$$V(q_1, q_2) = -\frac{G_1}{C_1^2}q_1{}^2 - \left(\frac{G_2}{C_2^2} - \frac{G_1 R G_3}{C_2^2}\right)q_2{}^2 - \left(\frac{G_1}{C_1 C_2} - \frac{G_1 R G_3}{C_1 C_2}\right)q_1 q_2 \tag{5.69}$$

If we forget temporarily that R is a function of $i_4 = G_3\, q_2/C_2$, the right-hand side of (5.69) is a quadratic form at q_1, q_2. It is a negative definite quadrative form, provided that

$$\left(\frac{G_1}{C_1C_2} - \frac{G_1RG_3}{C_1C_2}\right)^2 < 4\frac{G_1}{C_1^2}\left(\frac{G_2}{C_2^2} - \frac{G_1RG_3}{C_2^2}\right) \tag{5.70}$$

This condition can be transformed into

$$(1 + RG_3)^2 < 4G_2/G_1 \tag{5.71}$$

Let us remember here that R is not constant. As a consequence, if the nonlinear resistor is such that for all the currents $i_4 \neq 0$, we have

$$h(i_4)/i_4 < R_3(2\sqrt{R_1/R_2} - 1) \tag{5.72}$$

then $V(q_1,q_2) < 0$ if $q_2 \neq 0$. If $q_2 = 0$, we infer from (5.69) that $V(q_1,q_2) < 0$, except if $q_1 = 0$. In conclusion, if (5.72) is satisfied, then the stored energy is a strict Liapunov function of the circuit.

If the nonlinear resistor of the circuit is defined by the constitutive relation (5.1), an explicit expression can be given for the condition (5.72). We have

$$h(i_4)/i_4 = u_4/g(u_4) \tag{5.73}$$

and, therefore,

$$R = (a + bu_4^2)^{-1} \tag{5.74}$$

It follows that

$$R < 1/a \tag{5.75}$$

provided that $u_4 \neq 0$, or, which is equivalent, $i_4 \neq 0$. Consequently, the stored energy is a strict Liapunov function for this circuit, if

$$1/a \leqslant R_3(2\sqrt{R_2/R_1} - 1) \tag{5.76}$$

In the numerical example of subsection 5.1.8, with $R_3 = 10\,\text{k}\Omega$, the two members of (5.76) are equal. Conversely, (5.76) is not satisfied for $R_3 = 1\,\text{k}\Omega$.

A more detailed analysis of (5.69) would permit us to establish that the stored energy is a strict Liapunov function, even if (5.76) is not satisfied. However, it is only a strict Liapunov function in a certain interval $[E_1,\infty)$ with $E_1 < 0$.

5.2.23 Theorem: Bounded Solutions

Let us assume that the time-dependent sources of a circuit are bounded by A and that W is a Liapunov function of the circuit in the interval $[E_1,E_2)$. If $\xi(t)$ is a solution of the circuit, and if

$$E_0 = W(\xi(t_0)) < E_2 \tag{5.77}$$

then $\xi(t)$ is bounded. More precisely, if $K(A, E)$ is the bound introduced in definition 5.2.18, then

$$\|\xi(t)\| \leq \begin{cases} K(A, E_1) & \text{if} \quad E_0 \leq E_1 \\ K(A, E_0) & \text{if} \quad {}^\prime E_1 \leq E_0 < E_2 \end{cases} \tag{5.78}$$

Indeed, the function $f(t) = W(\xi(t))$ is continuous and differentiable, and its derivative is not positive when $E_1 \leq f(t) < E_2$. If, at an instant t, $f(t) < E_1$, then f can increase. However, its value cannot exceed E_1. Furthermore, if, at the instant t, we have $E_1 \leq f(t) < E_2$, then f cannot increase, this being true as long as its value remains greater than or equal to E_1. Consequently, if $f(t_0) \leq E_1$, we have $f(t) \leq E_2$, and if $E_1 \leq f(t_0) < E_2$, we have $f(t) \leq f(t_0)$ for all the $t \geq t_0$. This implies (5.78).

5.2.24 Corollary

Let us consider the case of a circuit where we do not know whether it satisfies the second hypothesis of subsection 5.2.1 or not. Let us assume that the circuit has a strict Liapunov function in the interval $[E_1, E_2]$. Let $\xi(t)$ be a solution, defined in the time interval $[t_0, t_0 + \epsilon]$, such that (5.77) holds. Then, $\xi(t)$ can be arbitrarily extended into the future, and the inequalities (5.78) hold.

This corollary is obtained by combining theorems 2.4.16 and 5.2.23.

5.2.25 Examples

In subsection 5.2.22, we showed a Liapunov function with $E_2 = \infty$ for the Wien-bridge oscillator. Also, global state equations were found for the same circuit in subsection 2.4.14. We concluded that all the solutions exist up to $t \to +\infty$. Note, however, that $E_1 < \infty$ if and only if the resistor characteristic satisfies $h(i)/i < R$ for a sufficiently large $|i|$. Therefore, we could also directly establish global existence in the future, using theorem 2.2.9, by showing that the right-hand side of the system of global state equations (2.160) and (2.161) is globally Lipschitz. However, the theorem 5.2.23 allows us to state, additionally that the solutions remain bounded.

An example, in which corollary 5.2.24 permits us to establish the global existence of the solutions in the future, whereas theorem 2.2.9 fails to do so, has already been given in subsection 2.2.13.

5.2.26 Comments

Theorem 5.2.23 and corollary 5.2.24 are powerful instruments to prove that solutions exist throughout the future, and that they remain bounded. To this end, we need only show, by the methods of chapter 3, that the associated resistive circuit has exactly one solution, and that there is a Liapunov function with $E_2 = \infty$. Normally the energy stored in the capacitors and the inductors will suffice for this function.

It is possible to generalize the Liapunov functions in various ways. Refer to [29, 30].

Theorem 5.2.23 also permits us to find explicit bounds for the solutions. Most of the time, however, we deal with strict Liapunov functions which allow us to find bounds that are even more interesting. In this context, the concept of *attractive set* must be introduced.

5.2.27 Definition

A *neighborhood* of a set $S \subset \mathbb{R}^n$ is an open set which contains S (fig. 5.39).

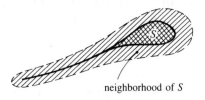

neighborhood of S

Fig. 5.39

5.2.28 Comment

The concept of the neighborhood of a set is a generalization of the concept of the neighborhood of a point introduced in subsection 2.3.17. Indeed, both concepts coincide if the set consists of only a single point.

5.2.29 Example

Any open solid sphere (2.119) is a neighborhood of itself because it is an open set. Such is not the case of a closed solid sphere (2.118), but any open sphere of radius $r' > r$ is a neighborhood of the solid sphere of radius r (fig. 5.40).

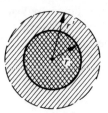

Fig. 5.40

5.2.30 Definitions

Let $S \subset \Lambda$ be a compact set in the generalized configuration space of a circuit. *Set S is globally attractive* if any solution of the circuit ends up in any neighborhood of S after a finite lapse of time. More precisely, if $\xi(t)$ is a solution of the circuit, and if $U \supset S$ is a neighborhood of S, then there exists a time T such that $\xi(t) \in U$ for $t > T$ (fig. 5.41).

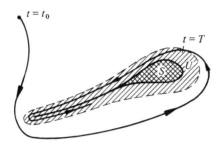

Fig. 5.41

Set S is an *attractive set* if all the solutions of the circuit which start from an initial point $\xi(t_0) \in U_0$, where U_0 is a neighborhood of S, end up in any neighborhood of S after a finite lapse of time (fig. 5.42).

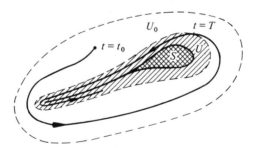

Fig. 5.42

5.2.31 Properties

If S is a (globally) attractive set, then any compact set which contains S is also (globally) attractive.

If S_1 and S_2 are two (globally) attractive sets, then their intersection is also an (globally) attractive set.

These properties are a direct consequence of definitions 5.2.30.

5.2.32 Comment

If we wish to obtain some idea about the order of magnitude of the currents and voltages in the circuit when $t \to \infty$, we should then try to find the smallest possible attractive sets. Is there a minimum attractive set? This question leads to the concept of *attractor*. We have already mentioned it without definition in subsection 5.1.14. An attractor is an attractive set which is minimum in some sense [28]. In the context of chaotic solutions, it is a very subtle concept.

5.2.33 Definition

The solutions of a circuit are *eventually uniformly bounded* if there is a constant K so that for any solution $\xi(t)$ of the circuit, there is a time T such that

$$\|\xi(t)\| \leq K \quad \text{for} \quad t \geq T \tag{5.79}$$

As a general rule, T depends on the solution.

5.2.34 Comment

The terminology of definition 5.2.33 is justified as follows. The solutions are said to be **uniformly** bounded because bound K in (5.79) does not depend on the solution. It is said to be an **eventual** bound because it is only valid for sufficiently long times.

5.2.35 Property

The solutions of a circuit are eventually uniformly bounded if and only if there is a globally attractive set.

Indeed, let S be a globally attractive set and U a bounded neighborhood of S. Such a neighborhood exists because S itself is bounded. If $\|\xi\| < K$, for all the $\xi \in U$, then, according to definition 5.2.30, the solutions of the circuit satisfy (5.79).

However, if the solutions satisfy (5.79), then the solid sphere of radius K is a globally attractive set. Indeed, because the solutions join this solid sphere in a finite time, they *a fortiori* reach any neighborhood of the solid sphere.

5.2.36 Theorem: Eventually Uniformly Bounded Solutions

Let us suppose that the time-dependent sources of a circuit are bounded by A and that W is a strict Liapunov function of the circuit in the interval $[E_1, E_2]$. Then, the set S of the generalized operating points ξ with $W(\xi) \leq E_1$ is an attractive set. More precisely, any solution $\xi(t)$ with $W(\xi(t_0)) < E_2$ ends up, after a finite lapse of time, in any neighborhood of S.

If $E_2 = \infty$, then S is globally attractive. In this case, the inequality (5.79) holds for any $K > K(A, E_1)$, where $K(A, E)$ is the constant introduced in definition 5.2.18.

We first show that S is compact. Set S is closed because W is continuous and Λ is closed (property 2.3.19). Set S is bounded due to the third condition of definition 5.2.18. To prove that S is attractive, we choose $U_0 = \{\xi | W(\xi) < E_2\}$ and an arbitrary neighborhood $U \subset U_0$ of S (fig. 5.43). Let us consider the set $R \subset \Lambda$ of points ξ such that $\xi \in U$ and $\xi \in U_0 = \{\xi | W(\xi) \leq E_2\}$. Then R is compact. In figure 5.43, R is cross-hatched. Then, the continuous function W possesses a minimum value E_3 in R [6]. Because S and R are disjoint, we necessarily have $E_1 < E_3$. Consequently, the neighborhood $U_3 = \{\xi | W(\xi) < E_3\}$ of S is contained in U and we need only show that any solution with the initial condition in U_0 within a finite time leads to U_3 and remains in U_3 thereafter.

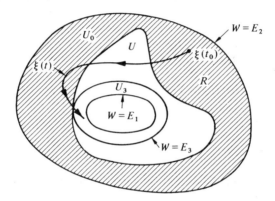

Fig. 5.43

Let $\xi(t)$ be a solution with $W(\xi(t_0)) \leq E_3$. According to theorem 5.2.23, we have $W(\xi(t)) \leq E_3$, which shows that once the solutions reach U_3, they remain there.

Let $\xi(t)$ be a solution with $E_3 < W(\xi(t_0)) = E_4 < E_2$. Then, if $f(t) = W(\xi(t))$, we have

$$df/dt = V(\xi(t), t) \leq -P(f(t)) \tag{5.80}$$

as long as $f(t) \geq E_1$. The functions V and P are part of definition 5.2.18. Let

$$P_0 = \min_{E_3 \leq E \leq E_4} P(E) \tag{5.81}$$

Because $P(E) > 0$ on the compact interval $[E_3, E_4]$, we have $P_0 > 0$. Consequently, provided that $f(t) \geq E_3$, we have $df/dt \leq -P_0$. Because this is possible during a finite time only, the solution necessarily reaches U_3.

5.2.37 Example

In subsection 5.2.22, we have shown that the energy stored in the two capacitors of the Wien-bridge oscillator is a strict Liapunov function if the characteristic curve of the nonlinear resistor satisfies inequality (5.72). In this case, the origin $u = i = q_C = 0$ is an attractive set, according to theorem 5.2.36. This is equivalent to saying that the constant solution frozen in the origin is globally asymptotically stable, which actually was noted in subsection 5.1.8 for the circuit with $R_3 = 10 \text{ k}\Omega$.

When inequality (5.72) is satisfied no longer, which is the case of the circuit with $R_3 = 1 \text{ k}\Omega$, the stored energy is not a Liapunov function in the entire interval $[0, \infty)$. Indeed, we noted in subsection 5.1.8 that the solutions tend toward a periodic solution. As mentioned in subsection 5.2.22, the stored energy still is a strict Liapunov function on an interval $[E_1, \infty)$, $E_1 > 0$. According to theorem 5.2.36, the set of the operating points ξ, such that $W(\xi) \leq E_1$, is globally attractive. It includes, in particular, the orbit of the periodic steady-state solution.

5.3 STEADY-STATE SOLUTIONS OF AUTONOMOUS AND PERIODIC CIRCUITS

5.3.1 Introduction

Let us consider a set of bounded and asymptotically stable solutions of a circuit, which have the same asymptotic behavior. How can we characterize such asymptotic behavior?

In the case of linear circuits, we use the steady-state solution. If the eigen-frequencies of the circuit are located in the left half-plane, and if the circuit only has one sinusoidal source, the steady-state solution is the unique sinusoidal solution of the circuit. Starting from this solution, we compose the steady-state solution from a circuit with an arbitrary number of time-dependent sources of any type, by using superposition.

This approach is no longer possible for nonlinear circuits. Among all the asymptotically stable solutions having the same asymptotic behavior, is there one that deserves to be designated a steady-state solution more than the others? It all depends on the aspect of the time-dependent sources. In the case of autonomous circuits and circuits with periodic sources, we may answer yes. The steady-state solution can possibly be defined in a satisfactory way for other types of time-dependent sources. However, it is difficult to imagine that a solution is privileged with respect to the others in the presence of sources with an absolutely irregular time dependence.

5.3.2 Definition

Let $\xi(t)$ be the solution of an autonomous circuit. An operating point ξ_∞ is a *limit point* of $\xi(t)$ if there is a time sequence $t_1, t_2, \ldots$, such that

$$\lim_{n \to \infty} \|\xi(t_n) - \xi_\infty\| = 0 \tag{5.82}$$

The *limit set* of $\xi(t)$ is the set of the limit points of $\xi(t)$ (fig. 5.44).

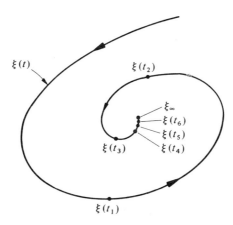

Fig. 5.44

5.3.3 Examples

In figures 5.44 and 5.45, we have represented two solutions with a different asymptotic behavior. In figure 5.44, all the sequences $\xi(t_n)$ converge toward the same point ξ_∞. Thus, the limit set consists of only the point ξ_∞. As opposed to this, the limit set of the solution of figure 5.45 is a complete closed curve. Two sequences $\xi(t_n)$ and $\xi(\tilde{t}_n)$ generally have limit points, ξ_∞ and $\tilde{\xi}_\infty$, which are different.

The solutions of the amplifier without signal (sec. 5.1.6) behave as the solution of figure 5.44. The dc operating point on its own constitutes the limit set of all the solutions.

All the solutions of the circuit with the tunnel diode of subsection 5.1.7 have only one limit point, but it is not the same for all the solutions. The solutions starting from the basin of attraction of the stable dc operating points $P_1(P_3)$ have $P_1(P_3)$ as limit point, and the solutions starting from the boundary between the two basins of attraction have the unstable dc operating point P_2 as limit point.

All the solutions of the oscillator of subsection 5.1.8, with $R_3 = 10$ kΩ, also have only one limit point, which is the same for all of them. This is the dc operating point $\xi = 0$. However, if $R_3 = 1$ kΩ, only the solution $\xi(t) \equiv 0$ has $\xi = 0$ as the unique limit point. All the other solutions behave as that of figure 5.45. Their limit set is the orbit of a periodic solution $\xi_\infty(t)$. Indeed, if T is the period of $\xi_\infty(t)$, we can

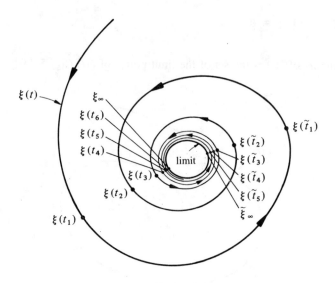

Fig. 5.45

see by way of numerical calculation that for a time t_0, we have

$$\xi(t_0 + nT) \rightarrow \xi_\infty(t_0) \quad \text{when} \quad t \rightarrow \infty \tag{5.83}$$

provided that $\xi(t) \not\equiv 0$.

Therefore, all the solutions, except $\xi(t) \equiv 0$, have an infinity of limit points.

5.3.4 Properties

Two solutions of an autonomous circuit with the same asymptotic behavior have the same limit set.

If a solution tends toward infinity when $t \rightarrow \infty$, it has no limit point.

If a solution is bounded, it has at least one limit point. In this case, the limit set is compact.

The first two properties are evident; the last property is proved in [4].

5.3.5 Definitions

A set S of generalized operating points for a circuit is *invariant* if for any solution $\xi(t)$ of the circuit with $\xi(t_0) \epsilon S$, we have $\xi(t) \epsilon S$ for any $t \geq t_0$.

5.3.6 Examples

The entire generalized configuration space is evidently invariant.

If $\xi(t)$ is a solution, defined in the time interval $[t_0, \infty)$, then its orbit $\gamma =$

$\{\xi(t)|t \geq t_0\}$ is an invariant set. Also, if $\xi(t_0)$ belongs to an invariant set S, the entire orbit γ belongs to S.

If W is a Liapunov function of a circuit, then the set of operating points with $W(\xi) \leq E$ is invariant, for any E.

5.3.7 Comment

The terms "positive limit point," "positive limit set," and "positively invariant set" (from [4]) are used to distinguish the time interval $[t_0, \infty)$ from $(-\infty, t_0]$. Here, because we are only interested in the future, we do not need any further specification.

5.3.8 Theorem: Invariance of Limit Sets

The limit set of each solution of an autonomous circuit is invariant.

The proof of this theorem is found in [4].

5.3.9 Comments

Theorem 5.3.8 means that if a solution $\xi_\infty(t)$, defined in the time interval $[t_0, \infty)$, is located at the instant $\xi(t)$ in the limit set of another solution $\xi_\infty(t)$, then it remains in that limit set for the entire future. Obviously, such a solution is a good candidate for a steady-state solution.

In fact, the theorem proved in [4] is stronger. It guarantees that the limit set not only is invariant toward the future, but also toward the past. If the starting solution $\xi(t)$ is bounded, so is the limit set. According to theorem 2.2.14, the solutions $\xi_\infty(t)$ in the limit set can be indefinitely continued toward the past.

5.3.10 Definition

Let $\xi(t)$ be a stable bounded solution of an autonomous circuit. If there is a stable $\xi_\infty(t)$ having an orbit located in the limit set of $\xi(t)$, and such that

$$\lim_{t \to \infty} \|\xi(t) - \xi_\infty(t)\| = 0 \tag{5.84}$$

then $\xi_\infty(t)$ is the *steady-state solution associated with* $\xi(t)$.

5.3.11 Comments

Let S be the limit set of a solution $\xi(t)$ and $\xi_\infty(t)$ a solution in S. In consideration of figure 5.45, we expect that the orbit of $\xi_\infty(t)$ fills the entire set S. Such is not always the case. In fact, S may comprise several orbits.

As a first example, we shall consider the system of differential equations in polar coordinates, according to [29]:

$$d\theta/dt = \sin^2\theta + (1 - r)^2 \tag{5.85}$$

314

$$dr/dt = r(1-r) \tag{5.86}$$

It follows from (5.86) that all the solutions tend toward the unit circle $r = 1$. Furthermore, if $r \neq 1$, we conclude from (5.85) that $d\theta/dt > 0$. Consequently, the limit set S of any solution $\xi(t)$, which starts outside the unit circle, is the entire unit circle (fig. 5.46). If $r = 1$, $d\theta/dt = \sin^2\theta$. There are thus two equilibrium points, $\theta = 0$ and $\theta = \pi$. Outside these points, we have $d\theta/dt > 0$. Consequently, a solution $\xi_\infty(t)$ that starts from a point ($r = 1$, $0 < \theta < \pi$) converges toward the point ($r = 1$, $\theta = \pi$) (fig. 5.46). Thus, the set S comprises four orbits: the two equilibrium points, as well as the two half-circles linking them.

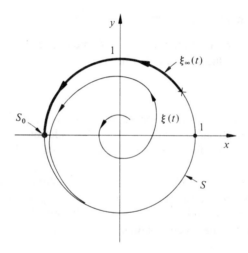

Fig. 5.46

If a solution $\xi(t)$ converges toward $\xi_\infty(t)$, it is evident that both solutions have the same limit set. Therefore, any solution of (5.85) and (5.86) which starts outside the unit circle cannot converge toward a solution on the unit circle because the limit set of the latter consists of only the point ($r = 1, \theta = \pi$) or ($r = 1, \theta = \pi$).

As a second example, we shall consider the following system of three differential equations. It is expressed in toroidal coordinates (fig. 5.47). We assume that it is only valid for $r < 2$.

$$dr/dt = r \cdot (1-r) \tag{5.87}$$
$$d\theta/dt = \omega_1 \tag{5.88}$$
$$d\varphi/dt = \omega_2 \tag{5.89}$$

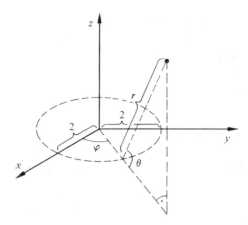

Fig. 5.47

The solutions of this system are easy to discuss because the three variables are uncoupled. It follows from (5.87) that a solution with an initial condition on the torus $r = 1$ remains on the torus (fig. 5.48). A solution with an initial condition outside the torus tends toward $r = 1$. More precisely, it converges toward the solution on the torus with the same initial angles θ_0 and φ_0. If the two angular velocities ω_1 and ω_2 have an irrational ratio, then the angles $\theta(t)$ and $\varphi(t)$ of the solution approach any pair of values θ and φ in the course of time. It follows that the limit set of the solution is the entire torus $r = 1$. However, although the solution $r = 1, \theta(t), \varphi(t)$ **passes arbitrarily close** to any point of the torus, it does not **exactly** pass through any point. In fact, there is an infinity of orbits on the torus. All the limit sets of the orbits are equal to the entire torus.

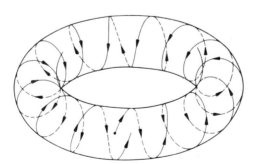

Fig. 5.48

316

The orbit of a solution from (5.87) to (5.89) on the torus is a curve which never closes on itself if ω_1/ω_2 is irrational. However, in the course of time, the solution returns **arbitrarily closely** to its starting point. Therefore, it is called an *almost periodic solution*. Thus, the solutions from (5.87) to (5.89), with ω_1/ω_2 being irrational, have an almost periodic steady-state solution. We shall return to the topic of almost periodic solutions in section 6.2.

We note as well that autonomous circuits may also exhibit chaotic solutions [33]. Because these solutions are unstable, definition 5.3.10 does not apply.

5.3.12 Theorem: Uniqueness and Nature of the Steady-State Solution

If a stable bounded solution $\xi(t)$ of an autonomous circuit has a steady-state solution $\xi_\infty(t)$, then it is unique. It is either constant, periodic, or almost periodic.

The theorem is a consequence of theorem 4.1.10 of [32].

5.3.13 Definition

Let Λ be the generalized configuration space of a circuit, and Λ_t the configuration space at the instant t. A pair of functions $F:\Lambda \rightarrow \Lambda$ and $f:\mathbb{R} \rightarrow \mathbb{R}$ constitutes a *symmetry of the circuit* if the following conditions are satisfied (fig. 5.49):

- The functions F and f are continuous and bijective. Their inverses are also continuous.
- The image of the interval $[t_0, \infty)$ under f is the interval $[f(t_0), \infty)$. If $t \rightarrow \infty$, then $f(t) \rightarrow \infty$. The same properties hold for f^{-1}.
- The function F transforms Λ_t into $\Lambda_{f(t)}$.
- If $\xi(t)$ is a solution of the circuit, defined in the interval $[t_0,\infty)$, then $\tilde{\xi}(\tilde{t}) = F(\xi(f^{-1}(\tilde{t})))$ is a solution defined in the interval $[f(t_0),\infty)$, and *vice versa*.

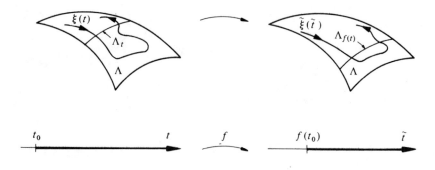

Fig. 5.49

5.3.14   Comment

In practice, the *solutions* of a nonlinear circuit are not known, and therefore the last condition cannot be directly checked against the example. Nonetheless, we can show that by replacing ξ with $\tilde{\xi} = F(\xi)$ and t with $\tilde{t} = f(t)$ in the **equation system** of the circuit, we return, possibly after some transformations, to the same system of equations.

5.3.15   Examples

An autonomous circuit is always symmetrical under an arbitrary *time translation*. A time translation by T consists of the pair of functions (F, f), where F is the identity and f is defined by

$$f(t) = t + \tau \tag{5.90}$$

A circuit with periodic time-dependent sources of period T is always symmetrical under a time translation by T.

An autonomous circuit whose elements are symmetrical has the symmetry (F, f), where $F(\xi) = -\xi$ and f is the identity. Indeed, by definition, a symmetrical element exhibits a symmetrical characteristic with respect to the origin and the Kirchhoff equations also remain invariant under the transformation $\xi \rightarrow -\xi$.

A circuit composed of symmetrical elements, except for the time-dependent sources having signals which satisfy $s(t + T/2) = -s(t)$ (fig. 5.50) has the following symmetry (F, f): $F(\xi) = -\xi$ and $f(t) = (t + T/2)$. Indeed, apart from the sources, the reasoning is the same as for the autonomous circuit with symmetrical elements. Regarding the sources, the transformation $s \rightarrow -s$ and $t \rightarrow t + T/2$ leaves the signal invariant. The ferroresonant circuit of subsections 5.1.10 and 5.1.11 is an example of such a symmetry.

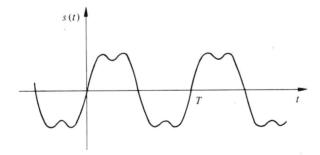

Fig. 5.50

5.3.16 Properties

If (F, f) is the symmetry of a circuit and $\xi(t)$ is an (asymptotically) stable solution, then the solution $\tilde{\xi}(\tilde{t}) = F\left(\xi(f^{-1}(\tilde{t}))\right)$ is also an (asymptotically) stable solution. The symmetry transforms the basin of attraction $\xi(t)$ into the basin of attraction $\tilde{\xi}(\tilde{t})$.

These properties are a consequence of the continuity of F and the invariance of the semi-infinite time-intervals under f. We will not give any detailed proofs here.

5.3.17 Property

If (F, f) is the symmetry of an autonomous circuit and $\tilde{\xi}(t)$ is a solution, then the limit set of $\xi(t)$ is transformed into the limit set of the solution $\tilde{\xi}(\tilde{t}) = F(\xi(f^{-1}(\tilde{t})))$ by F.

The proof is based on the same principle as that of property 5.3.16.

5.3.18 Corollary

If $\xi_\infty(t)$ is a steady-state solution of an autonomous circuit which has the symmetry (F, f), then $\tilde{\xi}_\infty(\tilde{t}) = F(\xi_\infty(f^{-1}(\tilde{t})))$ is also a steady-state solution. In particular, if the circuit has only one steady-state *solution*, then that solution is *symmetrical* relative to (F, f), i.e., we have

$$\xi_\infty(t) = F(\xi_\infty(f^{-1}(t))) \tag{5.91}$$

and its orbit is invariant under F.

Indeed, the first two conditions are a direct consequence of properties 5.3.16 and 5.3.17. The orbit invariance becomes evident if we rewrite (5.91) as

$$\xi_\infty(f(\tau)) = F(\xi_\infty(\tau)) \tag{5.92}$$

5.3.19 Examples

If an autonomous circuit has a unique steady-state solution, it is symmetrical with respect to an arbitrary time translation, which means it is constant. If a steady-state solution of an autonomous circuit is not constant, we find an infinity of other steady-state solutions with the same orbit by time translation. This phenomenon has already been shown in subsection 5.1.8 for the example of the Wien-bridge oscillator.

The Wien-bridge oscillator is also symmetrical with respect to the transformation $F(\xi) = -\xi$, $f(t) = t$ because it is composed of symmetrical elements. In the case where the oscillator exhibits steady-state solutions, we have observed that their orbits are always the same. It follows that it is symmetrical with respect to the origin, as confirmed by figure 5.5. In the case where there is only one steady-state solution, i.e., only one asymptotically stable dc operating point ξ, we necessarily have $\xi = -\xi$, i.e. $\xi = 0$, which can also be inferred from the circuit equations.

5.3.20 Poincaré Map

Until this point, we have limited ourselves in this chapter to autonomous circuits. Now, we will generalize the concepts developed to the case of periodic circuits.

An artificial construction permits us to reduce the case of a periodic source circuit of period T to that of an autonomous one. The idea is to limit the study of the solutions $\xi(t)$ to the discrete instants $t_0, t_0 + T, t_0 + 2T, \ldots$ For each initial condition ξ_0 at t_0, we thus find a sequence of points $\xi_n = \xi(t_0 + nT)$, $n = 0, 1, \ldots$ If we consider the transformation $\xi_0 \to \xi_1$, in the entire configuration space at t_0, it is identical to the transformation $\xi_1 \to \xi_2$, *et cetera*. Thus, the sequence ξ_n is obtained by an iteration of this transformation (fig. 5.51).

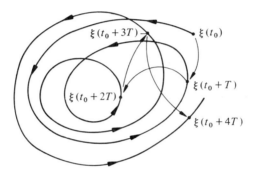

Fig. 5.51

5.3.21 Definition

Consider a circuit having time-dependent sources which are all periodic of period T. Let us assume that the hypotheses of subsection 5.2.1 on the solutions of the circuit hold. Let Λ_τ be the configuration space at the instant τ. Then, $\Lambda_\tau = \Lambda_{\tau+T}$, and we can define the function $F: \Lambda_\tau \to \Lambda_\tau$ by

$$F(\eta) = \xi(\tau + T) \tag{5.93}$$

where $\xi(t)$ is the solution with the initial condition:

$$\xi(\tau) = \eta \tag{5.94}$$

at the instant τ. Function F is called the *Poincaré map at the instant* τ (fig. 5.52).

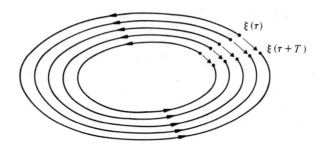

Fig. 5.52

5.3.22 Comment

The function F, which is part of a symmetry, and the Poincaré function F are of a different nature. Function F transforms *the entire generalized configuration space* into itself, and is normally a function given *explicitly*, whereas the Poincaré map *is not explicitly known* because it is defined starting from the solutions of the circuit. Furthermore, the Poincaré only transforms the *configuration space at the instant* τ into itself.

5.3.23 Properties

Let F be the Poincaré map of a circuit whose time-dependent sources are all periodic of period T. Let us assume that the hypotheses of subsection 5.2.1 concerning the solutions are satisfied. Thus, we have the following properties:

- Function F is injective and continuous.
- If the solutions of the circuit also exist for an arbitrarily long time toward the past, then F is bijective and F^{-1} is continuous.
- If the circuit is smooth, then F is smooth. If, in addition, the solutions also exist for an arbitrarily long time toward the past, F^{-1} is smooth.
- If the constitutive relations of the elements depend in a continuous way on a parameter λ, then F is continuous in λ.
- If the circuit is smooth and the constitutive relations of the elements depend in a smooth way on λ, then F is smooth in λ.
- If $\xi(t)$ is a solution of the circuit, defined in the time interval $[\tau, \infty)$, then

$$\xi(\tau + nT) = F^n(\xi(\tau)) \tag{5.95}$$

where

$$F^n(\eta) = F(F(... F(\eta) ...)) \tag{5.96}$$

Indeed, the first property results from the fact that the time-dependent sources are periodic of period T, and the other properties result directly from the hypotheses for the solutions of the circuit.

5.3.24 Definition

Let $F: \Lambda_\tau \to \Lambda_\tau$ be the Poincaré map of a circuit at the instant τ. The *orbit of a point F under* $\xi_0 \in \Lambda_\tau$ is the set of $\xi_n \in \Lambda_\tau$ with

$$\xi_n = F^n(\xi_0) \tag{5.97}$$

5.3.25 Example

The parameters of the configuration space Λ_τ of the junction diode circuit of subsections 5.1.13 and 5.1.14 is at any instant globally characterized by the state variables (q,i). The Poincaré map F can thus be represented as a transformation of the plane into itself, for any initial time τ. An orbit under F becomes a set of points (q_n, i_n) in the plane. In terms of the solution $(q(t), i(t))$ of the circuit with the initial condition $(q(\tau), i(\tau)) = (q_0, i_0)$, we can write

$$(q_n, i_n) = (q(t_0 + nT), \quad i(t_0 + nT)) \tag{5.98}$$

Two examples of orbits are shown in figure 5.26 and 5.27. Because they are generated by chaotic solutions of the circuit, they have a complicated structure.

At the other extreme, the orbit under F of a point ξ, which generates a periodic solution of period T, only comprises ξ. The orbit under F corresponding to the subharmonic solution of period $3T$ comprises the three points which are indicated by crosses in figure 5.27.

5.3.26 Property

The orbit under the Poincaré map of a point $\xi \in \Lambda_\tau$, which generates a periodic solution of period mT, consists of m points.

5.3.27 Analogy with Autonomous Circuits

The dynamical system realized by the iteration of the Poincaré map of a circuit is analogous to that of an autonomous circuit. The only difference is that time is discrete for the former and continuous for the latter. All of the concepts relating to autonomous circuits can be transferred in this way to periodic circuits. A *solution* of the discrete dynamic system is a sequence of points. The orbit has already been defined in subsection 5.3.24. *Stability* and *asymptotic stability* are defined by analogy with definitions 5.2.2 and 5.2.7, as are the concepts of *limit set* and *steady-state solution* by analogy with definitions 5.3.2 and 5.3.10. The theorem corresponding to theorem 5.3.12 permits us to conclude that constant, periodic, or almost periodic sequences are the only steady-state solutions of the discrete dynamic system.

5.3.28 Definition

Let $\xi(t)$ be a bounded asymptotically stable solution of a circuit having time dependent sources which are periodic of period T. The solution $\xi_\infty(t)$ is the *steady-state solution associated with* $\xi(t)$ if the solution $\xi_\infty(t_0 + nT)$ is the steady-state solution associated with the solution $\xi_\infty(t_0 + nT)$ of the discrete dynamic system generated by the Poincaré map at t_0, where t_0 is an arbitrary initial time.

5.3.29 Property

If ξ_∞ is the steady-state solution of a circuit having time-dependent sources which are periodic of period T, associated with the asymptotically stable solution $\xi(t)$, then

$$\lim_{t \to \infty} \ \|\xi(t) - \xi_\infty(t)\| \to 0 \qquad (5.99)$$

Indeed, by definition 5.3.28, the convergence (5.99) is guaranteed for the time sequence $t_0 + nT$. The continuous dependence of the solutions as functions of the initial solutions at $t_0 + nT$ for the interval $[t_0 + nT, t_0 + (n + 1)T]$ guarantees the convergence (5.99) for arbitrary time sequences.

5.3.30 Theorem: Uniqueness and Nature of the Steady-State Solution

The steady-state solution associated with a bounded stable solution of a circuit having time-dependent sources which are periodic of period T, is unique, if it exists. It is either periodic of period T or periodic of period mT, with $m > 1$ integer, or almost periodic.

This theorem is a direct consequence of the corresponding theorem for the discrete dynamical system generated by the Poincaré map.

5.3.31 Symmetry of Periodic Circuits

It is not difficult to show that corollary 5.3.18 also holds for circuits having periodic sources of period T. Consequently, if $\xi(t)$ is a periodic solution of period mT, then there are $m - 1$ other periodic solutions of the same period whose orbits are identical, as mentioned in section 5.1.

The ferroresonant circuit of subsections 5.1.9 and 5.1.10 is symmetrical under the transformation $\xi(t) \to -\xi$, $t \to t + T/2$. Consequently, if the periodic steady-state solution is unique, its orbit is symmetrical with respect to the origin. Figures

5.6, 5.11, and 5.13 illustrate this fact. Conversely, if a steady-state solution has an orbit that is not symmetrical with respect to the origin, then there is another, the orbit of which is the symmetrical image of the former with respect to the origin. An example is given in figure 5.12.

The subharmonic steady-state solutions of the ferroresonant circuit may or may not have a symmetrical orbit. Normally, the periodic solutions of period mT, with m being odd (even), have an orbit that is symmetrical (not symmetrical) with respect to the origin. The example of a periodic solution of period $3T$ in figure 5.14 has a symmetrical orbit.

5.4 STORED ENERGY

5.4.1 Introduction

In section 5.2, we introduced the Liapunov function concept and showed how it can be used to find attractive sets, whether they consist of an entire region of the configuration space or only a single point. We mentioned that the prototype of a Liapunov function is the energy stored in the capacitors and the inductors. The purpose of this section is to define and study this notion for nonlinear capacitors and inductors. Other sections of this chapter show the extent to which it is a Liapunov function of the circuit.

5.4.2 Energy Absorbed by a Charge-Controlled Capacitor

Consider a capacitor defined by the constitutive relation $u = h(q)$. If the charge depends on time according to the differentiable function $q(t)$, then the energy absorbed by the capacitor during the interval $[t_0, t_1]$ is

$$E = \int_{t_0}^{t_1} u(t)\, i(t)\, \mathrm{d}t = \int_{t_0}^{t_1} h(q(t))\, (\mathrm{d}q/\mathrm{d}t)\, \mathrm{d}t \qquad (5.100)$$

If $q(t)$ maps the time interval $[t_0, t_1]$ to the charge interval $[q_0, q_1]$ in a bijective way, then we can replace t by q as the integration variable, and we obtain

$$E = \int_{q_0}^{q_1} h(q)\, \mathrm{d}q = H(q_1) - H(q_0) \qquad (5.101)$$

where H is the indefinite integral of h. In (5.101), it is remarkable that E does not depend on the particular form of $q(t)$, but only on the initial point q_0 and the final point q_1. In the plane (q,u), energy E is represented by the area under the characteristic curve of the capacitor between q_0 and q_1 (fig. 5.53).

Let us consider a second function, $q(t)$, which transforms the time interval $[t_1, t_2]$ into the charge interval $[q_1, q_0]$ in a bijective way. By the same calculation as (5.100) and (5.101), we obtain the energy absorbed by the capacitor during interval $[t_1, t_2]$.

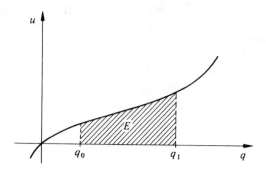

Fig. 5.53

$$E' = \int_{t_1}^{t_2} h\left(q'(t)\right)(\mathrm{d}q'/\mathrm{d}t)\,\mathrm{d}t = H(q_0) - H(q_1) = -E \qquad (5.102)$$

Thus, the energy E which has been absorbed by the capacitor on the way between q_0 and q_1 is entirely returned on the way back. This suggests the interpretation that energy E is stored rather than dissipated between q_0 and q_1, and then returned, rather than produced between q_1 and q_0.

Equation (5.102) also frees us from the hypothesis that $q(t)$ must be bijective. Indeed, if $q(t)$ is not bijective, we can remove, from interval $[t_0, t_1]$, a finite number of subintervals $[t'_k, t_k'']$ so that $q(t'_k) = q(t''_k)$, and in such a way that $q(t)$ transforms what is left of $q(t)$ into interval $[t'_k, t''_k]$ bijectively. These subintervals are shown in figure 5.54 for an example of $q(t)$. Because the integral (5.100) limited to $[t'_k, t''_k]$ is zero, the result (5.101) is still valid.

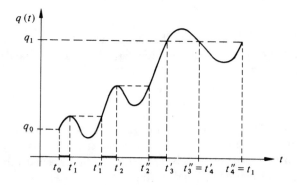

Fig. 5.54

5.4.3   Energy Absorbed by a Voltage-Controlled Capacitor

Consider a capacitor defined by the constitutive relation $q = g(u)$. If the voltage depends on time according to the differentiable function $u(t)$, then the energy absorbed by the capacitor during interval $[t_0, t_1]$ is

$$E = \int_{t_0}^{t_1} u(t)\, i(t)\, \mathrm{d}t = \int_{t_0}^{t_1} u(t)\, \frac{\mathrm{d}g(u(t))}{\mathrm{d}t}\, \mathrm{d}t \tag{5.103}$$

By partial integration, we obtain

$$E = u(t_1)\, g(u(t_1)) - u(t_0)\, g(u(t_0)) - \int_{t_0}^{t_1} g(u(t))\, (\mathrm{d}u/\mathrm{d}t)\, \mathrm{d}t \tag{5.104}$$

The integral in (5.104) is calculated again by changing the variable from t into u. If $u_0 = u(t_0)$ and $u_1 = u(t_1)$, and if $G(u)$ is the indefinite integral of $g(u)$, we find

$$E = u_1 g(u_1) - u_0 g(u_0) - G(u_1) + G(u_0) \tag{5.105}$$

We observe that E only depends on the initial voltage u_0 and the final voltage u_1. This suggests, as was the case in the preceding subsection, to consider energy E as being stored rather than dissipated.

Energy (5.105) can be represented as the difference of two areas in the plane (q,u). The first area, of value $u_1 g(u_1) - u_0 g(u_0)$, is shown in figure 5.55. The second area is that of figure 5.56, but the roles of q and u are interchanged. The result is represented in figure 5.57, where the vertically (horizontally) cross-hatched surfaces are to be taken positively (negatively).

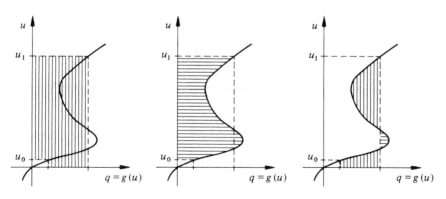

Fig. 5.55   Fig. 5.56   Fig. 5.57

326

Alternatively, the same result can be achieved by generalizing figure 5.53. Starting from the initial point $(g(u_0), u_0)$, we follow the characteristic with a point $(g(u'), u')$ until we reach the final point $(g(u_1), u_1)$. For each increase $\Delta u'$, we add the area, whether positive or negative depending on the sign of $\Delta g(u')$, between the characteristic and the charge axis. Three intermediate steps of this process are shown in figure 5.58. The final result coincides with figure 5.57.

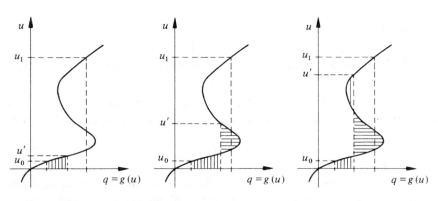

Fig. 5.58

5.4.4 Definitions

Consider a charge controlled capacitor, with the constitutive relation $u = h(q)$. The *stored energy* in the capacitor carrying charge q is defined by

$$W(q) = \int_{q_0}^{q} h(x)\,dx \qquad (5.106)$$

where q_0 is any admissible charge. If the capacitor is voltage controlled, with the constitutive relation $q = g(u)$, then the *stored energy* in the capacitor under voltage u is

$$W(u) = ug(u) - \int_{u_0}^{u} g(x)\,dx \qquad (5.107)$$

where u_0 is any admissible voltage.

In an analogous way, we define the *stored energy* in an inductor. If the inductor is flux controlled, with the constitutive relation $i = g(\varphi)$, then

$$W(\varphi) = \int_{\varphi_0}^{\varphi} g(x)\,dx \qquad (5.108)$$

whereas if it is current controlled, with the constitutive relation $\varphi = h(i)$, we define

$$W(i) = h(i)\,i - \int_{i_0}^{i} h(x)\,\mathrm{d}x \tag{5.109}$$

5.4.5 Properties

If the constitutive relation of a capacitor (inductor) is defined by a smooth function, then the stored energy is a smooth function, whether it is defined by (5.106) or (5.107) (by (5.108) or (5.109)).

If the constitutive relation of an inductor (capacitor) is defined by a piecewise-linear function, then the stored energy is a piecewise-quadratic function. If the inductor (capacitor) is charge-controlled (flux-controlled), then the stored energy has a continuous first derivative. Conversely, if it is voltage-controlled (current-controlled), then the first derivative generally has discontinuities. However, the function itself remains continuous.

5.4.6 Property

Let $W(t)$ be the energy stored in a capacitor or an inductor, evaluated along a solution of the circuit. Then

$$\mathrm{d}W/\mathrm{d}t = u(t)\,i(t) \tag{5.110}$$

which means that power is the derivative of stored energy.

Indeed, this property was the starting point of subsections 5.4.2 and 5.4.3.

5.4.7 Comment

As is the case for potential energy in mechanics, stored energy is defined only up to a constant. Indeed, in each one of the formulas (5.106) to (5.109), an arbitrary constant is present. This is no problem, because the fundamental relation is (5.110), which is independent of an additive constant in W. When the characteristic of the element passes through the origin, it is natural to choose the value zero for q_0, u_0, φ_0, i_0.

5.4.8 Example

Let us calculate the energy stored in a linear capacitor of value C. Its constitutive relation is

$$u = q/C \tag{5.111}$$

According to (5.106), we find

$$W(q) = \int_0^q (x/C)\,dx \tag{5.112}$$

$$= q^2/2C \tag{5.113}$$

Alternatively, we can apply (5.107):

$$W(u) = u \cdot Cu - \int_0^u Cx\,dx \tag{5.114}$$

$$= Cu^2/2 \tag{5.115}$$

Thus, our definitions correctly reproduce the known expressions for linear capacitors (vol. I, sec. 5.4.8; vol. IV, sec. 1.3.3).

The energy stored in an inductor with a ferromagnetic core, modeled by (1.29), is given, in the case $\varphi > \varphi_0$, by

$$W(\varphi) = (1/L_0) \int_0^{\varphi_0} \varphi\,d\varphi + (1/L_1) \int_{\varphi_0}^{\varphi} \varphi\,d\varphi - \left(\left(\frac{1}{L_1}\right) - \left(\frac{1}{L_0}\right)\right) \int_{\varphi_0}^{\varphi} \varphi_0\,d\varphi \tag{5.116}$$

$$= \varphi^2/2L_1 - (\varphi - \varphi_0/2) \cdot \varphi_0 \cdot \left(\left(\frac{1}{L_1}\right) - \left(\frac{1}{L_0}\right)\right) \tag{5.117}$$

For $0 \le \varphi \le \varphi_0$, $W(\varphi)$ is the energy of a linear inductor of value L_0. In addition, $W(\varphi)$ is an even function, because $i = g(\varphi)$ is odd. In short,

$$W(\varphi) = \begin{cases} \varphi^2/2L_0 & \text{for} \quad |\varphi| \le \varphi_0 \\ (\varphi^2/2L_1) - (\varphi - \varphi_0/2) \cdot \varphi_0 \cdot ((1/L_1) - (1/L_0)) & \text{for} \quad |\varphi| > \varphi_0 \end{cases} \tag{5.118}$$

We observe that W and $dW/d\varphi$ are indeed continuous at φ_0.

The energy stored in a superconducting junction, given by (1.30), is

$$W(\varphi) = (I_0/k_0) \cos k_0\varphi \tag{5.119}$$

5.4.9 Comment

How is stored energy to be defined when a capacitor is neither charge-controlled, nor voltage-controlled? An example of such a characteristic is given in figure 5.59. Let us apply the graphical method introduced at the end of subsection 5.4.3. Let (q_0, u_0) be a given initial point, and consider a point (q, u) moving clockwise on the characteristic until it is back at (q_0, u_0). For each increase $(\Delta q, \Delta u)$, we add the area between the segment of the characteristic and the charge axis. The sign is that of $q \cdot \Delta u$. In figure 5.60, we have represented the area for three intermediate values and for the final value of (q, u). We observe that the final area is not zero, but the total area inside the

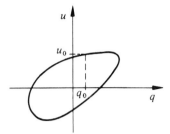

Fig. 5.59

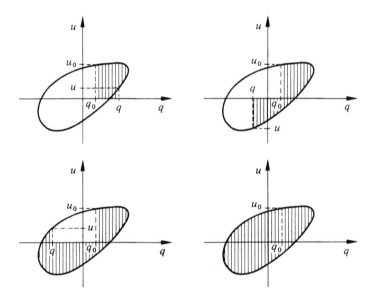

Fig. 5.60

characteristic. Consequently, if we define the area as being the stored energy, that energy is not exclusively a function of the point of the characteristic, but it also depends on the number of revolutions made by the orbit $(q(t), u(t))$ on the characteristic.

There are other capacitors which are neither charge controlled, nor voltage controlled, but for which this process still leads to a stored energy which is exclusively a function of the point of the characteristic. An example is given in figure 5.61. What makes the characteristic of figure 5.61 distinct from that of figure 5.59 is that the two points are linked by one path only.

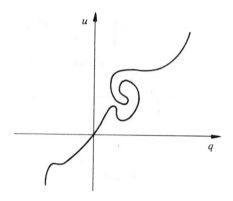

Fig. 5.61

In figure 5.62, we have represented the orbit in the plane (i, φ) of an *inductor with hysteresis* excited by a sinusoidal source. During one period of the source, point $(i(t), \varphi(t))$ makes one revolution on the orbit. The energy dissipated during that time is equal to the area contained within the orbit. The similarity between figures 5.62 and 5.59 suggests to interpret the area enclosed by the characteristic of figure 5.59 as the energy that is dissipated rather than stored during one revolution. However, if we adopt this point of view, then if $(q(t), u(t))$ makes one revolution on the characteristic in the counterclockwise direction, the one-port produces power. This surely is not the case of the inductor with hysteresis. Indeed, that inductor is not correctly modeled by a nonlinear inductor with the characteristic of figure 5.62. This figure only represents the orbit in the plane (φ, i) for a particular solution. In that case, the point $(\varphi(t), i(t))$, in fact, always turns in the clockwise direction. A more appropriate model would be a flux or current-controlled nonlinear inductor in parallel with a nonlinear resistor [34].

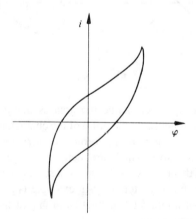

Fig. 5.62

Henceforth, we limit ourselves herein to capacitors (inductors) which are either charge-controlled (flux-controlled) or voltage-controlled (current-controlled). Consequently, *the energy will be exclusively a function* of the point of the characteristic, and, therefore, in the context of a circuit, a function *of the operating point.*

5.4.10 Passivity of Capacitors and Inductors

Is a capacitor passive? Definition 3.2.2 does not apply to this case, and the passivity concept must be defined in an appropriate manner.

It is not reasonable to call a capacitor active for being able to supply power. Indeed, any capacitor or inductor can return the energy previously stored. However, in this process of energy storage and recovery, no energy is created, and one is tempted to classify all capacitors and inductors as being passive.

If we connect a negative linear capacitor of value $-C$ and initial voltage u_0 to a linear resistor of value R (fig. 5.63), then the voltage follows the time evolution:

$$u(t) = u_0 \exp(t/RC) \tag{5.120}$$

and more and more power is dissipated in the resistor. Where does this energy come from?

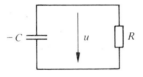

Fig. 5.63

According to (5.115), the energy stored in the capacitor is

$$W(u) = -Cu^2/2 \tag{5.121}$$

When we increase u, $W(u)$ becomes more and more negative, which amounts to extracting energy from the capacitor. Because W is not bounded below, the negative capacitor constitutes an infinite energy reservoir.

The stored energy concept has a somewhat unexpected meaning in the case of the negative capacitor. The term suggests that this energy is supplied in the past, and that no more than this energy can be recovered in the future. To be consistent with this point of view, one has to admit that an infinite energy has been stored in a negative capacitor, already at the instant the circuit was constituted. In actual fact, things are different. In figure 5.64, the negative capacitor is implemented by a positive capacitor and a negative impedance converter. The energy is produced by the dc sources of the operational amplifier as it comes out of the one-port.

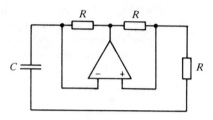

Fig. 5.64

In view of these considerations, we classify the capacitors and inductors having an infinite energy reservoir as being active. Definition 5.4.4 for the stored energy still applies, but the interpretation of this concept is no longer so evident.

5.4.11 Definition

A capacitor or an inductor is *passive* if its stored energy is bounded below. Otherwise it is *active*.

5.4.12 Generalization

The main property of stored energy is (5.110). Combined with the power balance equation (3.3), it permits to prove that stored energy is a Liapunov function for many circuits. As, in addition, an equation of incremental power balance is available, it would be interesting to generalize the concept of stored energy in such a way that an equation of type (5.110) holds with incremental powers.

5.4.13 Definitions

Consider a charge controlled capacitor defined by the constitutive relation $u = h(q)$. The *incremental energy that is stored* in the capacitor is a function $W(q_1, q_2)$, such that

$$(d/dt) W(q_1(t), q_2(t)) = \Delta u(t) \cdot \Delta i(t) \tag{5.122}$$

where

$$\Delta u(t) = u_2(t) - u_1(t) = h(q_2(t)) - h(q_1(t)) \tag{5.123}$$

$$\Delta i(t) = i_2(t) - i_1(t) = (dq_2/dt) - (dq_1/dt) \tag{5.124}$$

for any pair of differentiable functions $q_1(t)$ and $q_2(t)$.

In an analogous way, the incremental energy stored in a flux controlled inductor is a function $W(\varphi_1, \varphi_2)$ such that

$$(d/dt)\, W(\varphi_1(t),\ \varphi_2(t)) = \Delta u(t) \cdot \Delta i(t) \tag{5.125}$$

for any pair of differentiable functions $\varphi_1(t)$ and $\varphi_2(t)$.

5.4.14   Existence of Stored Incremental Energy

Let us assume that the stored incremental energy W of a capacitor exists. Then, according to (5.122),

$$(dW/dt) = (\partial W/\partial q_1)(dq_1/dt) + (\partial W/\partial q_2)(dq_2/dt) \tag{5.126}$$
$$= [h(q_1) - h(q_2)](dq_1/dt) + [h(q_2) - h(q_1)](dq_2/dt) \tag{5.127}$$

It follows that

$$(\partial/\partial q_2)\,[h(q_1) - h(q_2)] = (\partial/\partial q_1)\,[h(q_2) - h(q_1)] \tag{5.128}$$

and, therefore,

$$-(dh/dq)(q_2) = -(dh/dq)(q_1) \tag{5.129}$$

This equation must hold for arbitrary values q_1 and q_2, which is possible only if dh/dq is a constant, or, equivalently, if

$$u = h(q) = aq + b \tag{5.130}$$

Equation (5.130) stands for a linear capacitor in series with an independent voltage source.

In the case of a linear capacitor of value C, we easily find the incremental energy:

$$W(q_1, q_2) = (\Delta q)^2/2C = C(\Delta u)^2/2 \tag{5.131}$$

Similarly, the stored incremental energy of a linear inductor of value L is

$$W(\varphi_1, \varphi_2) = (\Delta \varphi)^2/2L = L(\Delta i)^2/2 \tag{5.132}$$

In short, the stored incremental energy of a capacitor or an inductor only exists if the element is linear.

5.4.15   Stored Incremental Energy for Autonomous Circuits

Definition 5.4.13 requires that equation (5.122) holds for any choice of $q_1(t)$ and $q_2(t)$. A more specific concept of stored incremental energy can be introduced by imposing one of the two functions. In the case of autonomous circuits, we choose a constant function, so as to satisfy equation (5.122) when one of the two solutions is frozen at a dc operating point.

This method applies even beyond linear capacitors and inductors, because, as incremental energy is now a function of only one variable, the pitfall of equation (5.129) is avoided.

5.4.16 Definition

Consider a charge controlled capacitor defined by the constitutive relation $u = h(q)$. The *stored incremental energy relative to a point* (q_0, u_0) of its characteristic is a function of $W(q)$, such that

$$dW(q(t))/dt = \Delta u(t) \cdot \Delta i(t) \tag{5.133}$$

where

$$\Delta u(t) = u(t) - u_{0} = h(q(t)) - h(q_0) \tag{5.134}$$

$$\Delta i(t) = i(t) - 0 = d(q(t) - q_0)/dt \tag{5.135}$$

for any differentiable function $q(t)$. If the capacitor is voltage controlled, with the constitutive relation $q = g(u)$, then the incremental energy relative to point (q_0, u_0) is a function $W(u)$, such that

$$(d/dt)W(u(t)) = \Delta u(t) \cdot \Delta i(t) \tag{5.136}$$

where

$$\Delta u(t) = u(t) - u_0 \tag{5.137}$$

$$\Delta i(t) = i(t) - 0 = d[g(u(t)) - q_0]/dt \tag{5.138}$$

In an analogous way, we define the stored incremental energy in an inductor, relative to a point (φ_0, i_0).

5.4.17 Calculation of Relative Stored Incremental Energy

Let us first consider a charge controlled capacitor defined by $u = h(q)$. We must find $W(q)$, such that

$$(dW/dq) \cdot (dq/dt) = (h(q) - h(q_0))(dq/dt) \tag{5.139}$$

By simplifying by dq/dt and integrating, we obtain

$$W(q) = \int_{q_0}^{q} (h(q') - h(q_0)) \, dq' \tag{5.140}$$

The free constant in the transition from (5.139) to (5.140) has been chosen such that $W(q_0) = 0$.

We can write (5.140) as

$$W(q) = H(q) - H(q_0) - (q - q_0) h(q_0) \tag{5.141}$$

where H is the indefinite integral of h. Remember that H is the stored nonincremental energy, introduced in subsection 5.4.4.

According to (5.110), we have

$$(dH/dt) = u \cdot i \tag{5.142}$$

from which we easily infer (5.133).

The case of a voltage-controlled capacitor, defined by $q = g(u)$ is similar. The stored incremental energy relative to point (q_0, u_0) of its characteristic is given by

$$W(u) = (u - u_0) g(u) - \int_{u_0}^{u} g(u') \, du' \tag{5.143}$$

Indeed,

$$(dW/dt) = \frac{d}{dt} \left[u g(u) - \int_{u_0}^{u} g(u') \, du' \right] - u_0 \frac{d}{dt} g(u) \tag{5.144}$$

The first term on the right-hand side of (5.144) is the derivative of the nonincremental energy. Then, according to (5.110):

$$(dW/dt) = u \cdot i - u_0 (dq/dt) = \Delta u \cdot \Delta i \tag{5.145}$$

In an analogous way, we find the incremental energy stored in an inductor, whether it is flux controlled or current controlled.

5.4.18 Definitions

A *capacitor is locally passive at a point* (u_0, q_0) of its characteristic if, for any other point (u, q) of its characteristic in a neighborhood of (u_0, q_0), the inequality

$$(u - u_0)(q - q_0) \geqslant 0 \tag{5.146}$$

is satisfied. More precisely, there is an $\epsilon > 0$ such that if

$$\| (u - u_0, q - q_0) \| < \epsilon \tag{5.147}$$

and if (u, q) belongs to the characteristic, we have (5.146). If this is not the case, the capacitor is *locally active at* (u_0, q_0). If a capacitor is locally passive at every point of its characteristic, we simply call it a *locally passive capacitor*.

If inequality (5.146) is strict, apart from point (u_0, q_0), it is a *strictly locally passive capacitor at* (u_0, q_0). If this is the case at every point, it is a *strictly locally passive capacitor*.

The same concepts are defined in an analogous way for an inductor.

5.4.19 Properties

Let us assume that a charge controlled capacitor is locally passive. The stored incremental energy relative to an arbitrary point (u_0, q_0) of its characteristic satisfies for any q and $\tilde{q}$:

$$W(q) \geqslant 0 \ , \quad W(q_0) = 0 \tag{5.148}$$

$$W(\tilde{q}) \geqslant W(q) \quad \text{if} \quad |\tilde{q} - q_0| > |q - q_0| \tag{5.149}$$

$$W(q) \rightarrow \infty \quad \text{when} \quad |q| \rightarrow \infty \tag{5.150}$$

The only exception to (5.150) is the case of a constitutive relation with $h(q) = h(q_0)$ for $(q \geqslant q_0)$ or for $q \leqslant q_0$. If the capacitor is strictly locally passive, then inequalities (5.148) and (5.149) are strict, except if $q = q_0$ and $\tilde{q} = q$, respectively. If the capacitor only is (strictly) locally passive in an interval $[q_1, q_2]$, then the (strict) inequalities at least hold for any q_0, q, and $\tilde{q}$ in that interval.

The analogous properties hold for the flux controlled inductors. Indeed, these properties result from formula (5.140) and the fact that, thanks to local passivity, $(h(q') - h(q_0))dq' \geqslant 0$.

5.4.20 Definitions

Let us assume that all the capacitors (inductors) of a circuit are either charge-controlled (flux-controlled) or voltage-controlled (current-controlled). The total energy stored in the capacitors and the inductors is the function $W : \mathbb{R}^{2b + N_C + N_L} \rightarrow \mathbb{R}$ defined by

$$W(\xi) = \sum_{k=1}^{N_C + N_L} W_k(x_k) \tag{5.151}$$

where

- $x_k = q_k$ and W_k is expression (5.106) if branch k carries a charge-controlled capacitor;
- $x_k = u_k$ and W_k is expression (5.107) if branch k carries a voltage-controlled capacitor;
- $x_k = \varphi_k$ and W_k is expression (5.108) if branch k carries a flux-controlled capacitor;
- $x_k = i_k$ and W_k is expression (5.109) if branch k carries a current-controlled capacitor.

The *total stored incremental energy relative to a dc operating point* is defined by (5.151) as well, but with expressions (5.140) and (5.143) instead of (5.106) and (5.107) for the W_k of the capacitors, and the corresponding expressions for the W_k of the inductors.

5.4.21 Properties

If all the capacitors (inductors) of a circuit are charge-controlled (flux-controlled), and if W is the total energy stored in the capacitors and the inductors, then, for any solution $\xi(t)$ of the circuit, we can write

$$dW(\xi(t))/dt = -\sum_{k=N_C+N_L+1}^{2b+N_C+N_L} u_k(t)\, i_k(t) \tag{5.152}$$

If W is the total stored incremental energy relative to a dc operating point $\bar{\xi}$, then

$$dW(\xi(t))/dt = -\sum_{k=N_C+N_L+1}^{2b+N_C+N_L} \Delta u_k(t)\, \Delta i_k(t) \tag{5.153}$$

where $\Delta u(t) = u(t) - \bar{u}$ and $\Delta i(t) = i(t) - \bar{\imath}$.

5.5 CONVERGENCE TOWARD A DC OPERATING POINT

5.5.1 Introduction

Let us assume that a circuit is composed of charge-controlled capacitors, flux-controlled inductors and resistors. Let $\bar{\xi} = (\bar{u}, \bar{\imath}, \bar{q}_C, \bar{\varphi}_L)$ be a dc operating point. If W is the total incremental energy stored in the capacitors and the inductors relative to $\bar{\xi}$, and if $\xi(t)$ is a solution of the circuit, equation (5.153) holds.

If the resistors are strictly locally passive, the right-hand side of (5.153) is negative. On the other hand, if the capacitors and inductors are strictly locally passive, function W is positive outside $\bar{\xi}$. Consequently, W is a strict Liapunov function, and the solutions tend toward $\bar{\xi}$ when $t \to \infty$. Thus, we conclude that the solution $\xi(t) \equiv \bar{\xi}$ is globally asymptotically stable. As we shall show hereafter, this reasoning has some weaknesses.

If the solution $\xi(t) \equiv \bar{\xi}$ is globally asymptotically stable, the dc operating point $\bar{\xi}$ is necessarily unique. The uniqueness of the dc operating point could have been also proved by theorem 3.4.2. In fact, we could have replaced, in the statement of theorem 3.4.2, the local passivity by the passivity relative to $\bar{\xi}$ without having to change the proof.

By limiting ourselves to resistors that are strictly locally passive or strictly passive relative to $\bar{\xi}$, we exclude the independent sources. Besides, in the presence of voltage source and inductor loops, and of current source and capacitor cut sets, the dc operating point cannot be unique, and, all the more so, the corresponding solution cannot be

338

globally asymptotically stable. Indeed, at a dc operating point, the capacitors (inductors) can be replaced by open circuits (shortcircuits), and we thus have a voltage source loop or a current source cut set. A circuit with such a loop or cut set has either zero or an infinity of solutions, whether it is linear or nonlinear. Consequently, if we want to find a class of circuits in which the independent sources are admitted, and in which all the solutions converge toward the same dc operating point, we have to exclude from the start the voltage source and inductor loops, as well as the current source and capacitor cut sets.

These considerations lead to the following conjecture, which will, however, be disproved by counterexamples.

5.5.2 Conjecture

Let $\bar{\xi}$ be a dc operating point of an autonomous circuit which satisfies the hypotheses of subsection 5.2.1, and which is composed of

- strictly locally passive charge-controlled capacitors;
- strictly locally passive flux-controlled inductors;
- strictly passive resistors relative to $\bar{\xi}$;
- constant voltage and current sources.

Let us assume that there is no capacitor and voltage source loop, nor any inductor and current source cut set. Then, any solution of the circuit converges toward $\bar{\xi}$, when $t \rightarrow \infty$.

5.5.3 Counterexample

A first counterexample to conjecture 5.5.2 is the circuit of figure 5.29. It actually only has one dc operating point, $u = i = 0$, but the constant solution frozen at that point is only stable, without being asymptotically stable.

Obviously, the solutions cannot tend toward zero because the energy stored in the capacitor and the inductor cannot be dissipated in the absence of resistors.

5.5.4 Counterexample

Let us consider the linear circuit of figure 5.65. Its configuration space is a plane of dimension 4 the parameters of which can be characterized by u_1, u_2, i_3, i_4. The only dc operating point is $u = i = 0$. The stored energy is

$$W(u_1, u_2, i_3, i_4) = (C_1 u_1^2/2) + (C_2 u_2^2/2) + (L_3 i_3^2/2) + (L_4 i_4^2/2) \tag{5.154}$$

and the power dissipated by the resistor R_5 is

$$P = u_5^2/R_5 = (u_1 - u_2)^2/R_5 \tag{5.155}$$

If $\xi(t)$ is a solution, we have, according to (5.152)

$$dW(\xi(t))/dt = -(u_1(t) - u_2(t))^2/R_5 \qquad (5.156)$$

Consequently, $dW/dt < 0$, except if $u_1 = u_2$, where $dW/dt = 0$. Otherwise, $W(\xi(t)) \rightarrow 0$, which implies $\xi(t) \rightarrow 0$, or $W(\xi(t)) \rightarrow E > 0$.

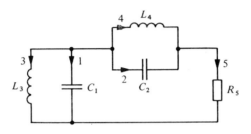

Fig. 5.65

Let us study the second possibility. In this case, the limit set S of the solution is part of the set of the operating points ξ with $W(\xi) = E$. According to property 5.3.4, S is not empty. If a solution $\xi_x(t)$ satisfies $\xi_x(t_0) \in S$, then $\xi_x(t) \in S$ for all the $t \geq t_0$. For such a solution, we thus have

$$u_1(t) = u_2(t) \qquad \text{for all the} \qquad t \geq t_0 \qquad (5.157)$$

The state equations of the circuit are

$$du_1/dt = i_1/C_1 = (-i_3 - (u_1 - u_2)/R_5)/C_1 \qquad (5.158)$$
$$du_2/dt = i_2/C_2 = (-i_4 + (u_1 - u_2)/R_5)/C_2 \qquad (5.159)$$
$$di_3/dt = u_3/L_3 = u_1/L_3 \qquad (5.160)$$
$$di_4/dt = u_4/L_4 = u_2/L_4 \qquad (5.161)$$

We then infer that

$$d(u_1 - u_2)/dt = -(i_3/C_1) + (i_4/C_2) - ((1/C_1) + (1/C_2))(u_1 - u_2)/R_5 \qquad (5.162)$$

If (5.157) holds, the left-hand side of (5.162) is identically null, and we find

$$0 = (i_3(t)/C_1) - (i_4(t)/C_2) \qquad \text{for} \qquad t \geq t_0 \qquad (5.163)$$

By differentiating (5.163) with respect to t, we obtain

$$u_1(t)/(C_1 L_3) = u_2(t)/(C_2 L_4) \quad \text{for} \quad t \geqslant t_0 \tag{5.164}$$

By combining the equations (5.157) and (5.164), we obtain

$$u_1(t) = u_2(t) = i_3(t) = i_4(t) \equiv 0 \tag{5.165}$$

unless

$$C_1 L_3 = C_2 L_4 \tag{5.166}$$

However, (5.165) is incompatible with $E > 0$. Consequently, if (5.166) is not satisfied, all the solutions converge toward zero and conjecture 5.5.2 holds.

Conversely, if (5.166) is satisfied, we find the following constant energy solutions.

$$u_1(t) = u_2(t) = a \cos \omega t + b \sin \omega t \tag{5.167}$$
$$i_3(t) = C_1 \omega b \cos \omega t - C_1 \omega a \sin \omega t \tag{5.168}$$
$$i_4(t) = C_2 \omega b \cos \omega t - C_2 \omega b \sin \omega t \tag{5.169}$$

where a and b are arbitrary constant and

$$1/\omega^2 = C_1 L_3 = C_2 L_4 \tag{5.170}$$

The stored energy is

$$E = ((C_1/2) + (C_2/2))(a^2 + b^2) \tag{5.171}$$

Since the circuit is linear, we could apply the Laplace transform method. We would observe that any solution of the circuit tends toward a solution of the form (5.167) to (5.169). This is the inverse Laplace transform of the term in $(p^2 + \omega^2)^{-1}$, resulting from the decomposition into simple fractions of the Laplace transform of the solution.

We conclude that conjecture 5.5.2 is wrong if and only if (5.166) holds. In that case, the solutions tend toward a solution of the form (5.167) to (5.169).

5.5.5 Comment

Contrary to counterexample 5.5.3, there is a resistor in the circuit of figure 5.65. Nevertheless, under certain conditions, the solution does not tend toward zero, because the resistor cannot dissipate the energy stored in the capacitors and the inductors. To avoid this case, we have to introduce enough resistors into the circuit.

5.5.6 Theorem

Let us consider an autonomous circuit with a dc operating point $\bar{\xi}$, which is composed of the following elements:

- strictly locally passive charge-controlled capacitors;
- strictly locally passive flux-controlled inductors;
- strictly passive resistors relative to $\bar{\xi}$;
- constant voltage and current sources.

Concerning the connections, let us assume that

- there is no loop that is exclusively composed of voltage sources, capacitors and inductors, nor any cut set that is exclusively composed of current sources, capacitors and inductors.

In addition, we assume

- that the hypotheses of subsection 5.2.1 hold.

Then all the solutions of the circuit converge toward $\bar{\xi}$ when $t \to \infty$.

Indeed, the total stored energy relative to $\bar{\xi}$ is a strict Liapunov function of the circuit. We check this fact by establishing the properties required for a strict Liapunov function in the order of definition 5.2.18.

The first condition holds thanks to properties 5.4.5, and the second one holds thanks to properties 5.4.19.

For the third condition, we have to show that if the total stored incremental energy W is bounded, all the variables of the circuit are bounded. We start with q_C and φ_L. According to subsection 5.4.19, the energy W_k of the kth capacitor (inductor) satisfies $W_k \to \infty$ if $|q_k| \to \infty$ ($|\varphi_k| \to \infty$). Consequently,

$$ W(\xi) \to \infty \quad \text{when} \quad |q_k| \to \infty \quad \text{or} \quad |\varphi_k| \to \infty \quad \text{for any} \quad k. \tag{5.172} $$

This implies that q_C and φ_L are bounded if W is bounded. Because q_C and φ_L are global parameters of the configuration space, all the other variables are functions of q_C and φ_L. Consequently, if q_C and φ_L are bounded, the other variables are also bounded.

The proof of the fourth condition is based on the relation (5.153). The function $-V$ of (5.48) thus is the total incremental power absorbed by the resistors,

$$ P_R(\xi) = \sum_{\text{resistors}} \Delta u_k \cdot \Delta i_k \tag{5.173} $$

since the constant sources satisfy $\Delta u_K \cdot \Delta i_K = 0$. The increments in (5.173) are relative to the dc operating point $\bar{\xi}$.

Thanks to property 5.2.21, we just have to show that $P_R(\xi) > 0$ if $\xi \neq \bar{\xi}$ to establish the last condition for a strict Liapunov function. Since the resistors are strictly passive with respect to $\bar{\xi}$, we have $P_R(\xi) \geq 0$, and $P_R(\xi) = 0$ is only possible when $\Delta u_R = \Delta i_R = 0$. The column matrices Δi_R and Δu_R designate the increments of currents and voltages in the resistors at the operating point ξ.

We show that $\Delta i_R = \Delta u_R = 0$ implies $\xi = \bar{\xi}$ by two applications of the colored-branch theorem. For the first one, we color all the capacitors, inductors and current sources green, and we orient them as their incremental voltages with respect to $\bar{\xi}$. The resistors and the voltage sources are colored red. Let us assume that all the incremental voltages of the resistors are null. Because there is no inductor-capacitor-current source cut set, each green branch is part of a uniform loop. The Kirchhoff equation of this loop implies

$$\sum_{\substack{\text{loop} \\ \text{LC,}}} \Delta u_k = -\sum_{\substack{\text{loop} \\ \text{resistors,}}} \Delta u_k = 0 \tag{5.174}$$

Because all the terms of the left side sum have the same sign, they are all zero. Consequently, the Δu_k of all the capacitors, all the inductors and all the sources are zero. By the dual application of the colored branch theorem, we prove that $\Delta i_R = 0$ implies that all the Δi_k are zero. Finally, $\Delta q_C = 0 (\Delta \varphi_L = 0)$ is a consequence of $\Delta u_C = 0 (\Delta i_L = 0)$ because the capacitors (inductors) are strictly locally passive. In short, we have shown that if $\Delta u_R = \Delta i_R = 0$, then $\xi = \bar{\xi}$. This completes the proof that W is a strict Liapunov function.

Then, theorem 5.2.36 guarantees that $\{\xi \,|\, W(\xi) = 0\}$ is an attractive set. Because it only contains point $\bar{\xi}$, all the solutions converge toward $\bar{\xi}$.

5.5.7 Example

Let us consider the circuit with the tunnel diode of section 2.1, but with a capacitor **and** an inductor (fig. 5.66). We easily find the global state equations:

$$L\,(di/dt) = -Ri - u + E \tag{5.175}$$

$$C\,(du/dt) = +i - g\,(u) \tag{5.176}$$

and the hypotheses of theorem 2.4.15 hold. We establish the second hypothesis of subsection 5.2.1 by showing with the methods of section 5.6 that the solutions of the circuit are bounded, and by applying theorem 2.2.14. The dc operating points of the circuit are the same as for the circuits discussed in section 2.1. The tunnel diode is passive relative to any dc operating point located in the part marked with a thick line on the characteristic in figure 3.21. For such a dc operating point, all the hypotheses of theorem 5.5.6 hold. This theorem then guarantees the convergence toward the dc operating point from any initial point (u,i).

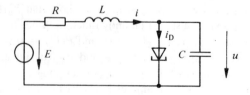

Fig. 5.66

5.5.8 Comment

Counterexample 5.5.3 does not satisfy the hypotheses of the theorem. Indeed, it includes a loop and a cut set without resistor.

Counterexample 5.5.4 also has loops and cut sets without resistors. Nevertheless, we observed that if $C_1L_3 = C_2L_4$, all the solutions converge toward zero. Therefore, the condition on the element connection is too strong. The next theorem gives a less restrictive condition, which is however more difficult to check. A rather complicated topological criterion has been given in [35].

5.5.9 Theorem

Let us assume that a circuit satisfies the same hypotheses as those of theorem 5.5.6, except for the fifth one, which is replaced by the following:

- there is no solution of the circuit such that the incremental energy stored in the capacitors and the inductors relative to $\bar{\xi}$ be constant and positive.

Then, the same conclusions as for theorem 5.5.6 are valid.

Indeed, by the same proof as for subsection 5.5.6, we prove that the incremental power absorbed by the resistors is non-negative. However, we cannot show any more that it is positive, and we must be satisfied with the fact that W simply is a Liapunov function, without being strict. Consequently, for any solution $\xi(t)$, we have

$$dW(\xi(t))/dt \leqslant 0 \qquad (5.177)$$

Then, either $W(\xi(t) \to 0$, and the theorem is proved, or $W(\xi(t)) \to E > 0$.

Let us consider the limit set of $\xi(t)$. Because of (5.177), $\xi(t)$ is bounded. According to subsection 5.3.4, it includes at least one point ξ_∞. However, if ξ_∞ is a limit point, we have $W(\xi_\infty) = E$. Let us consider the solution $\eta(t)$ with $\eta(0) = \xi_\infty$. According to theorem 5.3.8, it must remain in the limit set, and we thus have $W(\eta(t)) \equiv E$. This is in contradiction with the hypothesis. We conclude that $E > 0$ is not possible.

5.5.10 Comment

Theorems 5.5.6 and 5.5.9 do not apply in the presence of several dc operating points. However, it is easy to transform them so that they give some information on the basins of attraction of the various stable dc operating points.

5.5.11 Theorem

Let us consider an autonomous circuit with a dc operating point $\bar{\xi}$. Let W be the total incremental energy relative to $\bar{\xi}$, let Λ be the configuration space, and let the set S_E be defined by

344

$$S_E = \{\xi | W(\xi) < E\} \cap \Lambda \tag{5.178}$$

Let us assume that all the hypotheses of theorem 5.5.6 or theorem 5.5.9 are satisfied inside S_E. So, the capacitors and inductors must be strictly locally passive at any operating point in S_E, and

$$(u_k - \bar{u}_k)(i_k - \bar{i}_k) > 0 \tag{5.179}$$

must hold for any branch k which carries a resistor, if $u_k, i_k (\bar{u}_k, \bar{i}_k)$ are the voltage and the current of branch k at an operating point $\xi \in S_E$ (at $\bar{\xi}$). Outside S_E, no condition is imposed on the elements. Then, we have

$$\xi(t) \to \bar{\xi} \quad \text{for} \quad t \to \infty \quad \text{if} \quad \xi(t_0) \in S_E \tag{5.180}$$

Indeed, the same proof as in subsection 5.5.6 or 5.5.9 shows that W is a strict Liapunov function in $[0,E)$. It follows from theorem 5.2.36 that all the solutions which start in S_E converge toward $\bar{\xi}$.

5.5.12 Example

Let us return to the circuit of figure 5.66. And assume that the values E and R are such that there are three dc operating points, $P_1 = (u_1, i_1), P_2 = (u_2, i_2), P_3 = (u_3, i_3)$ (fig. 5.67). Note that at a dc operating point, the current in the diode, i_D, coincides with i. The incremental energy stored in the capacitor and in the inductor relative to these dc operating points is

$$W_j(u, i) = (C/2)(u - u_j)^2 + (L/2)(i - i_j)^2, \quad j = 1, 2, 3 \tag{5.181}$$

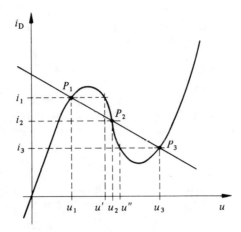

Fig. 5.67

Let $u'(u'')$ be the values of u where the current in the diode is the same as at $P_1(P_3)$. Then, we have

$$(u - u_1)(i_D - i_1) > 0 \quad \text{for} \quad u < u' \text{ and } u \neq u_1 \tag{5.182}$$

and an analogous inequality holds for $u > u''$. Let

$$E_1 = (C/2)(u' - u_1)^2 \tag{5.183}$$

and let

$$S_1 = \{(u, i) | W_1(u, i) \leq E_1\} \tag{5.184}$$

Then, according to (5.181), as soon as $(u,i) \in S_1$, we have

$$|u - u_1| < |u' - u_1| \tag{5.185}$$

and (5.182) holds. The hypotheses of theorem 5.5.11 are satisfied for $E = E_1$, and we conclude that S_1 is a basin of p_1. In an analogous way, we obtain the basin of attraction S_3 of P_3. However, P_2 is unstable. The basins of attraction S_1 and S_3 are ellipses in the plane (u,i) (fig. 5.68).

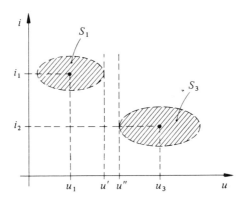

Fig. 5.68

5.5.13 Corollary

Let us consider a circuit composed of capacitors, inductors, resistors and constant sources, such that the condition on the element connection of theorems 5.5.6 or 5.5.9 and the hypotheses of subsection 5.2.1 are satisfied. Let $\bar{\xi}$ be a dc operating point of the circuit such that the capacitors, the inductors and the resistors are strictly locally passive at $\bar{\xi}$. Then, the solution $\xi(t) \equiv \bar{\xi}$ is asymptotically stable.

Indeed, there exists a $\epsilon > 0$ such that at any operating point ξ, with $\| \xi - \bar{\xi} \| < \epsilon$, the capacitors and the inductors are always strictly locally passive and the resistors satisfy (5.179). Let W be the total incremental energy stored in the capacitors and the inductors relative to $\bar{\xi}$, and let

$$E = \min_{\| \xi - \bar{\xi} \| = \epsilon} W(\xi) \tag{5.186}$$

Then, with S_E of (5.178), the hypotheses of theorem 5.5.11 are satisfied. Because the capacitors and the inductors are strictly locally passive at ξ, $E > 0$ and the set S_E is a neighborhood of $\bar{\xi}$ in Λ. According to theorem 5.5.11, the solutions that start from a point of S_E converge toward $\bar{\xi}$, which shows that the solution $\xi(t) \equiv \bar{\xi}$ is asymptotically stable.

5.6 ATTRACTIVE DOMAINS

5.6.1 Introduction

Let us assume that a circuit is composed of passive capacitors and inductors and strictly eventually passive resistors. Let W be the total stored energy. If $\xi(t)$ is a solution of the circuit, (5.152) holds.

Thanks to eventual passivity, the right-hand side of (5.152) is negative outside a bounded domain. If E is the maximum of W on this bounded domain, W is a strict Liapunov function in the interval $[E, \infty)$. Consequently, the set

$$D_E = \{ \xi \,|\, W(\xi) \leqslant E \} \cap \Lambda \tag{5.187}$$

is attractive, where Λ is the configuration space. This leads to the following conjecture. However, it turns out to be wrong, because the reasoning which suggested it is not complete.

5.6.2 Conjecture

Consider a circuit composed of

- charge-controlled passive capacitors;
- flux-controlled passive inductors;
- strictly eventually passive resistors.

Let us assume that there is no capacitor loop or inductor cut set. Then, there exists a bounded attractive set of the form (5.187).

5.6.3 Definition

Let us consider an autonomous circuit. A bounded attractive set D_E of the form (5.187), where Λ is the configuration space and $W: \Lambda \to \mathbb{R}$ is the stored energy, is called an *attractive domain*. If the set is globally attractive, it is a *globally attractive domain*.

5.6.4 Counterexamples

The counterexamples 5.5.3 and 5.5.4 for conjecture 5.5.2 are also counterexamples for conjecture 5.6.2. Indeed, these circuits have solutions with a stored energy that is arbitrarily large and constant. To avoid this phenomenon, one must make the same hypotheses as in section 5.5.

5.6.5 Counterexample

Consider a circuit composed of a nonlinear capacitor, defined by the constitutive relation $u = h(q)$, a linear resistor of value $R > 0$ and a voltage source of value E (fig. 5.69). The characteristic of the capacitor is represented in figure 5.70. The energy stored in the capacitor is represented in figure 5.71. Because it is non-negative, it indeed is a passive capacitor. The series connection of the source and the resistor possesses the characteristic of figure 5.72. We see that it is strictly eventually passive. Let us assume that the horizontal asymptote of the characteristic for $q \to \infty$ is located at a voltage $u_\infty < E$. The state equation of this circuit is

$$dq/dt = (E - h(q))/R \tag{5.188}$$

and because $h(q) < u_\infty$, we have

$$dq/dt > (E - u_\infty)/R > 0 \tag{5.189}$$

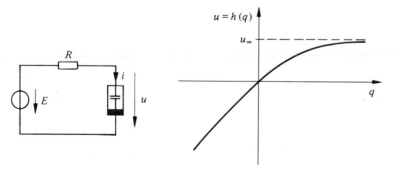

Fig. 5.69   **Fig. 5.70**

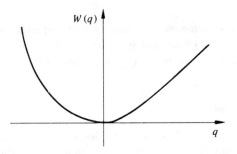

Fig. 5.71

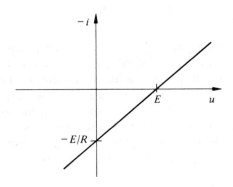

Fig. 5.72

It follows that for all the solutions we have $q(t) \to \infty$, which contradicts conjecture 5.6.2. Note, however, that voltage u and current i remain bounded.

To avoid this pitfall, we should require that $|q| \to \infty$ implies $|u| \to \infty$.

5.6.6 Comment

At first sight, it would seem that the circuit of figure 5.69 disproves theorem 5.5.6. Indeed, the capacitor is strictly locally passive and the resistor composed of the linear resistor and the voltage source is strictly passive relative to any point of its characteristic (fig. 5.72). Thus, we would expect all the solutions to tend toward a unique dc operating point. By definition, $i = 0$ at a dc operating point, hence $u = E$. However, there is no point on the capacitor characteristic with $u = E$ and, consequently, there is no dc operating point. Thus, this example does not contradict theorem 5.5.6, in which we assumed from the outset the existence of a dc operating point.

5.6.7 Theorem

Consider an autonomous circuit composed of

- charge-controlled passive capacitors, with $|u| \to \infty$ if $|q| \to \infty$;
- flux-controlled passive inductors, with $|i| \to \infty$ if $|\varphi| \to \infty$;
- strictly eventually passive resistors, with $u \cdot i \to \infty$, if $|u| \to \infty$ or $|i| \to \infty$.

Concerning the connections, we assume that

- there are no loops nor cut sets without resistors.

In addition, we assume that

- the hypotheses of subsection 5.2.1 hold.

Then, there exists a globally attractive domain.

5.6.8 Comments

The hypothesis that the points on the characteristic of a resistor must satisfy $u \cdot i \to \infty$ if $|u| \to \infty$ or $|i| \to \infty$ is not very restrictive. Thereby, we exclude for example a resistor with a constitutive relation $i = g(u)$ such that $u \cdot g(u) \to 0$ when $u \to \infty$ (fig. 5.73). This hypothesis is necessary to ensure that the set of the operating points where the resistors supply a total power that is positive to the capacitors and the inductors, is bounded.

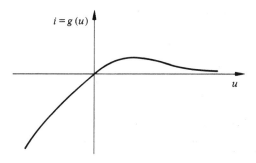

Fig. 5.73

Indeed, let us replace in the circuit of figure 5.66 the tunnel diode by a resistor with the characteristic of figure 5.73. At the operating points (u,i), where the voltage u of the nonlinear resistor is sufficiently large and positive, and when the current i in the inductor satisfies $0 < i < E/R$, the power produced by the series connection of the voltage source and the linear resistor is larger than the power absorbed by the nonlinear resistor.

The fact that resistors can supply a total power that is positive even at arbitrarily distant operating points does not necessarily imply that there is no attractive domain. Actually, it is the method used to establish attractive domains, as outlined in subsection 5.6.1, which fails.

The fourth hypothesis of theorem 5.6.7 is similar to the fifth hypothesis of theorem 5.5.6. As in subsection 5.5, it can be replaced by the corresponding hypothesis of theorem 5.5.9.

5.6.9 Proof

We start with the proof of theorem 5.6.7 by showing that the set S_R of the $\xi \in \Lambda$ such that the power P_R dissipated by the resistors satisfies

$$P_R(\xi) \leqslant 0 \tag{5.190}$$

is bounded. According to property 3.2.4, the power that can be supplied by a resistor is bounded. Let P_0 the maximum of these bounds. Then, if branch k carries a resistor, it satisfies

$$u_k \cdot i_k \geqslant -P_0 \tag{5.191}$$

Consequently, for any operating point ξ, which satisfies (5.190), and any branch k, which carries a resistor, we can write that

$$u_k \cdot i_k = P_R(\xi) - \sum_{\substack{j \neq k \\ \text{Resistors}}} u_j i_j \tag{5.192}$$

$$\leqslant (N_R - 1) P_0 \tag{5.193}$$

where N_R is the number of resistors. Because $u \cdot i \to \infty$ if $|u_k| \to \infty$ or $|i_k| \to \infty$, the set of the (u_k, i_k) which satisfies (5.193) and the constitutive relation of the resistor on branch k is bounded.

Since the variables u_R, i_R are bounded, we prove by the colored-branch theorem that the other variables are bounded. First, we color the capacitors and the inductors green, while orienting them in the same way as their voltage, and we color the resistors red. Because there is no cut set without resistor, each capacitor and each inductor are part of a uniform loop. According to the Kirchhoff equation for the loop, we have

$$\sum_{\substack{\text{loop} \\ L, C}} u_k = - \sum_{\text{loop } R} u_k \tag{5.194}$$

where all the terms of the left-hand sum have the same sign. Because the right-hand sum is bounded by $N_R U$, where U is the bound for the resistor voltages, each capacitor and inductor voltage also is bounded by $N_R U$. By dual application of the colored-branch theorem, we find bounds for the capacitor and inductor currents. Finally, since for each capacitor, $|q_k| \to \infty$ implies $|u_k| \to \infty$, q_C is bounded. In an analogous way, φ_L is bounded. Consequently, the set S_R of the $\xi \in \Lambda$, such that (5.190) holds, is bounded.

Since the stored energy W is a continuous function, it has a maximum E on S_R. Then, for any $\xi \in \Lambda$ with $W(\xi) > E$, we have $P_R(\xi) > 0$, and consequently, according to (5.152),

$$dW(\xi(t))/dt < 0 \qquad (5.195)$$

for any solution $\xi(t)$ and any instant t such that $W(\xi(t)) > E$.

To complete the proof that W is a strict Liapunov function in the interval $[E,\infty)$, we must still show that the set D_E', defined by (5.187), is bounded for any E'. The proof of this property is almost identical to that of subsection 5.5.6, and we will not repeat it here.

Finally, theorem 5.2.36 guarantees that D_E is an attractive set.

5.6.10 Comments

The theorem would be wrong if we admitted independent sources. The circuit of figure 5.74, which includes an exponential diode, can be used as a counterexample. Its state equation is

$$dq/dt = I + I_s(\exp(-q/(CnU_T)) - 1) \qquad (5.196)$$

and, therefore,

$$dq/dt > I - I_s \qquad (5.197)$$

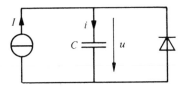

Fig. 5.74

Consequently, if $I > I_s$, we have $q(t) \to \infty$ when $t \to \infty$ for all the solutions.

However, if we include a passive linear resistor in series (parallel) with each voltage (current) source, we obtain eventually passive resistors. If the other hypotheses of theorem 5.6.7 are satisfied, the existence of an attractive domain is ensured. For instance, by adding to the circuit of figure 5.74 a resistor of value R in parallel with the current source, we obtain the circuit of figure 5.75. Its solutions satisfy the state equation:

$$\mathrm{d}q/\mathrm{d}t = I - q/(CR) + I_s(\exp(-q/(CnU_T)) - 1) \tag{5.198}$$

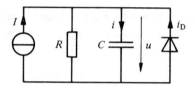

Fig. 5.75

The right-hand side of (5.198) is a strictly decreasing function of q. It passes through zero at a point $q_0 > 0$ (fig. 5.76). This is the dc operating point. Since the current i_D in the diode is negative when $q > 0$, we have $q_0 > ICR$. We conclude that the part of the configuration space which is characterized by $|q| \leq ICR$ is an attractive domain. We note that this domain becomes increasingly larger when the added resistor is increased, and becomes infinite when the resistor is removed.

We can even include bounded time-dependent sources, if they are provided with an internal resistor. Indeed, if a time-dependent voltage source with a signal $e(t)$, which satisfies

$$|e(t)| \leq A \tag{5.199}$$

is connected in series with a resistor of value R, then the points (u,i) admitted by the composite one-port are located in the cross-hatched area of figure 5.77. We see that

- outside a bounded region of plane (u,i), we have $u \cdot i > 0$;
- the power that can be produced by the one-port is bounded.

Consequently, the proof of theorem 5.6.7 remains valid in the presence of these time-dependent resistors.

Calculation of an attractive domain permits us to establish bounds on the solutions without knowing their nature. In practice, this is an important result because excessively high amplitudes may damage the circuit. However, the bounds obtained this way most of the time are too pessimistic. It looks as if the method is too simple to be accurate. Other Liapunov functions, apart from stored energy, might possibly improve the situation.

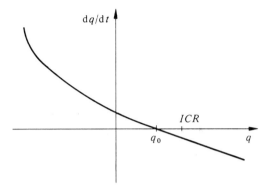

Fig. 5.76

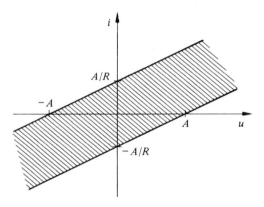

Fig. 5.77

5.6.11 Example

Let us return to the ferroresonant circuit of figure 5.52, with the characteristic of figure 1.37 for the inductor. We have seen in subsection 5.4.4 that q_1 and φ_2 are global state variables. Because the capacitor and the inductor possess bijective characteristics, u_1 and i_2 also are global state variables. For this example, we use u_1, i_2, and u_3 as the global parameters of the generalized configuration space. The variables i_1 and u_2 are expressed as functions of these parameters in (2.135) and (2.136).

The total power absorbed by the source and the two resistors is

$$P_R = u_3 i_3 + u_4 i_4 + u_5 i_5 \tag{5.200}$$

$$= -u_3 i_1 + R_4 i_1^2 + u_2^2/R_5 \tag{5.201}$$

By using (2.135) and (2.136), we obtain

$$
\begin{aligned}
P_R(u_1, i_2, u_3) &= -u_3(-u_1 + R_5 i_2 + u_3)/(R_4 + R_5) \\
&\quad + (-u_1 + R_5 i_2 + u_3)^2 R_4/(R_4 + R_5)^2 \tag{5.202} \\
&\quad + (-u_1 - R_4 i_2 + u_3)^2 R_5/(R_4 + R_5)^2 \\
&= u_1^2/(R_4 + R_5) + i_2^2 R_4 R_5/(R_4 + R_5) \\
&\quad - (u_1 + R_5 i_2) u_3/(R_4 + R_5) \tag{5.203} \\
&= (u_1 - u_3/2)^2/(R_4 + R_5) + (R_4 i_2 - u_3/2)^2 R_5/(R_4^2 + R_4 R_5) \\
&\quad - u_3^2/(4R_4) \tag{5.204}
\end{aligned}
$$

The set S_R of the generalized operating points, such that the power dissipated by the resistors and the source is negative or zero, is a cone in the space (u_1, i_2, u_3) (fig. 5.78). The sections of this cone, where u_3 is constant, are ellipses.

Let us assume that the source amplitude is limited to E:

$$|e_3(t)| \leqslant E \tag{5.205}$$

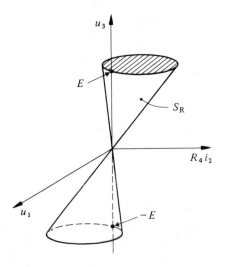

Fig. 5.78

Then, the solutions of the circuit are limited to the portion $|u_3| \leq E$ of Λ. In figure 5.78, we show the piece of cone inside of this portion.

Because the energy stored in the capacitor and the inductor does not depend on u_3, this piece of S_R must be projected on the plane (u_1, i_2). In other words, we must determine the set $\tilde{S}_R$ of the points (u_1, i_2) such that there exists a u_3 with $|u_3| \leq E$ and $P_R(u_1, i_2, u_3) \leq 0$. In figure 5.78, we see that $\tilde{S}_R$ is the union of the projections of the limit ellipses $u_3 = \pm E$ of the cone.

This fact can be easily proved by (5.203) as well. If $u_1 + R_5 i_2 \geq 0$ and $u_3 \leq E$, we have

$$P_R(u_1, i_2, u_3) \geq P_R(u_1, i_2, E) \tag{5.206}$$

If there is a $u_3 \leq E$ such that $P_R(u_1, i_2, u_3) \leq 0$, then according to (5.206), $P_R(u_1, i_2, E) \leq 0$. In other words, if $(u_1, i_2) \epsilon \tilde{S}_R$ and $u_1 + R_5 i_2 \geq 0$, then

$$P_R(u_1, i_2, E) \leq 0 \tag{5.207}$$

However, the points (u_1, i_2) which satisfy (5.207), and $u_1 + R_5 i_2 \geq 0$ are evidently part of $\tilde{S}_R$. By analogous reasoning, we find that the points $\tilde{S}_R$ in the half-plane $u_1 + R_5 i_2 \leq 0$ are characterized by

$$\cdot P_R(u_1, i_2, -E) \leq 0 \tag{5.208}$$

In figure 5.79, we have represented $\tilde{S}_R$. Now, we need to determine the maximum of W on $\tilde{S}_R$. By combining (1.29) and (5.118), we obtain the energy stored in the inductor as a function of i_2:

$$W_2(i_2) = \begin{cases} L_0 i_2^2/2 & \text{for} \quad |i_2| \leq i_0 \\ L_1 i_2^2/2 + (L_0 - L_1) i_0^2/2 & \text{for} \quad |i_2| > i_0 \end{cases} \tag{5.209}$$

The energy stored in the capacitor is

$$W_1(u_1) = C_1 u_1^2/2 \tag{5.210}$$

and the total stored energy is

$$W(u_1, i_2) = W_1(u_1) + W_2(i_2) \tag{5.211}$$

In figure 5.80, we have represented three level lines of W.

Since there is no local maximum of $W(u_1, i_2)$, the maximum of W on $\tilde{S}_R$ is taken on the boundary of $\tilde{S}_R$, where the inequality in (5.207) holds. Thus, searching for this

356

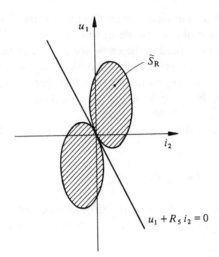

Fig. 5.79

maximum amounts to solving a problem of extremum with constraint, with the help of a Lagrange multiplier. Therefore, we introduce the two functions:

$$F_{\pm}(u_1, i_2, \lambda) = W(u_1, i_2) - \lambda P_R(u_1, i_2, \pm E) \tag{5.212}$$

in which we search for the local maximum. Be derivation with respect to u_1 and i_2, we obtain

$$0 = (\partial F_{\pm}/\partial u_1) = Cu_1 - 2\lambda(u_1 \mp E/2)/(R_4 + R_5) \tag{5.213}$$

$$0 = (\partial F_{\pm}/\partial i_2) = Li_2 - 2\lambda(R_4 i_2 \mp E/2)/(R_4 + R_5) \tag{5.214}$$

where the sign greater than (less than) is taken if $u_1 + R_5 i_2 \geq 0$ (≤ 0), and where $L = L_0$ ($L = L_1$) if $|i_2| \leq i_0$ ($|i_2| \geq i_0$). By eliminating λ, we obtain

$$u_1 = (Li_2 E)/(R_5 CE \mp 2(R_4 R_5 C - L)i_2) \tag{5.215}$$

The intersection of the curve (5.215) with the boundary of $\tilde{S}_R$ gives the location of the maximum E of W on $\tilde{S}_R$. The attractive domain D_E that we are searching for is limited by the level line $W = E$, which passes through this intersection point (fig. 5.81). If we attempt to calculate this intersection point analytically, we obtain an equation of degree 4. At that stage, we must resort to a numerical solution.

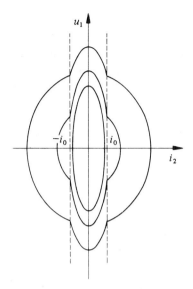

Fig. 5.80

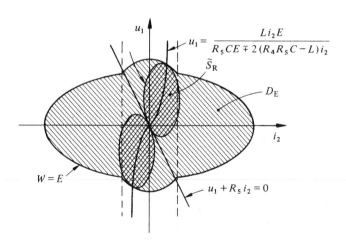

$$u_1 = \frac{Li_2 E}{R_5 CE \mp 2(R_4 R_5 C - L)i_2}$$

$\tilde{S}_R$

D_E

$W = E$

$u_1 + R_5 i_2 = 0$

Fig. 5.81

358

5.6.12 Numerical Example

Let us take again the parameters of the ferroresonant circuit of subsection 5.1.10, with $E = 160$ V. In figure 5.82, we have represented the boundaries of the regions $\tilde{S}_R$ and D_E, as in figure 5.81. At the scale of this drawing, the linear domain $|i_2| < i_0$ of the inductor practically coalesces with the vertical axis. Furthermore, the straight line $u_1 + R_5 i_2 = 0$ is almost vertical.

In figure 5.83, we have represented the globally attractive domain we obtained, as well as a solution which tends toward the smallest periodic solution of figure 5.7.

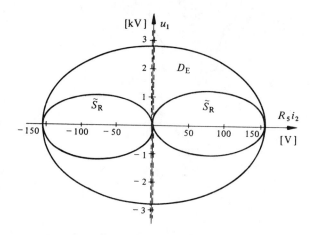

Fig. 5.82

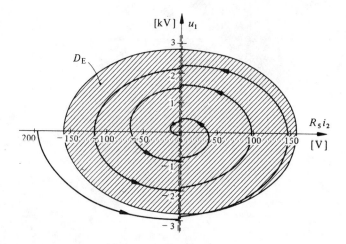

Fig. 5.83

The scale of figure 5.83 is much smaller than that of figure 5.7. In figure 5.84, we have added to the attractive domain a solution which converges toward the largest periodic solution of figure 5.7. Because the numerical experience has shown that any solution tends toward either one of these two periodic solutions, the orbit of the largest periodic solution delimits an attractive set. By comparing this attractive set with the one obtained by the calculation of subsection 5.6.11, we observe that the calculated version is about a factor of 5 too large.

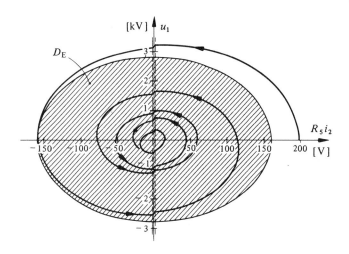

Fig. 5.84

5.7 CONVERGENCE TOWARD A TIME-DEPENDENT STEADY STATE

5.7.1 Introduction

In this section, we address the problem of whether a circuit with time-dependent sources has a unique asymptotic behavior. The Liapunov function method, as presented in section 5.2, does not apply because it involves only one solution, while it is necessary to prove the convergence of the difference of two solutions toward zero. There is a way to generalize the Liapunov function concept for this case [36].

5.7.2 Definition

Let $\Lambda \subset \mathbb{R}^{2b+N_C+N_L}$ be the generalized configuration space of a circuit. A function $W: \mathbb{R}^{2b+N_C+N_L} \times \mathbb{R}^{2b+N_C+N_L} \to \mathbb{R}$ is a *strict incremental Liapunov function* if it satisfies the following conditions:

- Function W is continuous and differentiable.
- Function W is non-negative on $\Lambda \times \Lambda$. More precisely, we have

$$W(\xi_1, \xi_2) \geqslant 0 \quad \text{if} \quad \xi_1, \xi_2 \in \Lambda \tag{5.216}$$

and

$$W(\xi_1, \xi_2) = 0 \quad \text{only if} \quad \xi_1 = \xi_2 \tag{5.217}$$

- There exists a continuous function $V: \Lambda \times \Lambda \to \mathbb{R}$ such that, for any pair of solutions $\xi_1(t)$, $\xi_2(t)$ of the circuit, we have

$$dW(\xi_1(t), \xi_2(t))/dt = V(\xi_1(t), \xi_2(t)) \tag{5.218}$$

- Function V is nonpositive on $\Lambda \times \Lambda$. More precisely, we have

$$V(\xi_1, \xi_2) \leqslant 0 \quad \text{if} \quad \xi_1, \xi_2 \in \Lambda \tag{5.219}$$

and

$$V(\xi_1, \xi_2) = 0 \quad \text{only if} \quad \xi_1 = \xi_2 \tag{5.220}$$

5.7.3 Example

In the circuit of figure 5.75, let us replace the constant current source by a time-dependent current source (fig. 5.85). As seen in subsection 5.6.10, the configuration space is globally parameterized by the charge q of the capacitor. Consequently, an incremental Liapunov function is defined as a function of two charges. Consider

$$W(q_1, q_2) = (q_1 - q_2)^2/(2C) \tag{5.221}$$

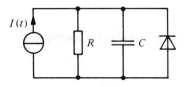

Fig. 5.85

Then, obviously, the first two conditions for an incremental Liapunov function are satisfied. In figure 5.86, we have represented three level curves of W in the plane (q_1, q_2). To check the third condition, we calculate the time derivative of W with the help of state equation (5.198):

$$dW(q_1(t), q_2(t))/dt = -(q_1 - q_2)[(q_1 - q_2)/C^2R - (g(q_1) - g(q_2)/C]$$

where $\tag{5.222}$

$$g(q) = I_s (\exp(-q/CnU_T) - 1) \tag{5.223}$$

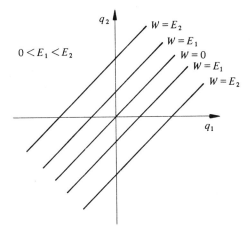

Fig. 5.86

The right-hand side of (5.222) is a function $V(q_1, q_2)$, as required by the third condition. It satisfies the last condition, because function g is a decreasing function. Three level curves of function V are represented in figure 5.87.

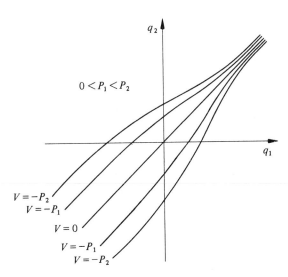

Fig. 5.87

362

If $q_1(t)$ and $q_2(t)$ are two solutions of (5.198), we thus have $dW(q_1(t), q_2(t))/dt$ < 0 as long as $q_1(t) \neq q_2(t)$. We conclude that the two solutions converge toward one another. The detailed arguments underlying this conclusion are given in subsection 5.7.4. In fact, it is necessary to assume that the two solutions are bounded. This is the case if $I(t)$ is bounded, as we have seen in subsection 5.6.10. It is surprising that no other condition has to be imposed to the signal $I(t)$ to establish a unique asymptotic behavior.

5.7.4 Theorem: Unique Asymptotic Behavior

Let us assume that the solutions of a circuit are bounded and the circuit has a strict incremental Liapunov function. Then the asymptotic behavior of the solutions is unique, i.e., if $\xi_1(t)$ and $\xi_2(t)$ are two solutions defined in the time interval $[t_0, \infty)$, then,

$$\lim_{t \to \infty} \|\xi_1(t) - \xi_2(t))\| = 0 \tag{5.224}$$

The proof of this theorem is analogous to that of theorem 5.2.36. Let $\xi_1(t)$ and $\xi_2(t)$ be two solutions of the circuit defined for $t_0 \leq t < \infty$. Because they are bounded, there exists a constant $K > 0$ such that $\|\xi_1\| \leq K$ and $\|\xi_2(t)\| \leq K$ for $t \geq t_0$. Consider the set B defined by

$$B = \{(\xi_1, \xi_2) | \xi_1 \in \Lambda, \, \xi_2 \in \Lambda, \, \|\xi_1\| \leq K \text{ and } \|\xi_2\| \leq K\} \tag{5.225}$$

Then, B is compact, and the solutions $\xi_1(t)$, $\xi_2(t)$ remain inside B. Because $K > 0$ is bounded, the incremental Liapunov function W is bounded on B. We designate this bound by E.

Let us consider the pairs of operating points in B which are distant from one another by at least $\epsilon > 0$. We thus introduce the set

$$R = \{(\xi_1, \xi_2) \in B | \, \|\xi_1 - \xi_2\| \geq \epsilon\} \tag{5.226}$$

Then, R also is compact. On R, function V takes negative values. Because R is compact, V has a maximum in R. This maximum is negative, and we designate it by $-P_0$. According to (5.218), we have

$$dW(\xi_1(t), \xi_2(t))/dt \leq -P_0 \tag{5.227}$$

Because W can only take values between 0 and E in R, inequality (5.227) means that the pair of solutions cannot remain longer than time E/P_0 in R. In other words, after a finite time, we have

$$||\xi_1(t) - \xi_2(t)|| \leqslant \epsilon \qquad (5.228)$$

Because ϵ can be arbitrarily small, this implies (5.224).

5.7.5 Example

The circuit of figure 5.88 is the same as that of figure 5.16, except that it includes a linear capacitor instead of a nonlinear capacitor. It is, in fact, the same circuit as in figure 5.66, except that it includes a sinusoidal source instead of a constant source and a strictly locally passive resistor instead of the tunnel diode.

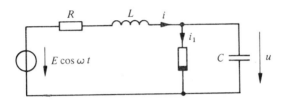

Fig. 5.88

Let us consider two solutions $\xi_1(t)$ and $\xi_2(t)$ of this circuit. The incremental stored energy with respect to these two solutions is

$$W(\Delta u, \Delta i) = C(\Delta u)^2/2 + L(\Delta i)^2/2 \qquad (5.229)$$

According to (5.133), we find

$$dW(\Delta u(t), \Delta i(t))/dt = -R(\Delta i(t))^2 - \Delta u(t)\Delta i_1(t) \qquad (5.230)$$

Because the nonlinear resistor is assumed to be strictly locally passive, the right-hand side is negative, and can only be zero if $\Delta u = \Delta i = 0$. Consequently, W is a strict incremental Liapunov function. Nonetheless, taking into account comments 5.6.10, theorem 5.6.7 guarantees that the solutions of the circuit are eventually uniformly bounded. Hence, according to theorem 5.7.4, the asymptotic behavior of the circuit is unique.

Remember that the circuit of figure 5.16 exhibits a lot of different asymptotic behaviors, multiple periodic solutions, subharmonic solutions and even some chaotic solutions. Therefore, for these phenomena to take place, it is indispensable for the junction capacitor of the diode to be nonlinear.

5.7.6   Theorem: Unique Asymptotic Behavior with Linear Reactances

Consider a circuit composed of

- positive linear capacitors;
- positive linear inductors;
- strictly locally passive, strictly eventually passive resistors;
- bounded time dependent and constant voltage (current) sources, with a positive linear resistor in series (parallel).

Concerning the connections, let us assume that

- there is no loop or cut set that is exclusively composed of capacitors and inductors.

In addition, we assume that

- the hypotheses of subsection 5.2.1 hold.

Then, the circuit possesses a unique asymptotic behavior.

For the proof, we first note that, due to theorem 5.6.7 and taking into account comment 5.6.10, the solutions are eventually uniformly bounded.

Then, we prove that the incremental energy stored in the capacitors and the inductors:

$$W(\Delta q_C, \Delta \varphi_L) = \sum_{k=1}^{N_C} (\Delta q_k)^2/(2C_k) + \sum_{k=N_C+1}^{N_C+N_L} (\Delta \varphi_k)^2/(2L_k) \qquad (5.231)$$

is a strict incremental Liapunov function. The first and the second conditions of definition 5.7.2 obviously are satisfied. The last two conditions are proved starting from equation (5.153):

$$dW/dt = -\sum \Delta u_k \cdot \Delta i_k \qquad (5.232)$$

where the sum involves the resistors and the sources. Because all the resistors, including the sources, are locally passive, the right-hand side of (5.232) is nonpositive. We still have to show that it can only be zero if $\Delta u = \Delta i = \Delta q_C = \Delta \varphi_L = 0$.

Let us thus assume that the right-hand side of (5.232) is zero. Because all the terms of the sum are nonpositive, we have $\Delta u_k \cdot \Delta i_k = 0$ if branch k carries a resistor or a source. Concerning the strictly locally passive resistors and composite source branches, it follows that $\Delta u_k = \Delta i_k = 0$. Starting from this fact, we prove by two applications of the colored branch theorem that all the increments are zero, in a way that is analogous to the proof of theorem 5.5.6.

Finally, theorem 5.7.4 guarantees that the circuit possesses a unique asymptotic behavior.

5.7.7 Comment

The proof of theorem 5.7.6 cannot be generalized to nonlinear capacitors and inductors, because there is no incremental stored energy in this case, as was seen in subsection 5.4.13.

However, it is contrary to intuition that the asymptotic behavior of a circuit has to abruptly change when a capacitor or an inductor becomes very slightly nonlinear. Indeed, the incremental Liapunov method can still be applied. However, relation (5.153) cannot be used any longer.

The idea is as follows [37]. We define function W by (5.231). If a capacitor (inductor) is linear, $C_k(L_k)$ designates its value. If it is nonlinear, $C_k(L_k)$ is a positive constant which we shall choose in an appropriate manner. Then,

$$
dW/dt = \sum_{k=1}^{N_C} \Delta q_k \Delta i_k / C_k + \sum_{k=N_C+1}^{N_C+N_L} \Delta \varphi_k \Delta u_k / L_k
\tag{5.233}
$$

We write

$$
dW/dt = -P + Q
\tag{5.234}
$$

where P is the incremental power absorbed by the resistors. In the case of linear capacitors and inductors, W is the incremental stored energy, and there isn't any corrective term Q. In the presence of nonlinear capacitors or inductors, we try to establish an inequality of the form

$$
|Q| \leqslant \lambda P
\tag{5.235}
$$

If (5.235) holds for $\lambda < 1$, W is a strict incremental Liapunov function and the circuit possesses a unique asymptotic behavior.

Strictly locally passive elements are not enough to guarantee a unique asymptotic behavior. Indeed, the ferroresonant circuit of subsection 5.1.10, as well as the circuit with a junction diode of subsection 5.1.12, only include strictly locally passive resistors, capacitors and inductors. Nevertheless, their asymptotic behavior often is not unique. Therefore we have to introduce stronger restrictions on the element characteristics.

5.7.8 Definitions

A voltage-controlled *resistor* is *uniformly locally passive* if two arbitrary points $(u_1, i_1) \neq (u_2, i_2)$ of its characteristic satisfy

$$
R_m \leqslant (u_1 - u_2)/(i_1 - i_2) \leqslant R_M
\tag{5.236}
$$

where R_m, R_M are two finite positive constants with the dimension of a resistance.

A charge-controlled *capacitor* is *uniformly locally passive* if two arbitrary points $(u_1, q_1) \neq (u_2, q_2)$ of its characteristic satisfy

$$C_m \leqslant (q_1 - q_2)/(u_1 - u_2) \leqslant C_M \qquad (5.237)$$

where C_m, C_M are two finite positive constants with the dimension of a capacitance.

A flux controlled *inductor* is *uniformly locally passive* if two arbitrary points $(\varphi_1, i_1) \neq (\varphi_2, i_2)$ of its characteristic satisfy

$$L_m \leqslant (\varphi_1 - \varphi_2)/(i_1 - i_2) \leqslant L_M \qquad (5.238)$$

where L_m, L_M are two finite positive constants with the dimension of an inductance.

5.7.9 Comments

If (u_1, i_1) is a point located on the characteristic of a uniformly locally passive resistor, then any other point (u_2, i_2) of its characteristic is located in the cross-hatched region of figure 5.89. Similarly, we need only change the label of the axes in figure 5.89 to obtain the admissible domain for the characteristic of a uniformly locally passive capacitor or inductor.

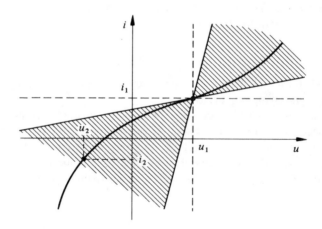

Fig. 5.89

We have limited the concept of uniform local passivity to resistors (capacitors, inductors) that are voltage (charge, flux) controlled, to avoid complications due to the characteristics which are of the same kind as that of figure 3.24. In fact, in the way this concept has been introduced, this is a restriction of the strict relative passivity rather than of the strict local passivity. However, in the case of resistors (capacitors, inductors) that are voltage (charge, flux) controlled, the two concepts of strict passivity for incremental variables are identical (subsec. 3.2.15). In that case, in fact, the strict local passivity, and, therefore, the uniform local passivity, imply that the resistors (capacitors, inductors) are also current (voltage, current) controlled.

If a uniformly locally passive resistor is defined by the constitutive relation $i = g(u)$ and if g is differentiable at u_0, then

$$1/R_M \leqslant (dg/du)(u_0) \leqslant 1/R_m \tag{5.239}$$

If the resistor is defined by $u = h(i)$, then

$$R_m \leqslant (dh/di)(u_0) \leqslant R_M \tag{5.240}$$

Analogous inequalities hold in the case of uniformly locally passive capacitors and inductors with differentiable characteristics.

If a uniformly locally passive resistor is defined by the constitutive relation $i = g(u)$, then function g and its inverse are Lipschitz on the entire space $\mathbb{R}$. On the other hand, if g and g^{-1} are Lipschitz on any $\mathbb{R}$, nothing guarantees that $\Delta u/\Delta i$ is positive. In fact, definition 2.2.2 only involves $|\Delta u/\Delta i|$.

5.7.10 Properties

Consider a voltage controlled resistor. Let (u_1, i_1), (u_2, i_2) be two points of its characteristic and $\Delta u = u_1 - u_2$, $\Delta i = i_1 - i_2$. The two following conditions are equivalent to the condition for uniform local passivity:

• There exist two finite positive constants $R_\mu > R_\Delta$, such that

$$|\Delta u/\Delta i - R_\mu| \leqslant R_\Delta \tag{5.241}$$

• There exist two finite positive constants R_n and $0 < \gamma \leqslant 1/2$, such that

$$\Delta u \cdot \Delta i \geqslant \gamma((u^2/R_n) + R_n i^2) \tag{5.242}$$

Analogous conditions characterize uniformly locally passive capacitors and inductors. The constants of (5.236), and those of (5.241) and (5.242) are linked by

$$R_\mu = (R_M + R_m)/2, \quad R_\Delta = (R_M - R_m)/2 \tag{5.243}$$

$$R_n = \sqrt{R_M R_m}, \quad \gamma = (\sqrt{R_M/R_m} + \sqrt{R_m/R_M})^{-1} \tag{5.244}$$

Indeed, inequality (5.241) is inferred from (5.236) by subtracting R_μ, and (5.242) is obtained by squaring (5.241).

5.7.11 Comments

If we rewrite (5.241) as

$$|(1/R_\mu) \, \Delta u/\Delta i - 1| \leqslant R_\Delta/R_\mu \tag{5.245}$$

we then see that $R_\Delta/R_\mu = (R_M - R_m)/(R_M + R_m)$ is a measure for the resistor nonlinearity. This ratio can vary between zero, in the case of a linear resistor, and one, in the limit case of a locally passive resistor having a characteristic which exhibits somewhere a slope that is zero or infinite, or a slope which tends toward zero or infinity at infinity. Hereafter, we shall use the corresponding measure for the nonlinearity of capacitors and inductors.

Inequality (5.242) expresses the fact that the incremental power absorbed by the resistor is at least $\gamma \, \| \, (\Delta u, \, \Delta i) \, \| ^ 2$, if the norm in the plane $(\Delta u, \, \Delta i)$ is defined with the normalization resistor R_n (subsec. 2.3.16). Hence, γ stands for a measure of local passivity. In the limit case, where $R_m = 0$ or $R_M = \infty$, we have $\gamma = 0$. The uniformly locally passive resistors have γ values between zero and one-half. This last case occurs if the resistor is linear.

5.7.12 Theorem: Unique Asymptotic Behavior with Nonlinear Reactances

Consider a circuit composed of uniformly locally passive resistors, inductors and capacitors, as well as bounded autonomous and time dependent voltage (current) sources, with a positive linear resistor in series (parallel). Let us assume that there exist constants $0 \leqslant \alpha_k < 1$, $k = 1, \ldots, N_C + N_L$ and $0 < \gamma \leqslant 12$, such that the following inequalities are satisfied:

- If branch k carries a capacitor, we have

$$C_{mk} \leqslant \Delta q_k/\Delta u_k \leqslant C_{Mk} \tag{5.246}$$

and

$$(C_{Mk} - C_{mk})/(C_{Mk} + C_{mk}) \leqslant \alpha \tag{5.247}$$

- If branch k carries an inductor, we have

$$L_{mk} \leqslant \Delta \varphi_k/\Delta i_k \leqslant C_{Mk} \tag{5.248}$$

and

$$(L_{Mk} - L_{mk})/(L_{Mk} + L_{mk}) \leqslant \alpha \tag{5.249}$$

- If branch k carries a resistor which is not a source, we have

$$R_{mk} \leqslant \Delta u_k/\Delta i_k \leqslant R_{Mk} \tag{5.250}$$

and

$$(\sqrt{R_{Mk}/R_{mk}} + \sqrt{R_{mk}/R_{Mk}})^{-1} \geqslant \gamma \qquad (5.251)$$

Concerning the connections, we assume that

- there are no loops or cut sets without resistors.

In addition, we assume that

- the hypotheses of subsection 5.2.1 hold.

Then, if

$$\alpha \cdot \sqrt{RG} < \gamma \qquad (5.252)$$

where

$$R = \sum_{\text{resistors}} \sqrt{R_{Mk}R_{mk}} \qquad (5.253)$$

$$G = \sum_{\text{resistors}} 1/\sqrt{R_{Mk}R_{mk}} \qquad (5.254)$$

it follows that the circuit possesses a unique asymptotic behavior.

For the proof, we first note that, thanks to theorem 5.6.7 and taking into account comment 5.6.8, the solutions are eventually uniformly bounded.

Then, we introduce the function:

$$W(\xi_1, \xi_2) = \sum_{k=1}^{N_C} (\Delta q_k)^2/(2C_{\mu k}) + \sum_{k=N_C+1}^{N_C+N_L} (\Delta \varphi_k)^2/(2L_{\mu k}) \qquad (5.255)$$

where $C_{\mu k} = (C_{Mk} + C_{mk})/2$ and $L_{\mu k} = (L_{Mk} + L_{mk})/2$. Let $\xi_1(t)$ and $\xi_2(t)$ be two solutions of the circuit. So,

$$dW(\xi_1(t), \xi_2(t))/dt = \sum_{k=1}^{N_C} \Delta q_k \cdot \Delta i_k/C_{\mu k} + \sum_{k=N_C+1}^{N_C+N_L} \Delta \varphi_k \cdot \Delta u_k/L_{\mu k} \qquad (5.256)$$

$$= Q - P \qquad (5.257)$$

where

$$Q = \sum_{k=1}^{N_C} (1/C_{\mu k})(\Delta q_k/\Delta u_k - C_{\mu k}) \cdot \Delta u_k \cdot \Delta i_k$$

$$+ \sum_{k=N_C+1}^{N_C+N_L} (1/L_{\mu k})(\Delta \varphi_k/\Delta i_k - L_{\mu k}) \cdot \Delta u_k \cdot \Delta i_k \qquad (5.258)$$

$$P = -\sum_{k=1}^{N_C+N_L} \Delta u_k \cdot \Delta i_k = \sum_{\text{resistors}} \Delta u_k \cdot \Delta i_k \qquad (5.259)$$

Equation (5.259) is a consequence of incremental power balance 3.1.8. Remember that the incremental power absorbed by a source, whether constant or time dependent, is zero at each instant. Thanks to inequalities (5.246) and (5.251) and to expressions that are analogous to (5.243) for $C_{\mu k}$ and $L_{\mu k}$, we find

$$|Q| \leq \alpha \cdot \left(\sum_{k=1}^{N_C + N_L} \alpha_k \right) \cdot (\max_{1 \leq k \leq N_C + N_L} |\Delta u_k \, \Delta i_k|) \tag{5.260}$$

However, we can bound Δu_k and Δi_k for each branch k which carries a capacitor or an inductor by applying corollary 3.1.15. We designate branch k as being branch v of the corollary. Set R is composed of resistors that are not sources, and set B is composed of all the other branches. Because there is no cut set without resistor, corollary 3.1.15 guarantees a loop exclusively composed of branch k and resistors. For that loop, we can write

$$\Delta u_k = - \sum_{\substack{j \neq k \\ \text{loop}}} \Delta u_j \tag{5.261}$$

and, therefore,

$$|\Delta u_k| \leq \sum_{\substack{j \neq k \\ \text{loop}}} |\Delta u_j| \leq \sum_{\text{resistors}} |\Delta u_j| \tag{5.262}$$

Using the Cauchy inequality [10], we find

$$\sum_{\text{resistors}} |\Delta u_k| = \sum_{\text{resistors}} \sqrt{R_{nk}} \cdot |\Delta u_k| / \sqrt{R_{nk}} \tag{5.263}$$

$$\leq \sqrt{\sum_{\text{resistors}} R_{nk}} \cdot \sqrt{\sum_{\text{resistors}} (\Delta u_k)^2 / R_{nk}} \tag{5.264}$$

$$\leq \sqrt{R} \cdot \sqrt{\sum_{\text{resistors}} ((\Delta u_k)^2 / R_{nk}) + R_{nk} (\Delta i_k)^2)} \tag{5.265}$$

where $R_{nk} = \sqrt{R_{Mk} R_{mk}}$. By analogous reasoning, we find

$$\sum_{\text{resistors}} |\Delta i_k| \leq \sqrt{G} \cdot \sqrt{\sum_{\text{resistors}} ((\Delta u_k)^2 / R_{nk}) + R_{nk} (\Delta i_k)^2)} \tag{5.266}$$

Thus, we have

$$|Q| \leq \alpha \cdot \sqrt{RG} \cdot \sum_{\text{resistors}} ((\Delta u_k)^2 / R_{nk}) + R_{nk} (\Delta i_k)^2) \tag{5.267}$$

Due to (5.242) and (5.244), we write

$$|Q| \leqslant \alpha \cdot \sqrt{RG} \cdot \sum_{\text{resistors}} \Delta u_k \Delta i_k (\sqrt{R_{Mk}/R_{mk}} + \sqrt{R_{mk}/R_{Mk}}) \qquad (5.268)$$

$$\leqslant \alpha \cdot \sqrt{RG} \cdot P/\gamma \qquad (5.269)$$

Finally, we obtain

$$dW/dt \leqslant (\alpha \sqrt{RG}/\gamma - 1) \cdot P \qquad (5.270)$$

By using the same reasoning as for the proof of theorem 5.7.6, we verify that the incremental power $P(\xi_1, \xi_2)$ absorbed by the resistors is negative, except when $\xi_1 = \xi_2$, where it is zero. It follows that W is a strict incremental Liapunov function if inequality (5.252) is not satisfied. In this case, theorem 5.7.4 guarantees a unique asymptotic behavior.

5.7.13 Comment

According to the interpretation of subsection 5.7.11, α is an upper bound for the nonlinearity of capacitors and inductors, and γ is a lower bound for the local passivity of resistors. Thus theorem 5.7.12 releases the restriction of exact linearity for capacitors and inductors in theorem 5.7.6.

5.7.14 Example

Let us consider the ferroresonant circuit of figure 2.52, with the characteristic of figure 1.37 for the inductor. As seen in subsections 5.1.10 and 5.1.11, this circuit does not always have a unique asymptotic behavior. Theorem 5.7.12 permits us to detect a region in the space of the circuit parameters, where the unique asymptotic behavior is guaranteed.

It is evident that the inductor is uniformly locally passive, with $L_M = 0$, $L_m = 1$. As the capacitor and the resistors are linear, the lower and upper bounds in (5.236) and (5.237) coincide with the element value. Consequently, $\gamma = 1/2$, $\alpha = (L_0 - L_1)/(L_0 + L_1)$, $R = R_4 + R_5$, $G = 1/R_4 + 1/R_5$. Condition (5.252) thus becomes

$$(L_0 - L_1)/(L_0 + L_1) \cdot \sqrt{2 + (R_4/R_5) + (R_5/R_4)} < 1/2 \qquad (5.271)$$

This inequality can also be written

$$L_0/L_1 < \frac{2\sqrt{2 + \dfrac{R_4}{R_5} + \dfrac{R_5}{R_4} + 1}}{2\sqrt{2 + \dfrac{R_4}{R_5} + \dfrac{R_5}{R_4} - 1}} \qquad (5.272)$$

If the inductor is linear, $L_0 = L_1$, inequality (5.272) always holds and the asymptotic behavior is unique. This fact already is a consequence of theorem 5.7.6. The right-hand side of (5.272) is a decreasing function of $R_4/R_5 + R_5/R_4$. If R_4 and R_5 tend toward zero or infinity, the right-hand side tends toward one, and the inductor must be closer and closer to a linear inductor for (5.272). Thus, inequality (5.272) clearly shows the limits imposed to the inductor nonlinearity by the resistive part of the circuit.

5.7.15 Comment

We observe that neither the amplitude of the sinusoidal source, nor the value of the capacitor are involved in criterion (5.272) for the uniqueness of the asymptotic behavior. However, example of subsection 5.1.10 shows that the asymptotic behavior may or may not be unique depending on the value of the source amplitude, and other numerical calculations could also show the importance of the capacitance. This clearly shows that theorem 5.7.12 can only partially solve the problem of unique asymptotic behavior. The condition it provides for the uniqueness of the asymptotic behavior is sufficient, but it is far from being necessary.

Apart from the fact that, in general, amplitudes of time-dependent sources do not play any role in the criterion of theorem 5.7.12, inequality (5.252) is not optimum. In fact, in its derivation, we have used rather rough bounds for the capacitor and inductor increments $|\Delta u_k|$ and $|\Delta i_k|$. Better bounds can be obtained for concrete circuits. For instance, condition (5.272) will be improved in subsection 5.7.16. In [37], a global measure of the local passivity of the resistive part of the circuit as an N-port has been used. Furthermore, a more powerful theorem has been stated for the special case of a linear resistive part.

5.7.16 Example

Let us try to improve the condition for the uniqueness of the asymptotic behavior in the case of the circuit of figure 2.52. Because (2.135) and (2.136) are linear, we can rewrite them with the incremental variables:

$$\Delta i_1 = (-\Delta u_1 + R_5 \Delta i_2)/(R_4 + R_5) \qquad (5.273)$$
$$\Delta u_2 = R_5(-\Delta u_1 - R_4 \Delta i_2)/(R_4 + R_5) \qquad (5.274)$$

from which we infer the incremental power absorbed by the resistors:

$$P(\Delta u_1, \Delta i_2) = (\Delta u_1)^2/(R_4 + R_5) + R_4 R_5 (\Delta i_2)^2/(R_4 + R_5) \qquad (5.275)$$

Conversely, we define function W by

$$W(\Delta u_1, \Delta i_2) = (\Delta q_1)^2/(2C_1) + (\Delta \varphi_2)^2/(2L) \qquad (5.276)$$

Here, we may choose constant L different from the value $(L_0 + L_1)/2$ used in (5.255). Then, (5.256) becomes

$$dW/dt = Q - P \qquad (5.277)$$

with

$$Q = (1/L) \cdot ((\Delta \varphi_2/\Delta u_2) - L) \cdot \Delta u_2 \cdot \Delta i_2 \qquad (5.278)$$

The term corresponding to the capacitor in Q is zero, because the capacitor is linear. In order to obtain a bound for $|Q|$, we look for lower and upper bounds for the factors $(1/L) \cdot ((\Delta \varphi_2/\Delta u_2) - L)$ and $\Delta u_2 \cdot \Delta i_2$. Concerning the first factor, we have

$$-(1 - L_1/L) \leqslant (1/L) \cdot ((\Delta \varphi_2/\Delta u_2) - L) \leqslant (1 + L_0/L) \qquad (5.279)$$

For the second factor, $\Delta u_2 \Delta i_2$ must be bounded by a multiple of P. To that effect, we calculate the extrema of $\Delta u_2 \Delta i_2/P$. According to (5.274), we have

$$\Delta u_2 \Delta i_2/P = -\frac{R_4 R_5 (\Delta i_2)^2 + R_5 \Delta u_1 \Delta i_2}{(\Delta u_1)^2 + R_4 R_5 (\Delta i_2)^2} \qquad (5.280)$$

$$= -\frac{R_4 R_5 \left(\dfrac{\Delta i_2}{\Delta u_1}\right)^2 + R_5 \dfrac{\Delta i_2}{\Delta u_1}}{1 + R_4 R_5 \left(\dfrac{\Delta i_2}{\Delta u_1}\right)^2} \qquad (5.281)$$

This function has two extrema at the points:

$$\Delta i_2/\Delta u_1 = (1 \pm \sqrt{1 + R_5/R_4})/R_5 \qquad (5.282)$$

which provides the inequalities:

$$-(\sqrt{1 + R_5/R_4} + 1)/2 \leqslant \Delta u_2 \Delta i_2/P \leqslant (\sqrt{1 + R_5/R_4} - 1)/2 \qquad (5.283)$$

Until now, L is an arbitrary positive constant. The best choice of L consists in making the product of the left-hand sides of (5.279) and (5.283) equal to the product of the right-hand sides. With this choice, we obtain, after some calculations

$$|Q|/P \leqslant \frac{(L_0 - L_1)R_5/4R_4}{\frac{1}{2}\left(\sqrt{1+\dfrac{R_5}{R_4}}-1\right)L_0 + \frac{1}{2}\left(\sqrt{1+\dfrac{R_5}{R_4}}+1\right)L_1} \qquad (5.284)$$

It follows that W is an incremental Liapunov function of the circuit and, consequently, that the asymptotic behavior is unique if the right-hand side is bounded by one. This inequality can further be transformed into

$$(L_0/L_1) < \left(\frac{\sqrt{1+\dfrac{R_5}{R_4}}+1}{\sqrt{1+\dfrac{R_5}{R_4}}-1}\right)^2 \qquad (5.285)$$

5.7.17 Comments

It is interesting to compare conditions (5.272) and (5.285) for the uniqueness of the asymptotic behavior of the circuit of figure 2.52. The right-hand side of (5.285) is larger than that of (5.272) because the function $f(x) = (x + 1)/(x - 1)$ is decreasing for $x > 1$. Thus, criterion (5.285) permits us to guarantee a unique steady-state for a larger region in the parameter space than does criterion (5.272), as would be expected. Furthermore, the right-hand side is a decreasing function of R_5/R_4, whereas the right-hand side of (5.272) is decreasing in $R_5/R_4 + R_4/R_5$. We, therefore, infer from (5.285) that a small value of R_4 and a large value of R_5 are favorable to a unique steady state. Such a conclusion cannot be drawn from criterion (5.272).

The problem of determining a region in the parameter space of a circuit where the steady state is unique, while there are several steady states elsewhere, is not easy to solve. In view of that, inequality (5.285), as is, in fact, inequality (5.272), is then surprisingly simple. However, the limit for the unique steady state suggested by (5.285) is still far too low compared to the actual limit estimated by numerical calculation [37]. In other words, there are still many values for the elements of the circuit such that the steady state is unique, while (5.285) is violated.

5.7.18 Conclusion

In this chapter, the concept of steady-state solution has been introduced, and we have discussed the Liapunov function method to prove the convergence of all the solutions toward a unique steady state solution. The same method also permits us to find attractive domains for the solutions. Other approaches to the same problem are known, such as, for instance, the describing function method [38], the Krylov-Bogolyubov method [39], or the Volterra series method [40]. However, these are limited to the case of weakly nonlinear circuits. The precise meaning of weak nonlinearity varies from one method to another. What is common to all is that they are based on a series expansion which no longer converges beyond certain limits. In most cases, these limits are not known, which makes the conclusions drawn from such methods uncertain. However, the results often contain more information concerning the steady-state solution than we were able to obtain in this chapter.

Although the Liapunov function method is applicable to highly nonlinear circuits, the class of circuits to which we could actually apply it is limited. Indeed, controlled sources and operational amplifiers have never been admitted. So, the results of this chapter are weak, if we compare them with those of chapter 3. This area of nonlinear circuit theory is still widely open for fundamental research.

Chapter 6

Analysis of Periodic and Almost Periodic Solutions

6.1 MEAN POWER IN A PERIODIC STEADY STATE AND THE PAGE THEOREM

6.1.1 Introduction

The purpose of this section is to study the energy balance in the particular case where the solution of the circuit is periodic. We shall not deal with the questions of whether a periodic solution exists, whether it is unique, stable, or even globally asymptotically stable. The results are valid for *any* periodic solution. However, it is evident that properties of a globally asymptotically stable solution are an essential aspect of circuit behavior, while properties of an unstable periodic solution are of no practical interest.

The study of periodic solutions is important because many nonlinear circuits are specially designed to transform power from one frequency to another. The general structure of such a circuit is a two-port that is excited by a sinusoidal source of angular frequency ω_1 and which supplies power at the angular frequency ω_2 to a load (fig. 6.1). More precisely, we wish that the circuit has a unique asymptotically stable periodic solution such that the total mean power supplied by the source at the angular frequency ω_1 is dissipated in the charge at the angular frequency ω_2. In general, such a solution does not exist, even if ideal components are available. It is necessary to dissipate part of the power inside the two-port, and part in the load at an angular frequency different from ω_2.

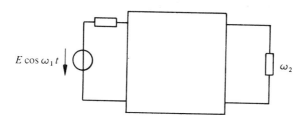

Fig. 6.1

378

We classify two-ports according to the relation between ω_1 and ω_2. If $\omega_1 \neq 0$ and $\omega_2 = 0$, we have a *rectifier*. If $\omega_1 = 0$ and $\omega_2 \neq 0$, we have an *oscillator*. If $\omega_2/\omega_1 = n$ ($\omega_1/\omega_2 = n$), where $n > 1$ is an integer, then the two-port is called a *frequency multiplier (divider)*.

6.1.2 Definition

Let $u(t)$ and $i(t)$ be the voltage and the current of a branch belonging to a periodic solution of period T for a circuit. The *mean power* $\overline{P}$ absorbed by the branch is defined by

$$\overline{P} = \frac{1}{T} \int_0^T u(t)\, i(t)\, \mathrm{d}t \tag{6.1}$$

6.1.3 Property: Mean Power Balance

Let $u_k(t)$, $i_k(t)$, $k = 1, \ldots, b$ be the voltages and the currents of all the branches of a circuit belonging to a periodic solution. Let $\overline{P}_k$ be the mean power absorbed by branch k. Then,

$$\sum_{k=1}^b \overline{P}_k = 0 \tag{6.2}$$

Indeed, according to the instantaneous power balance, we have at any instant

$$\sum_{k=1}^b u_k(t)\, i_k(t) = 0 \tag{6.3}$$

By taking the average of (6.3) over a period, we obtain (6.2).

6.1.4 Decomposition in Fourier Series

Let $u(t)$ and $i(t)$ be the voltage and the current of a branch belonging to a periodic solution with the fundamental angular frequency Ω. The decomposition in Fourier series:

$$u(t) = \sum_{m=-\infty}^{+\infty} U_m \exp(jm\Omega t) \tag{6.4}$$

$$i(t) = \sum_{n=-\infty}^{+\infty} I_n \exp(jn\Omega t) \tag{6.5}$$

leads to

$$u(t) \cdot i(t) = \sum_{m,n=-\infty}^{+\infty} U_m I_n \exp(j(m+n)\Omega t) \tag{6.6}$$

$$= \sum_{s=-\infty}^{+\infty} \exp(js\Omega t) \sum_{m+n=s} U_m I_n \tag{6.7}$$

By taking the average of (6.7) over a period, we eliminate all the terms with $s = 0$, because the average of $\exp(js\Omega t)$ is zero if $s \neq 0$. Consequently,

$$\bar{P} = \sum_{m=-\infty}^{+\infty} U_m I_{-m} \tag{6.8}$$

$$= U_0 I_0 + \sum_{m=1}^{\infty} (U_m I_{-m} + U_{-m} I_m) \tag{6.9}$$

Because $u(t)$ and $i(t)$ are real functions, we have $U_{-m} = U_m^*$ and $I_{-m} = I_m^*$; therefore,

$$\bar{P} = \sum_{m=0}^{\infty} \bar{P}_m \tag{6.10}$$

with

$$\bar{P}_0 = U_0 I_0 \tag{6.11}$$

$$\bar{P}_m = U_m I_m^* + U_m^* I_m \quad \text{for} \quad m > 0 \tag{6.12}$$

$\bar{P}_m$ is called *mean power*, or *active power*, at the angular frequency $m\Omega$ absorbed by the branch.

$\bar{P}_k$ of (6.2) should not be confused with $\bar{P}_m$ of (6.12). While the former is the total mean power absorbed by the kth branch, the latter stands for only a part of the mean power. Hereafter, $\bar{P}_{km}$ shall be used to designate the mean power absorbed by branch k at the angular frequency $m\Omega$.

6.1.5 Property. Mean Power Balance at Each Frequency

The Fourier coefficients of U_m and I_m in (6.4) and (6.5) are calculated by (vol. IV, sect. 7.4.6):

$$U_m = \frac{1}{T} \int_0^T u(t) \exp(-jm\Omega t)\, dt \tag{6.13}$$

$$I_m = \frac{1}{T} \int_0^T i(t) \exp(-jm\Omega t)\, dt \tag{6.14}$$

Equations (6.13) and (6.14) are linear functions of $u(t)$ and $i(t)$ respectively. If $u_k(t)$ and $i_k(t), i = 1, \ldots, b$ are the voltages and the currents of a periodic solution of a circuit, then the corresponding Fourier coefficients U_{km} and I_{km} satisfy the Kirchhoff equations, as do $u_k(t)$ and $i_k(t)$. Similarly, U_{km}^* and I_{km}^* satisfy the Kirchhoff equations because the conjugate complex is a linear operation.

By applying Tellegen's theorem, we find

$$\sum_{k=1}^b U_{km} I_{km}^* = 0 \tag{6.15}$$

and

$$\sum_{k=1}^b U_{km}^* I_{km} = 0 \tag{6.16}$$

therefore,

$$\sum_{k=1}^b \bar{P}_{km} = 0 \tag{6.17}$$

for all the $m = 0, 1, \ldots$

Thus, the mean power is zero, not only globally, as we have seen in subsection 6.1.3, but also separately at each frequency.

6.1.6 Mean Power Absorbed by a Capacitor or an Inductor

Consider a charge-controlled capacitor defined by the constitutive relation $u = h(q)$. Let us assume that the charge $q(t)$ is a periodic function of period T. Then, the current $i(t) = dq/dt$, the voltage $u(t) = h(q(t))$, and the stored energy $W(q(t))$ are also periodic functions of period T. According to (5.105), we have

$$u(t) \cdot i(t) = dW/dt \tag{6.18}$$

from which we infer the expression for the mean power:

$$\overline{P} = \frac{1}{T} \int_0^T \frac{dW}{dt} dt = [W(q(T)) - W(q(0))]/T = 0 \tag{6.19}$$

By using decomposition (6.10) of $\overline{P}$, we can also write

$$\sum_{m=0}^{\infty} \overline{P}_m = 0 \tag{6.20}$$

the same result is obtained analogously for voltage-controlled capacitors and flux- or current-controlled inductors.

6.1.7 Property

The mean power absorbed by a charge- or voltage-controlled capacitor (flux- or current-controlled inductor) is zero.

6.1.8 Comment

The fact that the mean power absorbed by a capacitor or an inductor is zero should not be surprising. However, sum (6.20) may include nonzero terms. For the sum to be zero, the positive terms must be compensated by negative terms. This means that the reactance absorbs power at certain frequencies, and returns it at other frequencies. This phenomenon does not happen, however, in the case of a linear reactance. Indeed, for a linear capacitor of value C, we have

$$I_m = jm\Omega C U_m \tag{6.21}$$

from which we infer $\overline{P}_m = 0$.

6.1.9 Example

Consider an inductor defined by the constitutive relation:

$$i = a\varphi + b\varphi^3 \tag{6.22}$$

Let us assume that φ depends on time in the following manner:

$$\varphi(t) = \varphi_1 \sin \Omega t + \varphi_3 \cos 3\Omega t \tag{6.23}$$

Then,

$$u(t) = d\varphi/dt = \varphi_1 \Omega \cos \Omega t - 3\varphi_3 \Omega \sin 3\Omega t \tag{6.24}$$

By introducing (6.23) into (6.22), we obtain, after some calculations,

$$4i(t) = 9b\varphi_3^3 \cos 9\Omega t + 3b\varphi_1\varphi_3^2 \sin 7\Omega t - 3b\varphi_1^2\varphi_3 \cos 5\Omega t \tag{6.25}$$
$$- 3b\varphi_1\varphi_3^2 \sin 5\Omega t + (4a\varphi_3 + 3b\varphi_3^3 + 6b\varphi_1^2\varphi_3) \cos 3\Omega t$$
$$- b\varphi_1^3 \sin 3\Omega t - 3b\varphi_1^2\varphi_3 \cos \Omega t + (4a\varphi_1 + 6b\varphi_1\varphi_3^2 + b\varphi_1^3) \sin \Omega t$$

By multiplying (6.24) with (6.25), we obtain a sum of terms of the form sin x · sin y, sin x · cos y, and cos x · cos y. The only terms having an average which is different from zero are sin x · sin y and cos x · cos y, when $x = y$. In our present case, they are the terms proportional to $(\cos \Omega t)^2$ and $(\sin 3\Omega t)^2$. Their contribution to (6.20) is, respectively,

$$\bar{P}_1 = -3b\Omega\varphi_1^3\varphi_3/8 \tag{6.26}$$

and

$$\bar{P}_3 = 3b\Omega\varphi_1^3\varphi_3/8 \tag{6.27}$$

If $b\varphi_1\varphi_3$ is positive, the inductor absorbs mean power at angular frequency 3Ω and returns it at angular frequency Ω.

If $b\varphi_1\varphi_3$ is negative, it is the opposite. Note, also, that if $b = 0$, there is no power transfer from one frequency to the other. Indeed, in this case, the inductor is linear.

6.1.10 Mean Power Absorbed by a Passive Resistor

Let $u(t)$ and $i(t)$ be the periodic voltage and current of a passive resistor. Passivity implies that

$$u(t) \cdot i(t) \geqslant 0 \tag{6.28}$$

It follows that the mean power satisfies:

$$\bar{P} \geqslant 0 \tag{6.29}$$

or, also, with decomposition (6.10),

$$\sum_{m=0}^{\infty} \bar{P}_m \geqslant 0 \tag{6.30}$$

If the resistor is strictly passive, inequalities (6.29) and (6.30) are strict, the only exception being $u(t) \equiv i(t) \equiv 0$.

6.1.11 Comment

If the passive resistor is linear, of value R, we have

$$U_m = RI_m \tag{6.31}$$

and, therefore, $\bar{P}_m \geq 0$ for all the m. In general, however, this does not hold for nonlinear resistors. As in the case of nonlinear reactances, nonlinear resistors are capable of transforming power from one frequency to another. However, only part of the absorbed power is returned, the rest is dissipated.

6.1.12 Example

The following example shows that negative terms $\bar{P}_m$ in (6.30) are actually possible. Consider the voltage at the terminals of a piecewise-linear diode, defined by (1.16), (fig. 6.2):

$$u(t) = u_1 \cos \Omega t - u_0 \tag{6.32}$$

with

$$0 < u_0 < u_1 \tag{6.33}$$

Current $i(t)$ in the interval $[-T/2, T/2]$, where $T = 2\pi/\Omega$, is given by (fig. 6.3):

$$i(t) = \begin{cases} g_0(u_1 \cos \Omega t - u_0) & \text{when} \quad u_1 \cos \Omega t < u_0 \\ g_s(u_1 \cos \Omega t - u_0) & \text{when} \quad u_1 \cos \Omega t > u_0 \end{cases} \tag{6.34}$$

The Fourier series of $i(t)$ includes all the harmonics. In particular, the constant component is

$$I_0 = -g_0 u_0 + (g_s - g_0) u_1 \sqrt{1 - (u_0/u_1)^2}/\pi \tag{6.35}$$

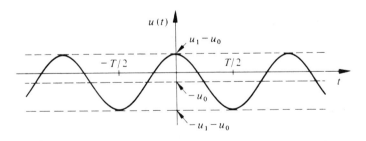

Fig. 6.2

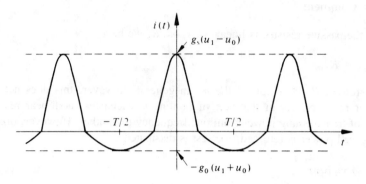

Fig. 6.3

It follows that the dc power absorbed by the diode is

$$\bar{P}_0 = g_0 u_0^2 - (g_s - g_0) u_0 u_1 \sqrt{1 - (u_0/u_1)^2}/\pi \tag{6.36}$$

This expression reaches a minimum when $(u_0/u_1)^2 = 1/2$ of value:

$$\bar{P}_0 = g_0 u_0^2 \left[1 - \left(\frac{g_s}{g_0} - 1 \right) \frac{1}{2\sqrt{2\pi}} \right] \tag{6.37}$$

If

$$g_s/g_0 > 1 + 2\sqrt{2\pi}, \tag{6.38}$$

which normally is amply satisfied, and for an appropriate ratio u_0/u_1, the diode supplies power at dc. This power arises from the power absorbed at the angular frequency Ω.

6.1.13 Page Theorem

The mean powers $\bar{P}_m$ (6.12) absorbed by a resistor which is passive with respect to any point of its characteristic satisfy the following relations:

$$\sum_{m=1}^{\infty} m^2 \bar{P}_m \geq 0 \tag{6.39}$$

and, for any real constant θ:

$$\sum_{m=1}^{\infty} (1 - \cos(m\theta)) \bar{P}_m \geq 0 \tag{6.40}$$

Indeed, if a resistor is strictly passive with respect to any point of its characteristic and if $u(t)$, $i(t)$ are periodic functions, we have, thanks to (3.15):

$$[u(t+\Delta t)-u(t)][i(t+\Delta t)-i(t)] \geqslant 0 \qquad (6.41)$$

and, consequently,

$$\frac{1}{T}\int_0^T [u(t+\Delta t)-u(t)][i(t+\Delta t)-i(t)]\,dt \geqslant 0 \qquad (6.42)$$

for an arbitrary t. By using the Fourier series (6.4), (6.5), we find

$$u(t+\Delta t)-u(t) = \sum_{m=-\infty}^{+\infty} \tilde{U}_m \exp(jm\Omega t) \qquad (6.43)$$

$$i(t+\Delta t)-i(t) = \sum_{n=-\infty}^{+\infty} \tilde{I}_n \exp(jn\Omega t) \qquad (6.44)$$

with

$$\tilde{U}_m = U_m[\exp(jm\Omega\Delta t)-1] \qquad (6.45)$$
$$\tilde{I}_n = I_n[\exp(jn\Omega\Delta t)-1] \qquad (6.46)$$

By analogous reasoning to that of subsection 6.1.4, we obtain, from (6.42):

$$0 \leqslant \tilde{U}_0\tilde{I}_0 + \sum_{m=1}^{\infty} (\tilde{U}_m\tilde{I}_m^* + \tilde{U}_m^*\tilde{I}_m) \qquad (6.47)$$

$$= \sum_{m=1}^{\infty} (U_m I_m^* + U_m^* I_m)(4 - 4\cos(m\Omega\Delta t)) \qquad (6.48)$$

$$= 4 \sum_{m=1}^{\infty} (1-\cos(m\Omega\Delta t))\bar{P}_m \qquad (6.49)$$

which is, up to factor 4, inequality (6.40). Because

$$\lim_{\theta \to 0} [1-\cos(m\theta)]/\theta^2 = m^2/2 \qquad (6.50)$$

we infer (6.39) from (6.40).

6.1.14 Comments

We infer from (6.39) and (6.40) the basic limitations for power conversion from one frequency to another in the case of circuits which only use locally passive resistors (e.g., diodes) as nonlinear elements.

The reasoning is as follows. Let us assume that a source supplies the mean power $\overline{P}_1$ at the angular frequency Ω to a locally passive resistor. Suppose that all the $\overline{P}_m = 0$ for $m > 1$, except $\overline{P}_n$, where we wish to draw power from the resistor. According to (6.39), we have

$$-\overline{P}_n \leqslant \overline{P}_1/n^2 \tag{6.51}$$

When n is large, only a very small part of the absorbed power is dissipated at the angular frequency $n\Omega$, the rest is dissipated by the resistor.

In the opposite situation, when power is supplied by the source at the angular frequency $n\Omega$ and where we want to draw power at the angular frequency Ω, we apply inequality (6.40) with $\theta = 2\pi/n$. Then, $1 - \cos(n\theta) = 0$, while $1 - \cos \theta > 0$. Consequently, through (6.40), we can conclude that

$$\overline{P}_1 \geqslant 0 \tag{6.52}$$

It is, therefore, impossible at the angular frequency to recover at Ω part of the power injected at $n\Omega$.

With the aid of Tellegen's theorem, we shall generalize the inequalities (6.51) and (6.52) of a resistor to any circuit in subsection 6.1.15.

6.1.15 Theorem

Consider a circuit composed of the following elements:

- a sinusoidal source of angular frequency ω;
- nonlinear resistors that are passive with respect to any point of their characteristic;
- passive linear resistors;
- linear capacitors and inductors.

A resistor, or, more generally, a one-port composed of linear elements is considered to be a load.

Let us first consider a periodic solution with the fundamental angular frequency $\Omega = \omega$. If $\overline{P}_s$ is the mean power supplied by the source, then the mean power at the angular frequency $n\omega$ that can be dissipated in the load is limited by $\overline{P}_s/n^2$.

Conversely, in the hypothetic case of a solution with the fundamental angular frequency $\Omega = \omega/n$, no mean power can be dissipated in the load at the angular frequency ω/n.

For the proof, we number the branches in such a way that the first branch is the source and the branches 2 to r are the nonlinear resistors. Let $\overline{P}_{km}$ be the mean power absorbed by branch k at angular frequency $m\Omega$. Then, $\overline{P}_{km} \geq 0$ for all m and all $k > r$. In addition, let us introduce

$$\overline{Q}_m = \sum_{k=2}^{r} \overline{P}_{km} \tag{6.53}$$

which is the mean power absorbed by all nonlinear resistors combined at the angular frequency $m\Omega$. According to (6.17), we have for all m

$$0 = \sum_{k=1}^{b} \overline{P}_{km} = \overline{P}_{1m} + \overline{Q}_m + \sum_{k=r+1}^{b} \overline{P}_{km} \geq \overline{P}_{1m} + \overline{Q}_m \tag{6.54}$$

Let us consider the first case where $\Omega = \omega$. Then, $\overline{P}_{11} = -\overline{P}_s$ and $\overline{P}_{1m} = 0$ for $m \neq 1$; therefore,

$$\overline{Q}_1 \leq \overline{P}_s \tag{6.55}$$

and

$$\overline{Q}_m \leq 0 \quad \text{for} \quad m \neq 1 \tag{6.56}$$

Equation (6.39) is written as

$$\sum_{m=1}^{\infty} m^2 \overline{P}_{km} \geq 0 \quad \text{for} \quad k = 2, ..., r \tag{6.57}$$

By adding all the inequalities (6.57), we obtain

$$\sum_{m=1}^{\infty} m^2 \overline{Q}_m \geq 0 \tag{6.58}$$

By combining inequalities (6.55), (6.56), and (6.58), we find for the mean power $\overline{P}_c$ absorbed by the load at the angular frequency $n\omega$

$$\overline{P}_c \leq \sum_{k=r+1}^{b} \overline{P}_{kn} = -\overline{Q}_n \tag{6.59}$$

$$\leq \frac{1}{n^2} \sum_{m \neq n} m^2 \overline{Q}_m < \overline{P}_s/n^2 \tag{6.60}$$

If, secondly, $\Omega = \omega/n$, then $\overline{P}_{1n} = -\overline{P}_s$ and $\overline{P}_{1m} = 0$ for $m \neq n$. Starting from (6.54), we find

$$\overline{Q}_n \leqslant \overline{P}_s \tag{6.61}$$

and

$$\overline{Q}_m \leqslant 0 \quad \text{for} \quad m \neq n \tag{6.62}$$

Equation (6.40) is written as

$$\sum_{m=1}^{\infty} (1 - \cos m\theta) \overline{P}_{km} \geqslant 0 \quad \text{for} \quad k = 2, ..., r \tag{6.63}$$

and, therefore,

$$\sum_{m=1}^{\infty} (1 - \cos m\theta) \overline{Q}_m \geqslant 0 \tag{6.64}$$

If $\theta = 2\pi/n$, the terms in which m is a multiple of n are zero in (6.64). For the other m, we have $1 - \cos m\theta > 0$ and, according to (6.62), $\overline{Q}_m \leqslant 0$ which is compatible with (6.64) if and only if $\overline{Q}_m = 0$. In particular, we have $\overline{Q}_1 = 0$. Then,

$$\sum_{k=r+1}^{b} \overline{P}_{k1} = -\overline{Q}_1 = 0 \tag{6.65}$$

which is not compatible with $\overline{P}_{k1} \geqslant 0$, $k = r + 1, \ldots, b$, unless

$$\overline{P}_{k1} = 0 \quad k = r+1, ..., b \tag{6.66}$$

In particular, no active power is absorbed by the charge at angular frequency ω/n.

6.1.16 Comments

Theorem 6.1.15 sets basic limitations for the efficiency of a diode frequency multiplier. Furthermore, we observe that it is impossible to construct a diode frequency divider. Nevertheless, no limitation arises from this theorem regarding the diode rectifier; an efficiency of 100% is possible in principle. We shall see in chapter 7 how and to what extent these theoretical limits can be reached by concrete circuits.

This theorem can easily be generalized to circuits which contain passive two-ports, such as the ideal transformer.

6.2 MEAN POWER IN AN ALMOST PERIODIC STEADY STATE AND THE MANLEY-ROWE EQUATIONS

6.2.1 Introduction

Some circuits have more than one time-dependent source, such as, in particular, the *modulator*. It is a three-port excited by two sources, one of which injects the signal and the other is a sinusoidal source called the *carrier*. The third port is terminated by the load, where the modulated signal is dissipated (fig. 6.4).

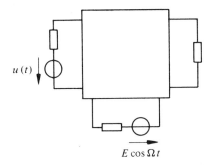

$$u(t) \qquad E \cos \Omega t$$

Fig. 6.4

Let us assume that in the absence of a signal, the modulator has a periodic solution $y_0(t)$. If the signal $u_s(t)$ is small with respect to the nonlinearities of the circuit at the operating point $y_0(t)$, the solution in the presence of the signal can be calculated by linearizing the circuit around $y_0(t)$. In such an approximation, the superposition principle is valid for the signal. Consequently, we need only study the response to a signal of the form exp $(j\omega t)$. The response to a small arbitrary signal is obtained, as in the case of linear circuits, via the Fourier transform. The circuits considered in this section thus have the structure of figure 6.5.

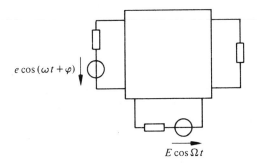

$$e \cos(\omega t + \varphi) \qquad E \cos \Omega t$$

Fig. 6.5

If a circuit has two sinusoidal sources of angular frequency Ω_1 and Ω_2, the ratio of which is a rational number, then we would expect a periodic steady state having a fundamental angular frequency that is the largest common denominator of Ω_1 and Ω_2. However, if Ω_1/Ω_2 is an irrational number, the circuit can no longer have a periodic solution. Nevertheless, the solutions may be stable. In that case, the asymptotic behavior of the solutions is not completely irregular; it is, in fact, a superposition of two periodic movements with different periods.

6.2.2 Definition

Let Ω_1 and Ω_2 be two angular frequencies, such that Ω_1/Ω_2 is an irrational number. A function is *almost periodic with basic angular frequencies Ω_1 and Ω_2* if it allows a representation of the form:

$$f(t) = \sum_{m=-\infty}^{+\infty} \sum_{n=-\infty}^{+\infty} F_{mn} \exp(j\Omega_{mn}t) \tag{6.67}$$

where

$$\Omega_{mn} = m\Omega_1 + n\Omega_2 \tag{6.68}$$

The numbers F_{mn} are called *Fourier coefficients*.

6.2.3 Comments

Definition 6.2.2 is easily generalized to almost periodic functions with basic angular frequencies $\Omega_1, \ldots, \Omega_N$, or even with an infinite sequence of basic angular frequencies. The essential property of almost periodic functions is the line spectrum of their Fourier transform.

An irrational number can be approximated by a rational number with arbitrary accuracy. Let f be an almost periodic function with basic angular frequencies Ω_1 and Ω_2. Let p/q be an approximation of Ω_1/Ω_2. Then,

$$\frac{\Omega_1}{p} \cong \frac{\Omega_2}{q} \tag{6.69}$$

and f is approximately periodic of angular frequency Ω_1/p. If we want to replace p/q by a more accurate approximation p'/q', then f will be all the more accurately periodic of angular frequency Ω_1/p'. In general $p' > p$ and the corresponding period is longer. In conclusion, if we wait long enough, function $f(t)$ is repeated almost exactly. This is the origin of the term *almost periodic*.

Periodic functions of period $2\pi/\Omega_1$, as well as periodic functions of period $2\pi/\Omega_2$, are special cases of almost periodic functions with the basic angular frequencies Ω_1 and Ω_2. They are obtained by setting $F_{mn} \neq 0$ for $n \neq 0$ and $F_{mn} = 0$ for $m \neq 0$, respectively.

6.2.4 Example

Consider

$$f(t) = \cos\pi t + \tfrac{1}{2}\sin 2\pi t + 2\sin t + \sin 3t \qquad (6.70)$$

Equation (6.70) defines an almost periodic function with the basic angular frequencies $\Omega_1 = \pi$, $\Omega_2 = 1$. The first rational approximation of $\Omega_1/\Omega_2 = \pi$ is 3. Consequently, f is approximately periodic with a fundamental angular frequency Ω between the values $\Omega_1/3 = 1.047$ and $\Omega_2 = 1$, which corresponds to a period T between the values $6\pi/\Omega_1 = 6$ and $2\pi\Omega_2 = 6.283$. We can observe from figure 6.6 that this approximation is very rough.

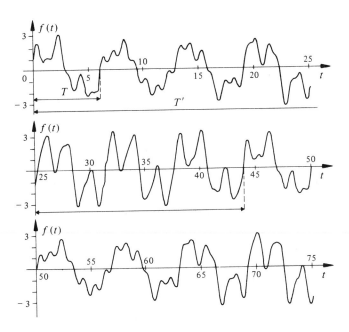

Fig. 6.6

A better rational approximation of π is $22/7 = 3.143$. According to this approximation, f is almost periodic with a fundamental angular frequency Ω' having a value that ranges between $\Omega_1/22 = 0.14280$ and $\Omega_2/7 = 0.14286$, which corresponds to a period T' between $44\pi/\Omega_1 = 44$ and $14\pi/\Omega_2 = 43.98$. Figure 6.6 confirms that this is already a good approximation.

6.2.5 Calculation of the Fourier Coefficients

Let f be an almost periodic function with the basic angular frequencies Ω_1 and Ω_2. Let T be an approximative period of f. Then, we can write

$$F_{mn} \cong \frac{1}{T} \int_0^T f(t) \exp(-j\Omega_{mn}t)\, dt \tag{6.71}$$

We can find more and more accurate periods which correspond to rational numbers that are closer and closer to Ω_1/Ω_2. However, these periods become longer and longer. This suggests the formula

$$F_{mn} = \lim_{T \to \infty} \int_0^T f(t) \exp(-j\Omega_{mn}t)\, dt \tag{6.72}$$

For the proof of this relation, refer to [41].

6.2.6 Property

Consider a capacitor defined by the constitutive relation $u = h(q)$. Let $q(t)$ be an almost periodic function with the angular frequencies Ω_1 and Ω_2. Thus, $u(t) = h(q(t))$ is also almost periodic with the angular frequencies Ω_1 and Ω_2. In addition, if Q_{mn} and U_{mn} are the coefficients of expansion (6.67) for $q(t)$ and $u(t)$, respectively, then U_{mn} depends on the set of the Q_{mn}, but is independent of Ω_1 and Ω_2. The same property holds for any element which is controlled by one of the two variables.

Indeed, according to definition 6.2.2, $q(t)$ can be represented in the form:

$$q(t) = \sum_{m,n=-\infty}^{+\infty} Q_{mn} \exp(j\Omega_{mn}t) \tag{6.73}$$

We associate with $q(t)$ the two-variable function:

$$\tilde{q}(t,s) = \sum_{m,n=-\infty}^{+\infty} Q_{mn} \exp j(m\Omega_1 t + n\Omega_2 s) \tag{6.74}$$

Equation (6.74) is periodic in t and s with the respective periods $2\pi/\Omega_1$ and $2\pi/\Omega_2$. We recover q from $\tilde{q}$ by

$$q(t) = \tilde{q}(t, t) \tag{6.75}$$

We define

$$\tilde{u}(t, s) = h(\tilde{q}(t, s)) \tag{6.76}$$

This function is also periodic in t and s with the same periods as q. Consequently, $\tilde{u}$ has a double Fourier series of the form:

$$\tilde{u}(t, s) = \sum_{m,n=-\infty}^{+\infty} U_{mn} \exp j(m\Omega_1 t + n\Omega_2 s) \tag{6.77}$$

Finally, we have

$$\tilde{u}(t, t) = h(\tilde{q}(t, t)) = h(q(t)) = u(t) \tag{6.78}$$

and, therefore,

$$u(t) = \sum_{m,n=-\infty}^{+\infty} U_{mn} \exp(j\Omega_{mn} t) \tag{6.79}$$

which proves that u is almost periodic with the basic angular frequencies Ω_1 and Ω_2.

The property still to be proved is as follows. If, in (6.73), the Q_{mn} remain unchanged and Ω_1 is replaced by Ω'_1, then, in (6.79), we need only replace Ω_1 by Ω'_1 without changing the U_{mn}. Let $q'(t)$ be the function obtained by this process. Thus, we must prove that $u'(t) = h(q'(t))$ has the expansion (6.79) with Ω_{mn} replaced by $\Omega'_{mn} = m\Omega'_1 + n\Omega_2$. To this end, we introduce $\tilde{q}'(t,s)$ by replacing Ω_1 with Ω'_1 in (6.74). Then,

$$\tilde{q}'(t, s) = \tilde{q}(\Omega'_1 t/\Omega_1, s) \tag{6.80}$$

If we define

$$\tilde{u}'(t, s) = h(\tilde{q}'(t, s)) \tag{6.81}$$

we find

$$\tilde{u}'(t, s) = h(\tilde{q}(\Omega'_1 t/\Omega_1, s)) = \tilde{u}(\Omega'_1 t/\Omega_1, s) \tag{6.82}$$

which amounts to saying that $\bar{u}'$ is also obtained by replacing Ω_1 with Ω'_1 in (6.77). In a way analogous to (6.78), we have

$$u'(t) = \tilde{u}'(t, t) \qquad (6.83)$$

which implies that (6.77)·with Ω'_{mn}, instead of Ω_{mn}, actually provides $u'(t)$.

6.2.7 Definition

A *solution* of a circuit is *almost periodic* with basic angular frequencies Ω_1 and Ω_2 if all the currents, all the voltages, all the fluxes, and all the charges of which it consists are almost periodic functions of time with the basic angular frequencies Ω_1 and Ω_2.

6.2.8 Comments

In view of property 6.2.6, we would expect a circuit such as that of figure 6.5 to have an almost periodic solution with the basic angular frequencies ω and Ω. Hereafter, we shall study the properties of such a solution, assuming its existence.

As noted in subsection 6.2.3, the periodic functions of period $2\pi/\omega$ and $2\pi/\Omega$ are special cases of almost periodic functions with the basic angular frequencies ω and Ω. Thus, some variables of an almost periodic solution can indeed be periodic. This is the case of the source voltages for the circuit in figure 6.5.

6.2.9 Definition

Let $u(t)$ and $i(t)$ be the voltage and the current of a branch belonging to an almost periodic solution of a circuit. The *mean power* absorbed by the branch is defined by

$$\bar{P} = \lim_{T \to \infty} \frac{1}{T} \int_0^T u(t)\, i(t)\, \mathrm{d}t \qquad (6.84)$$

6.2.10 Decomposition of the Mean Power

Let $u(t)$ and $i(t)$ be almost periodic functions given by

$$u(t) = \sum_{m,n=-\infty}^{+\infty} U_{mn} \exp(j\Omega_{mn} t) \qquad (6.85)$$

$$i(t) = \sum_{m',n'=-\infty}^{+\infty} I_{m'n'} \exp(j\Omega_{m'n'} t) \qquad (6.86)$$

Then,

$$u(t)\,i(t) = \sum_{m,n,m',n'=-\infty}^{+\infty} U_{mn}I_{m'n'}\exp\left(j\Omega_{m+n'\ n+n'}t\right) \tag{6.87}$$

$$= \sum_{\substack{m+m'=M\\n+n'=N}} \exp\left(j\Omega_{MN}t\right) \sum_{\substack{m+m'=M\\n+n'=N}} U_{mn}I_{m'n'} \tag{6.88}$$

According to (6.72), the mean power $\bar{P}$ is the Fourier coefficient of index $M = N = 0$ of $u(t)i(t)$. Hence,

$$\bar{P} = \sum_{m,n=-\infty}^{+\infty} U_{mn}I_{-m-n} \tag{6.89}$$

By transformations analogous to those of subsection 6.1.4, we obtain

$$\bar{P} = \sum_{m,n=0}^{\infty} \bar{P}_{mn} + \sum_{m,n=1}^{\infty} \bar{P}_{m-n} \tag{6.90}$$

where

$$\bar{P}_{00} = U_{00}I_{00} \tag{6.91}$$

$$\bar{P}_{mn} = U_{mn}I_{mn}^* + U_{mn}^*I_{mn} \qquad \text{for} \qquad (m,n) \neq (0,0) \tag{6.92}$$

$\bar{P}_{mn}$ is called the *mean power* or *active power, at the angular frequency* Ω_{mn}.

If several branches are considered simultaneously, $\bar{P}_{k,mn}$ is used to designate the mean power absorbed by the kth branch at the angular frequency Ω_{mn}.

6.2.11 Property: Mean Power Balance at Each Frequency

Let us consider an almost periodic solution with two basic angular frequencies of a circuit consisting of b branches. Then, for all the m, n, we have

$$\sum_{k=1}^{b} \bar{P}_{k,mn} = 0 \tag{6.93}$$

The proof is carried out as was done in subsection 6.1.5 by noting that the branch voltage (current) Fourier coefficients satisfy the Kirchhoff voltage (current)

396

equations. This is due to the fact that (6.72) is a linear expression at f. Consequently, Tellegen's theorem holds for the voltages $U_{k,mn}$ combined with the currents $I^*_{k,mn}$, and this implies (6.93).

6.1.12 Properties

Let us consider a solution of a circuit which is almost periodic with two basic angular frequencies. If a branch is occupied with a capacitor or an inductor, then the mean power absorbed by the branch is zero. By using (6.90), we can write

$$\sum_{m,n=0}^{\infty} \bar{P}_{mn} + \sum_{m,n=1}^{\infty} \bar{P}_{m-n} = 0 \tag{6.94}$$

If the capacitor or the inductor is linear, each term in (6.94) is individually zero.

The mean power absorbed in a branch occupied by a passive resistor is positive or zero. In that case, we thus have

$$\sum_{m,n=0}^{\infty} \bar{P}_{mn} + \sum_{m,n=1}^{\infty} \bar{P}_{m-n} \geqslant 0 \tag{6.95}$$

If the resistor is linear, each term of (6.95) is positive or zero.

The proofs of the different properties are analogous to those of subsections 6.1.6 and 6.1.10.

6.2.13 Manley-Rowe Equations

Consider a charge-controlled capacitor. Let $q(t)$ be an almost periodic function with the basic angular frequencies Ω_1 and Ω_2. Thus, we have

$$q(t) = \sum_{m,n=-\infty}^{+\infty} Q_{mn} \exp(j\Omega_{mn}t) \tag{6.96}$$

According to property 6.2.6, $u(t) = h(q(t))$ has an expansion of the same form:

$$u(t) = \sum_{m,n=-\infty}^{+\infty} U_{mn} \exp(j\Omega_{mn}t) \tag{6.97}$$

Furthermore, we have

$$i(t) = \frac{dq}{dt} = \sum_{m,n=-\infty}^{+\infty} j\Omega_{mn} Q_{mn} \exp(j\Omega_{mn} t) \tag{6.98}$$

The mean power absorbed by the capacitor at the angular frequency Ω_{mn} is written, according to (6.92):

$$\bar{P}_{mn} = U_{mn}^* I_{mn} + U_{mn} I_{mn}^* \tag{6.99}$$

$$= j\Omega_{mn}(U_{mn}^* Q_{mn} - U_{mn} Q_{mn}^*) \tag{6.100}$$

Let us now consider a new almost periodic function $q'(t)$ which has the same Fourier coefficients, but whose basic angular frequencies are Ω'_1 and Ω'_2. Consequently, $q'(t)$ has the expansion (6.116) with Ω_{mn} being replaced by

$$\Omega'_{mn} = m\Omega'_1 + n\Omega'_2 \tag{6.101}$$

According to property 6.2.6, the function $u'(t) = h(q'(t))$ also has the same Fourier coefficients as $u(t)$; only Ω_{mn} must be changed to Ω'_{mn} in (6.97) to obtain $u'(t)$. The mean power absorbed by the capacitor at the angular frequency Ω'_{mn}, when the time evolution of the charge is $q'(t)$, is thus written

$$\bar{P}'_{mn} = j\Omega'_{mn}(U_{mn}^* Q_{mn} - U_{mn} Q_{mn}^*) \tag{6.102}$$

Then,

$$\bar{P}'_{mn} - \bar{P}_{mn} = jm(\Omega'_1 - \Omega_1)(U_{mn}^* Q_{mn} - U_{mn} Q_{mn}^*) \tag{6.103}$$

and

$$\lim_{\Omega'_1 \to \Omega_1} \frac{\bar{P}'_{mn} - \bar{P}_{mn}}{\Omega'_1 - \Omega_1} = jm(U_{mn}^* Q_{mn} - U_{mn} Q_{mn}^*) \tag{6.104}$$

$$= jm\bar{P}_{mn}/\Omega_{mn} \tag{6.105}$$

Nonetheless, (6.94) holds for $\bar{P}_{mn}$ and $\bar{P}'_{mn}$, from which we infer

$$\lim_{\Omega'_1 \to \Omega_1} \frac{1}{\Omega'_1 - \Omega_1}\left[\sum_{m,n=0}^{\infty}(\bar{P}'_{mn} - \bar{P}_{mn}) + \sum_{m,n=1}^{\infty}(\bar{P}'_{m-n} - \bar{P}_{m-n})\right] = 0 \tag{6.106}$$

By combining (6.106) and (6.105), we obtain

$$\sum_{m,n=0}^{\infty} \frac{m}{m\Omega_1 + n\Omega_2} \bar{P}_{mn} + \sum_{m,n=1}^{\infty} \frac{m}{m\Omega_1 - n\Omega_2} \bar{P}_{m-n} = 0 \qquad (6.107)$$

By exchanging the roles of Ω_1 and Ω_2, we obtain

$$\sum_{m,n=0}^{\infty} \frac{n}{m\Omega_1 + n\Omega_2} \bar{P}_{mn} + \sum_{m,n=1}^{\infty} \frac{-n}{m\Omega_1 - n\Omega_2} \bar{P}_{m-n} = 0 \qquad (6.108)$$

We call (6.107) and (6.108) the *Manley-Rowe equations*.

6.2.14 Comments

The Manley-Rowe equations allow us to establish basic limitations for the efficiency of a modulator or a demodulator with nonlinear reactances.

The principle is as follows. Let us assume that a carrier (signal) source supplies the mean power $\bar{P}_\Omega(\bar{P}_\omega)$ at the angular frequency $\Omega_1 = \Omega(\Omega_2 = \omega)$ to the nonlinear reactance, which returns the power $\bar{P}_{\Omega+\omega} = \bar{P}_\Omega + \bar{P}_\Omega$ at the angular frequency $\Omega + \omega$. This amounts to setting $\bar{P}_{10} = \bar{P}_\Omega$, $\bar{P}_{01} = \bar{P}_\Omega$, and $\bar{P}_{11} = -\bar{P}_{\Omega+\omega}$. Therefore, we have assumed that nothing was lost in this transformation process, which means that all the other $\bar{P}_{mn}$ are zero. Equation (6.108) becomes

$$\bar{P}_{01}/\Omega_2 + \bar{P}_{11}/(\Omega_1 + \Omega_2) = 0 \qquad (6.109)$$

and, therefore,

$$\bar{P}_{\Omega+\omega} = -\bar{P}_{11} = P_{01}(\Omega_1 + \Omega_2)/\Omega_2 = \bar{P}_\omega(1+\Omega/\omega) \qquad (6.110)$$

We observe that the signal is not only modulated toward the upper sideband, but that it is also amplified by a factor $(1 + \Omega/\omega)$. The power required for this amplification comes from the carrier source, and this can be checked by writing (6.107). If (6.110) is applied to the demodulator, we observe that the signal is attenuated by the same factor $(1 + \Omega/\omega)$.

If the nonlinear reactance produces power at the angular frequency $\Omega - \omega$ (i.e. the lower sideband), we have $\bar{P}_{\Omega-\omega} = P_{1-1}$ and, instead of (6.109), we find

$$\bar{P}_{01}/\Omega_2 - \bar{P}_{1-1}/(\Omega_1 - \Omega_2) = 0 \qquad (6.111)$$

What is surprising in (6.111) is that $\overline{P}_{01}$ and $\overline{P}_{1-1}$ may have the same sign. We establish the signs by writing (6.107):

$$\overline{P}_{10}/\Omega_1 + \overline{P}_{1-1}/(\Omega_1 - \Omega_2) = 0 \tag{6.112}$$

therefore,

$$\overline{P}_{\Omega - \omega} = \overline{P}_{\Omega}\,(1 - \omega/\Omega) \tag{6.113}$$

and, by (6.111):

$$-\overline{P}_{\omega} = \overline{P}_{\Omega}\,\omega/\Omega \tag{6.114}$$

Consequently, the signal source does not supply any power to the nonlinear reactance. On the contrary, it receives some from it. This phenomenon is called the *Hartley effect*. In the usual case, where $\Omega \gg \omega$, however, practically all the power of the carrier is transformed into the lower sideband, and only a small part goes to the signal angular frequency. Nevertheless, this small part can cause the instability of the circuit which prepares the signal.

The modulation gain suggests that a signal could be amplified by alternating modulation and demodulation. However, what is gained by modulation is lost through demodulation. In *magnetic amplifiers*, the signal is modulated by a nonlinear inductor and demodulated by a diode. This method makes it possible, in principle, to keep the modulation gain.

Another type of amplifier exploits the Hartley effect. The power developed by the reactance at the angular frequency of the signal is considered as the superposition of a power $\overline{P}_s$ supplied by the signal source and of a power $\overline{P}_r$ returned by the reactance. Such a decomposition is obtained by introducing an incident wave and a reflected wave (vol. VI, sec. 6.3; vol. XIX, sec. 2.3). Because the balance $\overline{P}_{\omega} = \overline{P}_s - \overline{P}_r$ is negative, the reflected wave constitutes an amplification of the incident wave. This purely formal interpretation is put into practice by separating the incident waves and the reflected waves by a nonreciprocal linear three-port. In particular, the *circulator* of figure 6.7 is a nondissipative linear three-port, such that the incident wave $\xi_1(\xi_2, \xi_3)$ is equal to the reflected wave $\eta_2(\eta_3, \eta_1)$.

The basic circuit of an amplifier based on this principle, which is called a *parametric amplifier*, is represented in figure 6.8. The signal source supplies the power $\overline{P}_s$ to port 1 of the circulator, and the charge dissipates the power $\overline{P}_r$ it receives from port 3. The nonlinear capacitor connected to port 2 transforms the power $\overline{P}_{\Omega}$ developed by the angular frequency Ω. According to (6.114), we have

$$\overline{P}_r - \overline{P}_s = \overline{P}_{\Omega}\,\omega/\Omega \tag{6.115}$$

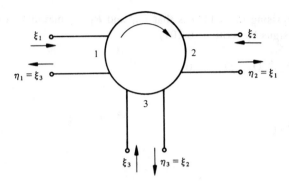

Fig. 6.7

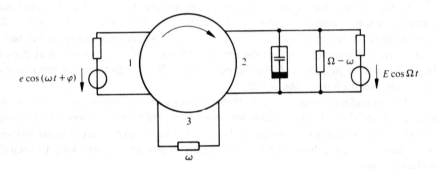

Fig. 6.8

otherwise

$$\overline{P}_r = \overline{P}_s + \overline{P}_\Omega \, \omega/\Omega \tag{6.116}$$

In order to satisfy the power balance, we must still provide some power dissipation at the angular frequency $\Omega - \omega$ in the second load, even if we do not use it in this case. Of course, the amplifier gain (6.116) constitutes a limit, which is only reached when all the other powers $\overline{P}_{k,mn}$ vanish. The only way to approximate this ideal situation is to incorporate filters in the two loads in order to eliminate undesirable angular frequencies. Parametric amplifiers are used in microwave circuits where non-reciprocal multiports can easily be implemented (vol. XIII, sec. 4.10).

Also, it should be noted that for a nonlinear reactance, we have $\overline{P}_{00} = 0$. This is derived from (6.100), as $\Omega_{00} = 0$. Consequently, it is out of the question to synthesize a rectifier or an oscillator with operation based on nonlinear reactances.

Relations (6.107) and (6.108), which have been derived for a nonlinear reactance, can be generalized to the case of a whole circuit by making use of property 6.2.11.

Chapter 7

Optimization of Diode Circuits

7.1 IDEAL FILTER METHOD

7.1.1 Introduction

As mentioned in the introduction to this book, the frequency method, which is ideal for linear circuits, is failing with nonlinear circuits because neither the isomorphic response theorem, nor the superposition theorem are valid. The ideal filter method is meant to extend as much as possible the linear circuit approach that remains valid for an important class of practical nonlinear circuits [42].

We are concerned particularly with the circuits defined in sections 3.3 and 3.4, having solutions that are periodic or almost periodic, and the function of which is always to effect a frequency change: rectifiers, oscillators, frequency multipliers, and modulators. These circuits are inserted into systems which include linear parts to simulate a linear system globally. In particular, even if the superposition and the isomorphic response theorems are not valid for certain subsystems, they must be valid for the global system, which is considered to be a relation between input and output.

7.1.2 Example

Let us consider the link between two telephone subscribers in its simplest case, where it has only one modulation (fig. 7.1). In cascade connection, we find: a lowpass filter, a modulator with a local oscillator, a line, a passband filter, and a detector. Although the modulator with its oscillator and the detector are nonlinear circuits, a good performance of the link supposes that a sinusoidal signal of the source in the interval from 0 to 3.4 kHz generates a sinusoidal frequency in the load at the same frequency, even if the amplitude and the phase have been modified. If two sinusoidal signals of angular frequencies ω_1 and ω_2 are simultaneously applied, the response should be the sum of the responses to the two signals separately applied; in other words, the superposition principle must apply.

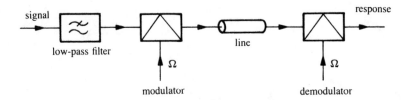

Fig. 7.1

Therefore, the modulator and the detector must belong to a special class of nonlinear circuits. Such circuits may and should perform certain frequency transposition functions. The response to a sinusoid is not a sinusoid of the same frequency, but it is still a sinusoid. The simultaneous excitation by two sinusoids of angular frequencies ω_1 and ω_2 cannot generate *intermodulation products* of angular frequencies $m\omega_1 + n\omega_2$, where m and n are integers different from zero. Were this the case, the signal would be affected by a *nonlinear distortion*, which would deteriorate the spectrum irreversibly.

7.1.3 Describing Function Principle

According to the foregoing discussion, certain circuits, while being nonlinear, still have a solution which is sinusoidal or almost sinusoidal, wherein the deviation with respect to the purely sinusoidal solution constitutes an imperfection to be kept within strict limits. Thus, at the terminals of a nonlinear one-port of that type, a sinusoidal voltage will generate a sinusoidal current, and *vice versa*. The difference with respect to the linear case lies in the fact that the current amplitude is no longer proportional to the voltage amplitude, but rather depends on it through a more complicated function, which is the *describing function*.

7.1.4 Example

Let us consider the circuit of figure 7.2 in which the source creates a voltage $e = E\cos\omega t$, the resonant circuit is tuned on the angular frequency $\omega = 1/\sqrt{LC}$, and the resistor is defined by the relation $u = ai + bi^3$. Because of its resonance, the LC circuit exhibits an impedance Z that is zero at the angular frequency ω. Thus, assuming that $i = I\cos\omega t$, we have

$$u = (aI + (3/4)bI^3)\cos\omega t + (bI^3/4)(\cos 3\omega t) \tag{7.1}$$

Both u and i have a term at ω, and the ratio of the two amplitudes defines a describing function $R(I)$, which has the dimensions of a resistor:

$$R = a + (3/4)bI^2 \tag{7.2}$$

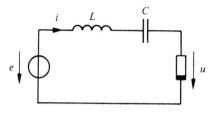

Fig. 7.2

Voltage u also has a term at the angular frequency 3ω of amplitude $bI^3/4$. Its amplitude depends on b: if this coefficient is not too large, this term can be disregarded.

Even if b is important, we can still disregard this term, provided that the filtering is selective enough. Indeed, a small current term at angular frequency 3ω will balance the voltage $bI^3/4$ if the value of $Z(3\omega)$ is sufficient. We find $Z(3\omega) = (8/3)jL\omega$, and by choosing L large enough, we can thus assign as small a value as desirable to the current amplitude at the angular frequency 3ω. It is, therefore, legitimate to assume, as we have done above, that, as an approximation, $i = I\cos\omega t$.

7.1.5 Ideal Filters

The preceding example shows that there are two instances in which the describing function method is applicable: when nonlinearity is low; when nonlinearity is high and the circuit is filtered. While the former instance is trivial, the latter is less so. It is commonplace to state that an almost linear circuit has an almost linear behavior, but it is more surprising to note that harmonic analysis applies to a highly nonlinear circuit if it is well filtered. We are primarily interested in the second case. Indeed, the circuits used in practice must be simultaneously highly nonlinear for the frequency change function to be effective and well filtered for the frequency analysis to make sense.

To avoid repeatedly returning to the discussion relating to filtering, we shall introduce two ideal one-ports: the *ideal current filter* and the *ideal voltage filter*, the impedances of which are defined, respectively, by the equations:

$$\begin{cases} Z(\omega_0) = 0 \\ Z(\omega) = \infty \quad \omega \neq \omega_0 \end{cases} \tag{7.3}$$

and

$$\begin{cases} Z(\omega_0) = \infty \\ Z(\omega) = 0 \quad \omega \neq \omega_0 \end{cases} \tag{7.4}$$

Hereafter, we shall represent the two filters by the one-ports of figures 7.3 and 7.4.

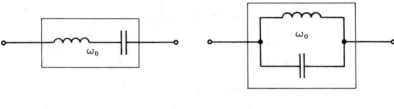

| Fig. 7.3 | Fig. 7.4 |

Of course, such impedances cannot be synthesized because the impedance of a circuit with lumped constants is a real rational function. We can, however, approximate these definitions as closely as desired by choosing a large enough value for the inductor of a resonant circuit. This last consideration is, in fact, nothing more than a formal clause which delimits, within the ideal element definition, the approximation part relating to the describing function method.

7.2 DIODE CIRCUITS BETWEEN RESISTIVE TERMINATIONS

7.2.1 Reminder

The definition of piecewise-linear diodes and ideal diodes was given in subsection 1.3.9. If we start with the passing $g_s(r_s)$ and blocking $g_o(r_o)$ conductances (resistances), we define two parameters:

$$\rho = \sqrt{r_s r_o} \tag{7.5}$$

$$\alpha = \sqrt{r_s/r_o} \tag{7.6}$$

The ideal diode has an indeterminate ρ and $\alpha = 0$. A linear resistor ($r_s = r_o$) would be characterized by $\alpha = 1$. A good diode has an $\alpha \ll 1$.

A circuit consisting of piecewise-linear diodes, capacitors, inductors, resistors, and sources satisfies the conditions of theorem 5.7.6, provided that there is not any pathological loop or cut set. It follows that the solutions of the circuit are globally asymptotically stable. There is only one steady-state solution. In particular, filtered circuits, to be studied in the following sections, have a harmonic response toward which all solutions tend, whatever the initial conditions. Hence, the solution found by the describing function method is always stable and unique.

Indeed, a pure harmonic solution is obtained by considering a circuit in which inductors tend toward infinity and capacitors tend toward zero, or *vice versa*. Insofar as the solutions are continuous functions of the inductor value or the capacitor value, the limit obtained when inductors and capacitors tend toward infinity and toward zero enjoys the properties of the family of the solution.

7.2.2 Simple Diode

Consider the circuit of figure 7.5, where $e = \sqrt{2E} \cos \omega t$. We have $i = e/(R_1 + R_2 + r)$, where r is the variable resistance of the diode given by

$$r = (r_s + r_0)/2 + (\text{sgn}(u))(r_s - r_0)/2 = a + b \, \text{sgn}(u) \tag{7.7}$$

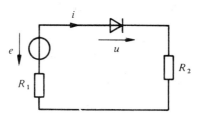

Fig. 7.5

By setting $A = R_1 + R_2 + a$, while taking into account $(\text{sgn}(u))^2 = 1$, we find

$$i = e/(A + b \cdot \text{sgn}(u)) = e \cdot (A - b \cdot \text{sgn}(u))/(A^2 - b^2) \tag{7.8}$$

Thus, the current has a term $e \cdot A/(A^2 - b^2)$ of the same frequency as the source, which is of no use. The term sgn (u) is the rectangular wave sgn(cosωt) because current i of this purely resistive circuit is in phase with e and that $u = ri$. We have

$$\text{sgn}(\cos \omega t) = (4/\pi) \sum_{n=0} (-1)^n \cos((2n+1)\omega t)/(2n+1) \tag{7.9}$$

and

$$\cos \omega t \, (\text{sgn} \cos \omega t) = (2/\pi) \sum_{n=0}^{\infty} (-1)^n [\cos 2n\omega t + \cos(2n+2)\omega t]/(2n+1) \tag{7.10}$$

$$= -(2/\pi) \sum_{n=0}^{\infty} (-1)^n \cos 2n\omega t/(4n^2 - 1) \tag{7.11}$$

If we multiply this last expression by the amplitude $-\sqrt{2E}b/(A^2 - b^2)$, we find the values of the current terms, which are different from the fundamental one, and we discover that they are all even harmonics of ω.

7.2.3 Simple Diode Rectification

The dc term is

$$I_0 = -2\sqrt{2}bE/(\pi(A^2 - b^2)) \tag{7.12}$$

and we easily find that $A^2 - b^2 = (R_1 + R_2 + r_o)(R_1 + R_2 + r_s)$. The *dc conversion transmittance* can be written with $V_0 = R_2 I_0$:

$$s_{12} = \frac{2V_0\sqrt{R_1}}{E\sqrt{R_2}} = \frac{\sqrt{2}}{\pi}\frac{2(r_o - r_s)\sqrt{R_1 R_2}}{(R_1 + R_2 + r_s)(R_1 + R_2 + r_o)} \tag{7.13}$$

This expression for the transmittance has been chosen by analogy with the off-diagonal term of the distribution matrix: its square is the ratio of the continuous power dissipated in the charge to the maximum active power, which can be drawn from the source at angular frequency ω.

Formula (7.13) is symmetrical in R_1 and R_2: it is maximum for $R_2 = R_1 = R$. The maximum is obtained for $R = \rho/2$. We thus obtain

$$s_{12} = \frac{\sqrt{2}}{\pi}\frac{(r_o - r_s)\rho}{(\rho + r_s)(\rho + r_o)} = \frac{\sqrt{2}}{\pi}\frac{1 - \alpha}{1 + \alpha} \tag{7.14}$$

If we calculate the attenuation:

$$A = \ln(1/|s_{12}|) = \ln(\pi/\sqrt{2}) + \ln((1 + \alpha)/(1 - \alpha)) \tag{7.15}$$

and if we use the equivalent formulas:

$$e^{2x} = (1 + \alpha)/(1 - \alpha) , \quad \text{th}\, x = \alpha \tag{7.16}$$

we then find

$$A = \ln(\pi/\sqrt{2}) + 2\,\text{arcth}\,(\alpha) \tag{7.17}$$

7.2.4 Comments

Formula (7.17) shows two terms in the attenuation. The first one is independent of the diode quality: it is 0.8 Np or 6.93 dB. It measures the dc to ac *conversion attenuation* by means of the circuit in figure 7.5 and it is the only attenuation if the diode is ideal, that is, if $\alpha = 0$. The second term is the *dissipation attenuation* due to the diode; it is 2 arcth $(\alpha) \doteq 2\alpha$ Np or 17.4α dB.

If we take ρ as the reference resistor, the dual of the diode is the diode itself connected in the opposite direction, that is, in such a way that the periods where it is passing become blocking, and *vice versa*. The dual of the circuit of figure 7.5 is the circuit of figure 7.6. The optimum terminations are 2ρ, and the attenuation is always (7.17).

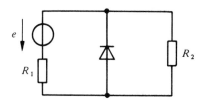

Fig. 7.6

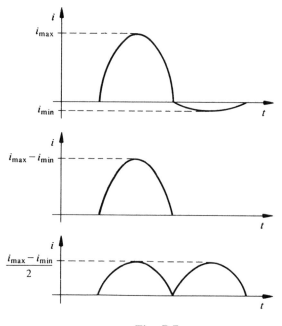

Fig. 7.7

Note that for both figure 7.5 and figure 7.6, the linear resistors at the diode terminals are in the optimum case ρ as a whole, i.e., the geometric average between the resistors r_s and r_o, each of which being the diode resistor half of the time. For most of the circuits studied hereafter, we shall find a similar rule.

The waveform of a current flowing in the circuit of figure 7.6 is represented in figure 7.7 for a period of the source. During the positive (negative) half-waves of the source, the current has an amplitude $i_{max} = \sqrt{2}E/(r_s + R_1 + R_2)$ and $i_{min} =$

$\sqrt{2}E/(r_o + R_1 + R_2)$, respectively. This current can be decomposed by subtracting a sinusoid of amplitude i_{min}, which is not of any interest for frequency multiplication or rectification. Hence, we find the half-sinusoid in the second diagram of figure 7.7. Additionally, this wave contains a term at the angular frequency ω of amplitude $i_{min} + (i_{max} - i_{min})/2 = (i_{max} + i_{min})/2 = \sqrt{2}E(r_s + r_o)/[2(r_s + R_1 + R_2)(r_o + R_1 + R_2)]$. After subtraction of this term, we find the rectified full-wave sinusoid of the last diagram, which no longer has any fundamental. As a whole, the eliminated fundamental term has an amplitude $eA/(A^2 - b^2)$, which indeed corresponds to the term, eliminated from the beginning in subsection 7.2.2.

7.2.5 Simple Diode Frequency Multiplication

By (7.11) we find, for the mth-order even harmonic, the amplitude:

$$\frac{2\sqrt{2}}{\pi}\ \frac{Eb}{A^2 - b^2}\ \frac{(-1)^{m/2}}{m^2 - 1} \tag{7.18}$$

Therefore, the conversion transmittance is written

$$s_{12} = \frac{4}{\pi(m^2 - 1)}\ \frac{(r_o - r_s)\sqrt{R_1 R_2}}{(R_1 + R_2 + r_s)(R_1 + R_2 + r_o)} \tag{7.19}$$

and the attenuation is

$$A = \ln(\pi(m^2 - 1)/2) + 2\,\text{arcth}\,(\alpha) \tag{7.20}$$

We find again, as in formula (7.17), a conversion attenuation which increases here as the square of the mth-order harmonic, and the same dissipation attenuation between the same optimum terminations.

7.2.6 Diode with Capacitive Polarization

The circuit of figure 7.5 is limited to an even harmonic generation. This results from figure 7.7 (see also the rule in vol. IV, sec. 7.4.12). We can also understand this result by considering that the characteristic of a piecewise linear diode can be written

$$g = (g_s + g_o)/2 + (g_s - g_o)\,\text{sgn}\,(u/2) \tag{7.21}$$

We can thus decompose the diode, according to figure 7.8, into two conductances in parallel, one of which is linear and the other nonlinear, such that the current flowing through it is $(g_s - g_o)|u|/2$ and an even function of u. As a necessary result, the current flowing through this nonlinear branch will satisfy the relation $i(t) = i(t + T/2)$ with $T = 2\pi/\omega$, if u is a periodic function of an angular frequency ω, such that

$$u(t) = -u(t + T/2) \tag{7.22}$$

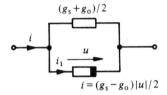

Fig. 7.8

The only way to generate odd harmonics is by either altering relation (7.21) so that its odd part would not be linear any more, or not applying a voltage satisfying (7.22). In the case with which we are concerned, both methods coalesce when the diode is polarized according to figure 7.9. The simplest way is to have an automatic polarization constituted by a resistor R_0 connected in parallel with a capacitor which is large enough to exhibit a negligible impedance for the alternative terms present in the current. The characteristic of the resulting one-port is

$$i = (g_s + g_0)(u - U_0)/2 + (g_s - g_0)|u - U_0|/2 \tag{7.23}$$

which is represented in figure 7.10. The odd part of this characteristic is calculated by

$$i_i = [i(u) - i(-u)]/2 \tag{7.24}$$

and is obviously not linear, as can be seen in the graphic construction of figure 7.11.

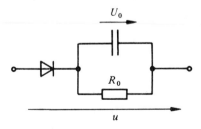

Fig. 7.9

410

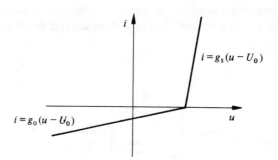

Fig. 7.10

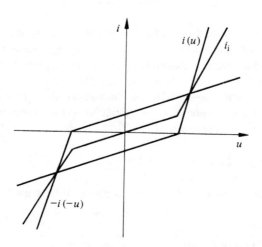

Fig. 7.11

If we apply a voltage $e = \sqrt{2}E \cos\omega t$ to the terminals of the one-port of figure 7.9, the diode will be passing for $e - U_0$ positive, i.e., from $-\theta$ to $+\theta$ during the period running from $-\pi$ to $+\pi$, with the *conduction angle* θ being given by

$$\sqrt{2}E \cos\theta = U_0 \tag{7.25}$$

In the calculation of harmonics, we disregard the source and the polarization leakage current $g_0(e - U_0)$. We must then analyze the current $g_s(e - U_0)$ from $-\theta$ to $+\theta$, and zero elsewhere. The amplitude of the harmonic n is

$$\sqrt{2}I_n = (2/\pi)(g_s - g_o) \cdot \int_0^\theta (e - U_0)\cos(n\omega t)\, d(\omega t) \tag{7.26}$$

according to the Fourier expansion of the wave. By making the calculation and eliminating U_0 with (7.25), we have

$$I_n = \frac{g_o - g_s}{n\pi} E \left[\frac{\sin(n+1)\theta}{n+1} - \frac{\sin(n-1)\theta}{n-1} \right] = \frac{g_o - g_s}{n\pi} ED(\theta) \tag{7.27}$$

By an analogous calculation, we find the dc term and the angular frequency term ω of the analyzed wave, to which we must add the leakage current term:

$$I_1 = E\left[\frac{g_s - g_o}{\pi} \left(\theta - \frac{\sin 2\theta}{2} \right) + g_o \right] = E\left[\frac{g_s - g_o}{\pi} C(\theta) + g_o \right] \tag{7.28}$$

$$I_0 = U_0\left[\frac{g_s - g_o}{\pi} (\operatorname{tg}\theta - \theta) - g_o \right] = U_0\left[\frac{g_s - g_o}{\pi} A(\theta) - g_o \right] \tag{7.29}$$

7.2.7 Odd Harmonic Generation by the Polarized Diode

Consider the circuit of figure 7.12 in which we assume, by analogy with the result of subsection 7.2.3, that the terminations are $R_1 = R_2 = R$ and equal. The current and its Fourier expansion are calculated, as in subsection 7.2.6, by replacing g_s with $(r_s + 2R)^{-1}$, and g_o with $(r_o + 2R)^{-1}$. From (7.27), we have

$$\left| \frac{I_n}{E} \right| = \frac{(r_o - r_s)}{n\pi(r_o + 2R)(r_s + 2R)} D(\theta) \tag{7.30}$$

and

$$s_{12} = 2\left| \frac{RI_n}{E} \right| = \frac{2}{n\pi} \frac{(r_o - r_s)R}{(r_o + 2R)(r_s + 2R)} D(\theta) \tag{7.31}$$

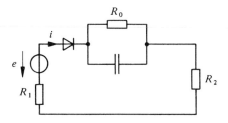

Fig. 7.12

The function $D(\theta)$ is maximum for the value of θ, which cancels

$$dD/d\theta = 2 \sin(n\theta) \sin \theta = 0 \tag{7.32}$$

We eliminate the trivial solution $\theta = 0$, which would correspond to a conduction angle zero, and therefore to an harmonic which is also zero. For $\sin n\theta = 0$, or $\theta = k\pi/n$, we have

$$D(\theta) = (-1)^k (2n/(n^2 - 1)) \sin(k\pi/n) \tag{7.33}$$

and the absolute value of $D(\theta)$ will increase as $k\pi/n$ becomes closer to $\pi/2$. Therefore, the best choice of k is $(n - 1)/2$, which gives $\theta = \pi/2 - \pi2n$ and $\sin(k\pi/n) = \cos(\pi/2n)$.

Also, the factor accounting for $R, r_o,$ and r_s in (7.31) is identical to the one involved in (7.13), and will be maximized by the identical choice $R = \rho/2$. In this case, we have

$$A = \ln \frac{\pi(n^2 - 1)}{2 \cos \pi/2n} + 2 \operatorname{arcth}(\alpha) \tag{7.34}$$

Finally, we must choose R_0 so that it gives the appropriate value of U_0. We shall calculate by formula (7.29), where g_s and g_o are replaced by the values mentioned at the start of this subsection, and where $U_0 = R_0 I_0$. By disregarding the terms at α, we find, to the first order:

$$R_0 = \pi\rho/[\cotg(\pi/2n) - \pi/2 - \pi/(2n)] \tag{7.35}$$

The current wave flowing in the circuit consists of sinusoid peaks, as represented in the lower diagram of figure 7.16. The circuit of figure 7.12 is a *limiter*.

7.2.8 Inductive Polarization Diode

While the dual of a single diode is the same diode, this is no longer true for the dual of the one-port of figure 7.9, which is represented in figure 7.13, and has a

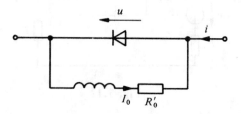

Fig. 7.13

characteristic as shown in figure 7.14. We assume that the inductor is large enough to carry a constant current I_0. The characteristic of figure 7.14 permits, as does that of figure 7.10, the generation of odd harmonics.

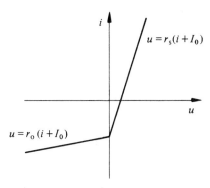

Fig. 7.14

The simplest circuit is that of figure 7.15. Its optimum operation is identical to that of figure 7.12, of which it is the dual. Of course, in a technology where inductors are made out of coils, it is better to use a capacitive polarization circuit for economical reasons.

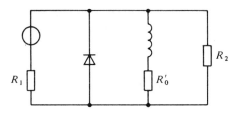

Fig. 7.15

The one-port of figure 7.13 can also be inserted in series between the source and the charge. In that case, the current flowing in the charge of figure 7.17 has the form of a peakless sinusoid, as represented on the second line of figure 7.16. During the blocking period, the only current which flows is the constant current I_0 of the inductive polarization. During the passing period, the current is a sinusoid. Although i is positive, the current $i - I_0$ in the diode is negative, which indeed corresponds to the state of the diode.

By considering figure 7.16, we can see that the sum of the currents in the circuits 7.12 and 7.17 is a sinusoid at the angular frequency ω, provided that the conduction

414

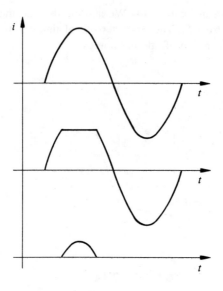

Fig. 7.16

angle θ is the same. The Fourier analysis of the two lower waves of figure 7.16 necessarily gives amplitudes which are equal for each harmonic, except for the sign, because these harmonics must vanish in the sum of these two waves. Hence, if each termination of circuit 7.17 is also equivalent to $\rho/2$, the circuit attenuation will be given by (7.34). The reader may wish to calculate, for practice, $R'_0 = \rho^2/R_0$ in this optimum case. It is understood that the circuit of figure 7.17 has a dual with capacitive polarization (to be drawn for practice).

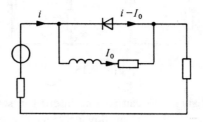

Fig. 7.17

7.2.9 Full-Wave Diode Circuits

The useful part of the voltage on the upper line of figure 7.7 is the half-sinusoid of the second line. This half-sinusoid also contains another term ω, which is of no interest for frequency multiplication or rectification. The purpose of a full-wave circuit is to obtain directly an output current, which is a sinusoid with two rectified half-waves.

The circuits of figures 7.18 and 7.19, respectively called the *Graetz bridge* and the *Jaumann structure*, and located between resistive terminations, precisely implement this function. This is evident for the second circuit, which is simply the combination of two circuits, according to figure 7.5, placed side by side and fed by sources with phases that differ by 180°, due to the ideal transformer. This version of the full-wave circuit clearly shows that what we want to keep is the even part of the diode characteristic only, by making the addition $i(u) + i(-u)$, which, of course, will only generate the even harmonics. In the circuit of figure 7.18, which we shall analyze in detail, the two diodes $D_1(D_2)$ are passing during the positive (negative) half-waves of the source. We naturally assume that all the diodes have the same characteristic.

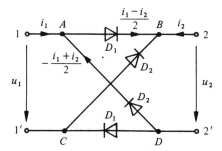

Fig. 7.18

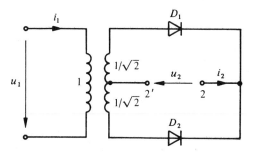

Fig. 7.19

According to the Kirchhoff voltage lemma, we find

$$u_{AB} = u_{DC} = (u_1 - u_2)/2 \tag{7.36}$$

$$u_{CB} = u_{DA} = -(u_1 + u_2)/2 \tag{7.37}$$

From the Kirchhoff current lemma, we find that the currents in D_1 and D_2 are respectively equivalent to $(i_1 - i_2)/2$ and $-(i_1 + i_2)/2$. The two-port equations are thus written

$$u_1 - u_2 = r_1 (i_1 - i_2) \tag{7.38}$$
$$u_1 + u_2 = r_2 (i_1 + i_2) \tag{7.39}$$

where r_1 and r_2 denote the values of (7.7) for D_1 and D_2.

Let us assume that the two-port is placed between a source such that $u_1 = e - R_1 i_1$ and a charge $u_2 = - R_2 i_2$.

During the positive half-waves, $r_1 = r_2$ and $r_2 = r_o$, and during the negative half-waves the reverse is true. Whatever the half-wave, the Graetz bridge is equivalent to a lattice having branch impedances which are r_o and r_s. The image impedances of this lattice are $\sqrt{r_o r_s} = \rho$ (according to vol. XIX, sec. 3.1.11), and the image attenuation is 2 arcth $(r_s/r_o)^{1/2} = 2$ arcth(α). Thus, the dissipation attenuation remains what it was in formulas (7.17) and (7.20).

Furthermore, the branch conductances change value for each half-wave, and the two-port of figure 7.18 can thus be modeled between resistive terminations by the circuit of figure 7.20, which includes a periodic switch, i.e., a two-port that is equivalent to an ideal transformer with a ratio alternating between $+1$ and -1.

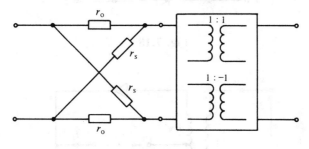

Fig. 7.20

Thus, the output current indeed has the aspect represented on the last line in figure 7.7. The term at the source frequency has disappeared, and all the even harmonics, including the dc term, have doubled in value. The conversion attenuations of (7.17) and (7.20), therefore, must be reduced by 6 dB.

7.2.10 Full-Wave Polarized Diode Circuits

The problem here is to isolate with a circuit the only odd term given by (7.23) and represented in figure 7.11. We simply connect two one-ports in parallel according to figure 7.9, the diodes of which are oriented in opposite directions. We obtain the circuit of figure 7.21, implementing the function $i(u) - i(-u)$, that is an odd function of u.

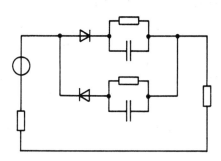

Fig. 7.21

If the leakage current is not taken into account, the current has the form of figure 7.22, consisting of the two peaks of the sinusoid. It only includes odd harmonics, and when comparing it with the last line of figure 7.16, it is obvious that the amplitude of each one of these harmonics has been multiplied by a factor 2, which again amounts to a 6 dB reduction of the conversion attenuation (7.34).

Fig. 7.22

The dissipation attenuation must be calculated with respect to the circuit, as well as the optimum terminations and the polarization resistors. We shall simply give here the results of a calculation the reader may wish to make for practice. The optimum terminations are $\rho/(2\sqrt{2})$, the polarization resistor is R_0, given by (7.35), divided by $\sqrt{2}$, and the dissipation attenuation equals $2\sqrt{2}$ arcth (α).

The systematic occurrence of this factor $\sqrt{2}$ at the outcome of an explicit calculation can be understood by comparison with the foregoing circuits. Circuits of figures 7.5 and 7.12 include a diode having an internal resistance which oscillates between the values r_s and r_o, and a resistance at the terminals which is optimal when equivalent to the geometric average of these values. The circuit of figure 7.21 includes a nonlinear one-port having an internal resistance that oscillates between $r_o/2$ when

418

both diodes are blocking, and r_s (while disregarding r_o in parallel) when one is passing. Here, the geometric average is $\rho/\sqrt{2}$.

7.2.11 Double Polarization Diode Circuits

Instead of combining the currents coming from two identical one-ports, as was done in subsection 7.2.10, we can combine the currents coming from the one-ports of figures 7.9 and 7.19. Figure 7.16 shows that these currents have complementary forms, provided that the conduction angles are 2θ and $2\pi - 2\theta$. The harmonic amplitudes are equal, except for the sign. To add them up, a bridge connection is required, as represented in figure 7.23, which constitutes a half-wave double-polarization circuit. Apart from leakage currents, the output current should have the aspect represented in figure 7.24

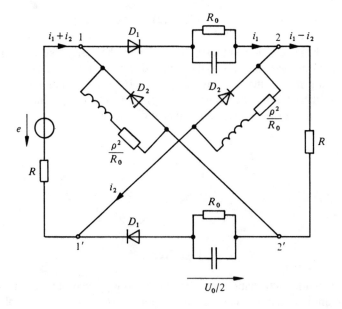

Fig. 7.23

Correct bridge operation implies that the diodes switch over simultaneously and in the opposite direction. Rather than undertaking a general analysis of the circuit, we shall assume that the optimum circuit is transformed into itself by duality except for the inversion of the straight and crossed branches of the lattice. This assumption is based on the idea that in this case the optimum filter and its dual only differ by the sign of the output current, which does not affect the attenuation. In this hypothesis, the polarization resistances respectively are R_0 and ρ^2/R_0, while the terminations are ρ. We shall check that this hypothesis indeed generates the desired switching.

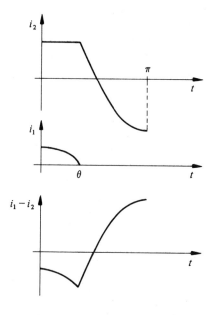

Fig. 7.24

The equations of the two loops, respectively consisting of the two straight branches with terminations and the two crossed branches with terminations of the lattice, are expressed as

$$e - U_0 = 2(\rho + r_1) i_1 \tag{7.40}$$
$$e = 2(\rho + r_2) i_2 - r_2 I_0 \tag{7.41}$$

From which we can infer the current expressions:

$$i_1 = (e - U_0)/(2(\rho + r_1)) \tag{7.42}$$
$$i_2 = (e + r_2 I_0)/(2(\rho + r_2)) \tag{7.43}$$

The switching of the first diode is performed for the angle θ defined by

$$\sqrt{2} E \cos \theta = U_0 \tag{7.44}$$

starting from (7.42). To obtain the same switching angle of the diodes D_2, the current:

$$i_2 - (I_0/2) = (e - \rho I_0)/(2(\rho + r_2)) \tag{7.45}$$

must change sign for the same argument. It is, therefore, necessary that

$$U_0 = \rho I_0 \tag{7.46}$$

which is in accordance with the argument that the circuit is transformed into itself by duality.

By adding (7.42) and (7.43), while taking into account (7.46) and $r_1 r_2 = \rho^2$, as a result of the switching in the opposite direction of the two pairs of diodes, we find

$$i_1 + i_2 = \frac{e(2\rho + r_1 + r_2)}{2(\rho + r_1)(\rho + r_2)} = \frac{e(2\rho + r_o + r_s)}{2(\rho + r_s)(\rho + r_o)} \tag{7.47}$$

thus, the current in the source is sinusoidal. This is the first example of a nonlinear two-port exhibiting a terminated input impedance which is linear. Our interest in this situation stems from the fact that no harmonic is flowing through the source impedance and is dissipated in it. Indeed, we observe that (7.47) is equivalent to the formula:

$$e = 2\rho(i_1 + i_2) \tag{7.48}$$

which expresses that the input impedance of the two-port is precisely equivalent to ρ, in other words, that the matching is carried out at the input.

If the sum of i_1 and i_2 gives a sinusoid, then their difference, i.e., the output current, indeed contains twice as many odd harmonics as the circuit studied in subsection 7.2.7. The attenuation (7.34) is replaced by

$$A = \ln \frac{\pi(n^2 - 1)}{4\cos(\pi/2n)} + 2 \operatorname{arcth}(\alpha) \tag{7.49}$$

then 6 dB have been gained, as expected.

The calculation of R_0, which is given by formula (7.35), is left to the reader for practice.

7.2.12 Exercise

Draw the generating circuit of a double-polarization, full-wave, odd-harmonic multiplier. What is the output waveform? What is the diode switching diagram? Show that the attenuation is 6 dB lower than the attenuation given by (7.49).

7.3 RECTIFIERS

7.3.1 Half-Wave Rectification

The simplest rectifying circuit is that of figure 7.25. Because R_2 only carries a constant current and R_1 carries a current at the source angular frequency ω, there is no power dissipation, except on the diode. These are the optimum conditions mentioned in subsection 6.1.15 to obtain an *ideal rectification*. This means that the maximum active power that can be drawn from the source, i.e., $E^2/4R_1$, is converted into dc power on the charge R_2 except for the losses on the diode, which are to be minimized. We consider that the source and the diode are given and that R_1 and R_2 must be determined to effect the input matching and minimize losses on the diode.

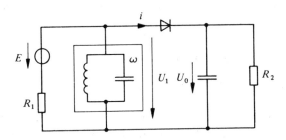

Fig. 7.25

The analysis of the circuit is immediate from the formulas (7.28) and (7.29), which give the dc term and the angular frequency term ω for a diode subjected to a voltage $\sqrt{2}U_1\cos\omega t - U_0$ if U_1 is substituted for E. The input matching is obtained if $E - R_1I_1 = R_1I_1$, or, also, if $E = 2U_1$ and $U_1 = R_1I_1$. By (7.28), written with U_1 instead of E, we thus have

$$R_1[((g_s - g_o)/\pi)\,C(\theta) + g_o] = 1 \tag{7.50}$$

This formula gives the value of R_1 to be chosen to obtain the matching as a function of the conduction angle θ. That angle is determined by the choice of R_2. Indeed, we have $U_0 = R_2I_0$, which gives, with (7.29):

$$R_2[((g_s - g_o)/\pi)\,A(\theta) - g_o] = 1 \tag{7.51}$$

Also, the conduction angle determines a relation:

$$\sqrt{2}U_1\cos\theta = U_0 \tag{7.52}$$

The transmission coefficient is written

$$s_{12} = \frac{2U_0}{E}\sqrt{\frac{R_1}{R_2}} = \frac{2\sqrt{2}U_1\cos\theta}{E}\sqrt{\frac{R_1}{R_2}} = \cos\theta\sqrt{\frac{2R_1}{R_2}} \qquad (7.53)$$

By (7.51) and (7.50), we find

$$s_{12}^2 = \frac{2\cos^2\theta[(g_s - g_o)A(\theta) - \pi g_o]}{(g_s - g_o)C(\theta) + \pi g_o} \qquad (7.54)$$

We can reasonably assume that θ is small. Indeed, from 0 to θ, the difference $\sqrt{2}U_1\cos\omega t - U_o$ is equal to the voltage drop $r_s i$, where r_s is small. Therefore, the conduction angle must not be too large. To optimize the function $s_{12}(\theta)$, we can thus replace it by its MacLaurin expansion in series starting from the formulas:

$$\cos\theta = 1 - (\theta^2/2!) + (\theta^4/4!) - (\theta^6/6!) \ldots \qquad (7.55)$$

$$A(\theta) = (\theta^3/3) + (2\theta^5/15) + (17\theta^7/315) \ldots \qquad (7.56)$$

$$C(\theta) = (2\theta^3/3)[1 - (\theta^2/5) + (2\theta^4/105) + \ldots] \qquad (7.57)$$

In the expansion of (7.54), we must take into account another large quantity $\alpha^2 = g_o/g_s$. By keeping the first term at θ, which is of the order of θ^2, and the first term at α^2 which is of the order of α^2/θ^3, we obtain

$$s_{12}^2 \cong 1 - 2\theta^2/5 - 9\pi\alpha^2/2\theta^3 \qquad (7.58)$$

the maximum of which is obtained for

$$\theta^5 \cong 135\pi\alpha^2/8 \qquad (7.59)$$

Then, we find

$$s_{12} \cong 1 - (45/8)(\pi\alpha^2/\theta^3) \qquad (7.60)$$

and

$$A = 14,2\,\alpha^{4/5}\ \text{dB} \qquad (7.61)$$

The terminations are respectively equivalent to

$$R_2 = 2R_1 = 2(4\pi r_s/9)^{2/5}(r_o/5)^{3/5} \qquad (7.62)$$

7.3.2 Comments

The circuit of figure 7.25 indeed constitutes an ideal rectifier because the attenuation (7.61) can be made as small as desired by choosing a diode with a small enough α. It should be noted, however, that A cannot be made equal to zero, which means that the diode cannot be ideal. In this case, indeed, θ must be strictly equal to zero because $\sqrt{2}\, U_1 \cos \omega t - U_0$, i.e., the voltage at the terminals of the short circuit consisting of the passing diode, can be zero at only one instant.

In the case of the piecewise-linear diode, θ is small and the diode is blocking for much longer than it is passing. This makes it possible to understand why the terminations are not of the order of $\rho = r_o^{1/2} r_s^{1/2}$, as previously, but on the order of $r_s^{2/5} r_o^{3/5}$. This also explains why the dissipation attenuation is on the order of $\alpha^{4/5}$ instead of α.

7.3.3 Full-Wave Rectification

We could, of course, place the Graetz bridge of figure 7.18 between a source and a charge, both being in parallel with an ideal voltage filter, as in figure 7.25, but nothing really new would be obtained. However, the circuit of figure 7.26, which consists of a Graetz bridge placed between inverse filters, exhibits an interesting behavior. The variables at the bridge ports are partly defined by the ideal filters. We have

$$i_1 = \sqrt{2} I_1 \cos \omega t, \quad u_2 = U_0 \tag{7.63}$$

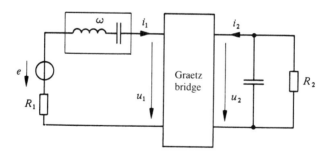

Fig. 7.26

Starting from the system of (7.38) and (7.39), we can express the two variables, which are not filtered, in the form:

$$i_2 = [(r_1 - r_2)\, i_1 + 2u_2]/(r_1 + r_2) \tag{7.64}$$

$$u_1 = [2r_1 r_2 i_1 + (r_2 - r_1)\, u_2]/(r_1 + r_2) \tag{7.65}$$

Equations (7.38) and (7.39) also show that the value of r_1 and r_2 depends on the sign of $(i_1 - i_2)$ and $-(i_1 + i_2)$. If the first (second) current is positive, $r_1(r_2)$ is r_o. Starting from (7.64), we can write

$$-(i_1 + i_2) = -2(u_2 + r_1 i_1)/(r_1 + r_2) \tag{7.66}$$

$$i_1 - i_2 = -2(u_2 - r_2 i_1)/(r_1 + r_2) \tag{7.67}$$

As a result, we have the following rule: in a plane (u_1, i_1), $r_1 = r_s$ in the area where $u_2 - r_2 i_1 < 0$, and $r_2 = r_s$ in the area where $u_2 + r_1 i_1 < 0$. Note that the passing or blocking state of a diode depends on a function in which the resistance of another diode is involved.

The diagram of figure 7.27 summarizes the conclusions that can be drawn from the study of (7.66) and (7.67). The equation $u_2 - r_2 i_1 = 0$ is represented by two half lines passing through the origin and located in the first and third quadrants. It is clear that $u_2 + r_1 i_1$ is negative in the third quadrant and $r_2 = r_s$, whereas it is positive in the first quadrant and $r_2 = r_o$. This explains the choice made for r_2 when writing the equations of the two half lines. An identical analysis allows us to explain the aspect taken by the boundary $u_2 + r_1 i_1$ between the areas where $r_2 = r_s$ and r_o. The four half lines determine four areas: I (D_1 passing, D_2 blocking), II (D_1 and D_2 blocking), III (D_1 blocking, D_2 passing), and IV (D_1 and D_2 passing).

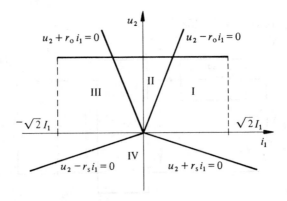

Fig. 7.27

The relation (7.63) constitutes the parametric equations of the locus that is representative of the bridge operation. This locus consists of a segment of a line parallel to the axis i_1, the ordinate of which at the origin is U_0. By setting

$$U_0 = \sqrt{2} r_o I_1 \cos\theta \tag{7.68}$$

we may determine the conduction angle θ. The representative point will then run through: state I from 0 to θ; state II from θ to $\pi - \theta$; state III from $\pi - \theta$ to π; the subsequent half-wave brings the representative point back to the starting point, while going through the same states in the opposite direction. Because the resulting waves are symmetrical with respect to π, the Fourier analysis will be made on the half period from 0 to π.

By (7.64) and (7.65), the unknown quantities, in state II, take the values:

$$i_2^{II} = u_2/r_o \qquad u_1^{II} = r_o i_1 \tag{7.69}$$

States I and III respectively generate the forms:

$$i_2 = [2u_2 \pm (r_s - r_o) i_1]/(r_o + r_s) \tag{7.70}$$

$$u_1 = [2r_o r_s i_1 \pm (r_o - r_s) u_2]/(r_o + r_s) \tag{7.71}$$

where the upper (lower) sign corresponds to I (III).

The dc term I_0 is calculated by

$$\pi I_0 = \int_0^\theta i_2^I \, d\omega t + \int_\theta^{\pi - \theta} i_2^{II} \, d\omega t + \int_{\pi - \theta}^\pi i_2^{III} \, d\omega t; \tag{7.72}$$

and is equivalent to

$$I_0 = \frac{U_0}{r_o} \left[1 - \frac{2}{\pi} \frac{r_o - r_s}{r_o - r_s} A(\theta) \right] \tag{7.73}$$

By an analogous calculation, we find

$$U_1 = r_o I_1 \left[1 + \frac{2}{\pi} \frac{r_o - r_s}{r_o + r_s} C(\theta) \right] \tag{7.74}$$

Because $U_0 = -R_2 I_0$, we find from (7.73):

$$\frac{r_o}{R_2} = \frac{2}{\pi} \frac{r_o - r_s}{r_o + r_s} A(\theta) - 1 \tag{7.75}$$

If the input matching is required, $U_1 = R_1 I_1$, $E = 2U_1$, and (7.74) gives

$$\frac{R_1}{r_o} = 1 + \frac{2}{\pi} \frac{r_o - r_s}{r_o + r_s} C(\theta) \tag{7.76}$$

The transmission coefficient is written

$$s_{12} = \frac{2U_0}{E}\sqrt{\frac{R_1}{R_2}} = \frac{\sqrt{2}r_0\cos\theta}{\sqrt{R_1R_2}} \tag{7.77}$$

Diagram 7.27 shows that θ is close to $\pi/2$. Hence, we can set $\theta = \pi/2 - \epsilon$, with ϵ being small. The expressions (7.75) and (7.76) become, to the first order,

$$R_1/r_0 = (4\epsilon/\pi)(1 + \pi\alpha^2/2\epsilon) \tag{7.78}$$

$$r_0/R_2 = (2/\pi\epsilon)(1 - \pi\epsilon) . \tag{7.79}$$

and (7.77) becomes

$$s_{12}^2 = 1 - \pi\epsilon - \alpha^2\pi/2\epsilon \tag{7.80}$$

which is maximum for $\epsilon = \alpha\sqrt{2}$. The attenuation is $\alpha\pi/\sqrt{2}$ Np, or 19.2 α dB. The terminations are $R_1 = 2\rho\sqrt{2}/\pi$ and $R_2 = \pi\rho/2\sqrt{2}$.

7.4 FREQUENCY MULTIPLICATION

7.4.1 Half-Wave Frequency Multiplier

The circuit which is most tempting for generating even harmonics is that of figure 7.28. We know from subsection 7.2.5 that circuit 7.5 generates all the harmonics with some conversion attenuation. At first sight, it would seem that circuit 7.28 ensures a zero conversion attenuation because the source only carries a current of angular frequency ω, while the charge carries a current of angular frequency $n\omega$ (even n). The only remaining attenuation would be a dissipation attenuation on the diode.

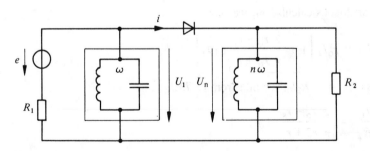

Fig. 7.28

In fact, this circuit is cleverly designed to violate the Page theorem according to which a diode cannot convert more than $1/n^2$ of the fundamental power into power at harmonic n. The detailed calculation of the circuit of figure 7.28 actually confirms this theorem. We discover that the frequency doubler exhibits an attenuation $A = 13.56 + 17\alpha^2$dB for optimum terminations $R_1 = 2.18r_s$ and $R_2 = 1.80r_s$. This result is very disappointing because the unfiltered circuit of figure 7.5 generates through (7.20) a 3.46 dB attenuation for $m = 2$. Therefore, far from being of any use, filtering actually deteriorates the conversion attenuation. This paradoxical result is easily explained: the voltage at the diode terminals is $\sqrt{2}U_1\cos\omega t - \sqrt{2}U_n\cos n\omega t$. This voltage can only be zero for some isolated time values during each period. On some intervals, the voltage will be positive with the diode passing. At that time, the voltage balance requires that the terms ω and $n\omega$ of i generate voltages of the same magnitude as $r_s i$. That is why the terminations are on the order of r_s, and the dissipations in the diode and the terminations are on the same order of magnitude. Thus, the 13.56 dB attenuation represents a dissipation attenuation which is different from α. Because a fraction $(1 - 1/n^2)$ of the power at angular frequency ω cannot be dissipated at the angular frequency $n\omega$, it is dissipated in the diode. The Page theorem thus constitutes an impassable limit and it is risky to try to circumvent it at any price.

The circuit of figure 7.29 remedies these disadvantages. Nevertheless, the fundamental power fraction $(1 - 1/n^2)$ which cannot be dissipated on R_2 will be dissipated on R_0. However, the voltage:

$$u = \sqrt{2}U_1 \cos\omega t - \sqrt{2}U_n \cos n\omega t - U_0 \qquad (7.81)$$

can be made negative for most of the period by choosing U_0 to be on the same order of magnitude as $\sqrt{2}U_1$. Then, there will be only a short conduction interval in the neighborhood of the origin. Most of the time, the diode is blocking, and we thus avoid the necessity of choosing terminations that are on the order of r_s, with the resulting attenuation problems of dissipation on the diode.

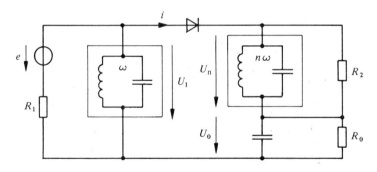

Fig. 7.29

The conduction angle θ is given by

$$\sqrt{2}U_1 \cos\theta - \sqrt{2}U_n \cos n\theta - U_0 = 0 \qquad (7.82)$$

We calculate the terms I_1, I_n, and I_0 of current i by a Fourier expansion in series. We have, for example,

$$(\pi/2)\sqrt{2}I_1 = g_s \int_0^\theta u \cos x \, dx + g_0 \int_\theta^\pi u \cos x \, dx \tag{7.83}$$

The results for I_1 and I_n are as follows:

$$I_1 = [g_0 + (g_s - g_0) C(\theta)/\pi]U_1 + [(g_s - g_0) D(\theta)/\pi] nU_n \tag{7.84}$$

$$-nI_n = [(g_s - g_0) D(\theta)/\pi]U_1 + [g_0 + (g_s - g_0) E(\theta)/\pi] nU_n \tag{7.85}$$

In these formulas, U_0 no longer appears because it has been systematically eliminated, due to (7.82). Formulas (7.84) and (7.85) have the form of the short-circuit admittance equations of a two-port. This two-port has been shown in figure 7.30. The functions $C(\theta)$ and $D(\theta)$ have been defined in subsection 7.2.6. The function $E(\theta)$ is given by

$$E(\theta) = \theta - (\sin 2n\theta)/(2n) \tag{7.86}$$

$$= (2n^2\theta^3/3)\cdot(1 - (n^2\theta^2/3) + (2n^4\theta^4/105) + ...) \tag{7.87}$$

The two-port of 7.30 can be represented by its matrix:

$$\pi Y/g_s = \begin{pmatrix} \pi\alpha^2 + (1-\alpha^2)C & (1-\alpha^2)D \\ (1-\alpha^2)D & \pi\alpha^2 + (1-\alpha^2)E \end{pmatrix} \tag{7.88}$$

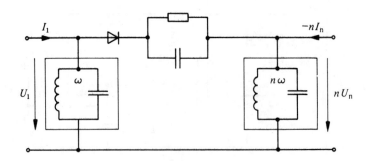

Fig. 7.30

As a result of the choice of output variables nU_n and $-nI_n$, (7.88) has the same form as the admittance matrix of a reciprocal linear two-port, and all the formalism developed for the analysis of such two-ports is thus valid. The existence of a nonlinearity is only detectable in the input angular frequency ω and the output angular frequency $n\omega$.

The image attenuation a of a two-port is given by

$$\coth a = \sqrt{Z_{11} Y_{11}} = \sqrt{Y_{11} Y_{22}/(Y_{11} Y_{22} - Y_{12}^2)} \tag{7.89}$$

according to formula ((3.13), vol. XIX). Because $\cosh^2 a = (1 - \th^2 a)^{-1}$, we also have

$$\cosh^2 a = \frac{Y_{11} Y_{22}}{Y_{12}^2} \tag{7.90}$$

The MacLaurin expansions in series of C, D, and E show that these three functions are on the order of θ^3. In the expressions (7.88), we shall keep for each element a term in α^2, and disregard the terms $\alpha^2 D$, $\alpha^2 C$, and $\alpha^2 E$. Accordingly, (7.89) is written

$$\cosh^2 a = (CE/D^2)[1 + \alpha^2 \pi (1/C + 1/E)] \tag{7.91}$$

$$= 1 + (n^2 - 1)^2 \theta^4/525 + [3(n^2 + 1)/(2n^2)](\alpha^2/\theta^3) \tag{7.92}$$

The image attenuation will be minimum for

$$\theta^7 = 4.725 \pi (n^2 + 1)\alpha^2/(8n^2(n^2 - 1)^2) \tag{7.93}$$

By using the expansion $\cosh^2 a = 1 + a^2$, we find

$$a = 1/15(n^2 - 1)^{3/7} [4.725 \pi (n^2 + 1)/(8n^2)]^{2/7} \alpha^{4/7} \text{ Np} \tag{7.94}$$

This is the minimum dissipation attenuation. Of course, we must add the conversion attenuation resulting from the fact that the chosen output quantities are equal to nU_n and $-nI_n$. Thus, the power at the angular frequency $n\omega$ is indeed equivalent, except for the dissipation on the diode, to $1/n^2$ of the power drawn from the source $U_1 I_1$, i.e., the upper limit predicted by the Page theorem. The more α tends toward zero, the closer we shall be to that limit. The input image impedance is

$$R_i = \coth a/Y_{11} \cong 1/a Y_{11} = 3\pi r_s/2a\theta^3 \tag{7.95}$$

$$= 0,326(n^2 - 1)^{3/7} (1 + 1/n^2)^{5/7} r_0^{5/7} r_s^{2/7} \tag{7.96}$$

Similarly, we find $R_2 = R_1/n^2$.

Because $U_n^2/R_2 = U_1^2/(R_1 n^2)$, except for the dissipation on the diode, we find $U_n/U_1 = 1/n^2$. However, (7.82) must hold for a small θ, and therefore $U_0 = \sqrt{2}(U_1 -$

430

U_n) $= \sqrt{2}U_n (n^2 - 1)$. Because the power which is not transformed into the harmonic n is transformed into dc power, we also have $U_0^2/R_0 = (n^2 - 1)U_n^2/R_2$. Finally, we find

$$R_0 = 2(n^2 - 1)R_2 \qquad (7.97)$$

7.4.2 Comments

The circuit studied in subsection 7.4.1 can also be considered as a filtered version of the polarized diode, which permits generation of odd harmonics. The circuit of figure 7.29 thus constitutes a universal frequency multiplier.

7.4.3 Frequency Multiplication by the Graetz Bridge

Based on our experience of subsections 7.3.3 and 7.4.1, the circuit which we will choose is that of figure 7.31. Because the filtering is carried out as in figure 7.26, the relations (7.64) and (7.65) are still valid, and the passing and blocking areas of the diodes are still those of figure 7.27.

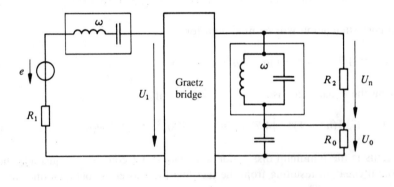

Fig. 7.31

We shall make the following hypothesis: in order to attain the optimum operation, the locus of the representative point must run exclusively through areas I and III. By avoiding areas II and IV, we avoid the configurations in which both pairs of diodes are simultaneously passing and blocking. In that case, the bridge is balanced, and no power can pass between input and output. It is, therefore, suitable to avoid these configurations. To this end, we shall choose $U_0 = \sqrt{2}U_n$ so that the output voltage will never be negative. Also, (7.11) shows that the amplitude of the (even) harmonic n is of sign $-(-1)^{n/2}$. Because $n\pi/2 = (-1)^{n/2}$, it follows that all the odd harmonics coming from the rectification of the two cosinusoid half-waves take the value $-\sqrt{2}U_n$

for $\omega t = \pi/2$. Thus, the representative curves pass through the origin at that point, which ensures switching between areas I and III. Finally, the output voltage is, thus,

$$u_2 = \sqrt{2}U_n[1 - (-1)^{n/2}\cos n\omega t] \tag{7.98}$$

According to the switching diagram, the Fourier analysis gives

$$(\pi/2)I_n = \int_0^{\pi/2} i_2^I \cos nx \, dx + \int_{\pi/2}^{\pi} i_2^{III} \cos nx \, dx \tag{7.99}$$

By a similar calculation, we can find U_1 as a function of I_1 and U_n as well. Furthermore, we have, at the output, $U_n = -R_2I_n$, and the input matching requires $U_1 = R_1I_1$. By combining these relations with the expressions of I_n and U_1, we find

$$\frac{U_n}{U_1} = \frac{R_2}{R_1}\frac{4}{\pi(n^2-1)}\left(1 - \frac{2R_2}{r_o}\right). \tag{7.100}$$

$$\frac{U_1}{U_n} = \frac{4}{\pi}\frac{n}{n^2-1}\left(1 + \frac{2r_s}{R_1}\right) \tag{7.101}$$

The transmission coefficient:

$$s_{12} = (U_n/U_1)\sqrt{(R_1/R_2)} = (1/n)(1 - (r_s/R_1) - (R_2/r_o)) \tag{7.102}$$

The conversion attenuation again reaches the minimum predicted by the Page theorem. The dissipation attenuation is minimum for

$$R_1R_2 = \rho^2 \tag{7.103}$$

By multiplying (7.101) and (7.100), we find, when disregarding the terms in R_2/r_o or r_s/R_1,

$$R_1/R_2 = 4n^2/(\pi^2(n^2-1)^2) \tag{7.104}$$

and, therefore, $R_1 = 4n\rho/(\pi(n^2-1))$ and $R_2 = \pi(n^2-1)\rho/(4n)$.
We can show that

$$R_0 = 2R_n/(n^2-1) \tag{7.105}$$

7.4.4 Exercise

Analyze circuit 7.21 between two ideal voltage filters so as to obtain an optimum odd-harmonic generator.

432

7.5 DIODE CIRCUIT MODULATION

7.5.1 Principle

A general definition of the modulator has been given in subsection 6.2.1. The special requirements relating to the absence of nonlinear distortion have been mentioned in subsection 7.1.2. To meet these requirements, the diode modulators are fed by a carrier which is of large amplitude compared to that of the signal so that the passing state or the blocking state of a diode is exclusively determined by the carrier. This is, of course, an approximation, because the state of a diode depends on the voltage applied to it, part of which comes from the signal. Nevertheless, for the objectives of subsection 7.1.2 to be met, this approximation must be satisfied or the carrier sufficiently prevailing over the signal. The real modulator will depart from the ideal modulator by a certain nonlinear distortion, which must be kept below a certain threshold. We shall not calculate this nonlinear distortion here and will instead be satisfied with a study of the ideal behavior of the modulator.

As long as we limit ourselves to this type of analysis, the signal may be represented by a simple sinusoid. Of course, a purely sinusoidal signal has no meaning, but we have good reasons for analyzing the real signal in these sinusoidal terms because the ideal modulator must satisfy the superposition principle for the signal terms.

7.5.2 Single Diode Modulator

Let us consider the circuit of figure 7.32. The signal and carrier voltages $e_s = \sqrt{2}E_s\cos\omega t$ and $e_p = \sqrt{2}E_p\cos\Omega t$ are applied in series to the diode. We assume that $E_p \gg E_s$ and that

$$r = a + b\,\mathrm{sgn}(\cos\Omega t) \tag{7.106}$$

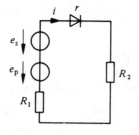

Fig. 7.32

This relation would be exact if e_p were a rectangular wave whose amplitude is simply higher than that of e_s. It is approximated in the case considered, the larger is E_p, the better we can, by increasing E_p, approximate the ideal situation as closely as desired.

The current is written with the notations of subsection 7.2.2.

$$i = \frac{e_s(A - b\,\text{sgn}(\cos\Omega t))}{A^2 - b^2} \qquad (7.107)$$

This current includes a term $e_s A/(A^2 - b^2)$, which constitutes a leakage of the signal. The term $\text{sgn}(\cos\Omega t)$ supplies by (7.9) all the odd harmonics of angular frequency Ω with an amplitude that is inversely proportional to their order. Consequently, the term $be_s\,\text{sgn}(\cos\Omega t)/(A^2 - b^2)$ supplies all the terms $\cos[(2n + 1)\Omega \pm \omega]t$. The *main sidebands*, of angular frequencies $\Omega \pm \omega$, have an efficient amplitude $-[E_s b/(A^2 - b^2)]2/\pi$ by (7.8). The transmission coefficient of the signal source toward charge R_1 is written

$$s_{12} = \frac{1}{\pi} \frac{2(r_o - r_s)\sqrt{R_1 R_2}}{(R_1 + R_2 + r_s)(R_1 + R_2 + r_o)} \qquad (7.108)$$

The dissipation attenuation will be minimum for $R_1 = R_2 = \rho/2$, as in subsection 7.2.3. We obtain

$$A = \ln\pi + 2\,\text{arcth}\,\alpha \qquad (7.109)$$

i.e., 9.94 dB for an ideal diode.

7.5.3 Comment

The voltages e_s and e_p have been chosen with the same phase because their angular frequencies are not in a rational ratio. The relative phase difference of the two waves varies at each period of e_p. In fact, in the course of time, we may approximate any phase difference. Consequently, nothing is lost if we assume that the phase difference is zero at the time origin.

7.5.4 Modulator with Balanced Carrier

In the analysis of subsection 7.5.2, the presence of e_p has been disregarded in the calculation of i, apart from the fact that the (large) current at the angular frequency of the carrier entirely determines the state of the diodes. However, this current also flows on the charge, where it is of no interest. On the contrary, if we wish to keep only one sideband, the presence of the carrier will impose filtering requirements that are very difficult to meet.

The three-port of figure 7.33 includes a 33' port, to which is connected the carrier source which controls the passing, or blocking state of the two diodes connected in series. The signal source is connected to the 11' port and the charge at 22'. The carrier is not transmitted toward the source nor the charge, each one being connected between equipotential points with respect to the source of the carrier. The equivalent circuit of the two-port for the signal transmission is that of figure 7.34, with a resistor

434

2r(t) coming from the connection in series of two diodes. This is the only modification with respect to circuit 7.32. Thus, the optimum terminations are ρ, and the attenuation is given by (7.109).

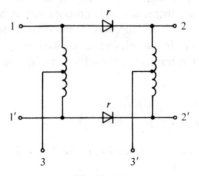

Fig. 7.33

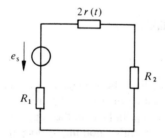

Fig. 7.34

An equivalent circuit is that of figure 7.35, where the series resistor met by the current of the source is $r(t)$, and the optimum terminations are ρ/2. The dual circuit is that of figure 7.36, the terminations of which are 2ρ by duality. This circuit is called the *Cowan modulator*.

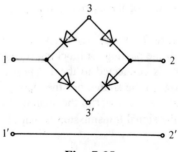

Fig. 7.35

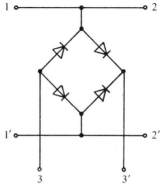

Fig. 7.36

In the case of ideal diodes, the circuits considered in subsections 7.5.2 and 7.5.4 are equivalent to either the circuit of figure 7.37 or figure 7.38, which contain a periodically activated switch, with a period $T = 2\pi/\Omega$, during which they are open and closed for a half period.

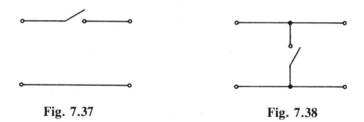

Fig. 7.37   **Fig. 7.38**

7.5.5 Doubly-Balanced Modulators

All the modulators previously studied let a current leakage pass from the signal to the charge, which is one of the causes of the conversion attenuation calculated in (7.109). It is possible to eliminate this leakage by adding the output signals of two identical modulators, one of which includes a phase difference π between input and output, which means simply exchanging the wires. The carrier is simultaneously applied with a phase difference π between the two modulators.

The principle of such a circuit is represented in figure 7.39, and a practical implementation, the *ring modulator* is represented in figure 7.40. Regarding this circuit, note that the diodes **are not** oriented as in the Graetz bridge.

On figure 7.39, it appears that the leakage from the source signal to the charge is the sum zero of two terms $(e_s - e_s) A/(A^2 - b^2)$. The modulation products appear in the terms $-be_s \text{sgn}(\cos\Omega t)/(A^2 - b^2)$ and $-b(-e_s)\text{sgn}(-\cos\Omega t)/(A^2 - b^2)$. Therefore, they add up, and the main sidebands are twice what they were in subsection

436

7.5.2. However, if we consider piecewise-linear diodes, the ring modulator is a lattice of impedances r_o and r_s, the image impedance of which is ρ. Finally, the attenuation is written

$$A = \ln \pi/2 + 2 \operatorname{arcth}(\alpha) = 3{,}94 + 17{,}4\,\alpha \text{ dB} \qquad (7.110)$$

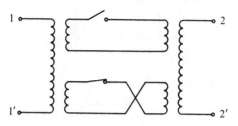

Fig. 7.39

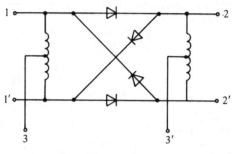

Fig. 7.40

7.5.6 Equations of the Ring Modulator

The hypothesis made in subsection 7.5.1 on the state of the diode, which is exclusively determined by the carrier, and the way it has been used in subsections 7.5.2, 7.5.4, and 7.5.5, amounts to considering time-dependent linear circuits in terms of signal transmission and modulation. The purpose of the following subsections is to show the describing function method, embodied by the concept of the ideal filter, which can also be applied to this type of circuit.

As a preliminary, we shall set a formal equation of the ring modulator. If the diodes of figure 7.40 are ideal, the two-port between port 1 and port 2 is equivalent to an ideal transformer with the ratio:

$$n(t) = \operatorname{sgn}(\cos \Omega t) \qquad (7.111)$$

which alternatively takes the values $+1$ and -1, and which we shall call the *ideal transformer*. If we take into account the piecewise-linear diodes, the two-port is equivalent to a lattice of impedances r_o and r_s connected in cascade with the ideal switch. If A_0, with th $(A_0/2) = \alpha$ designate the image attenuation of the lattice having an impedance of ρ, then the impedance matrix is written by (vol. XIX, sec. 3.1.10)

$$\begin{pmatrix} \rho \coth A_0 & \rho/\mathrm{sh}A_0 \\ \rho/\mathrm{sh}A_0 & \rho \coth A_0 \end{pmatrix} \tag{7.112}$$

The cascade connection of the ideal switch causes Z_{11} and Z_{22} to be multiplied by $n^2(t) = 1$, and Z_{12} by $n(t)$. Thus, the time-dependent linear equations of the two-port are

$$u_1 = \rho \coth A_0 \cdot i_1 + (\rho/\mathrm{sh}A_0) \, n(t) \cdot i_2 \tag{7.113}$$

$$u_2 = (\rho/\mathrm{sh}A_0) \, n(t) \cdot i_1 + \rho \coth A_0 \cdot i_2 \tag{7.114}$$

7.5.7 Ring Modulator Between Similar Filters

Let us consider the case of the ring modulator placed between two ideal current filters, respectively tuned on ω and $\Omega + \omega$. The currents are thus written $i_1 = \sqrt{2}I_1\cos\omega t$ and $i_2 = \sqrt{2}I_2\cos(\Omega + \omega)t$. If we introduce these currents into (7.114) and (7.113), we obtain the relations between the amplitudes:

$$U_1 = \rho I_1 \coth A_0 + 2\rho I_2/(\pi \cdot \mathrm{sh}A_0) \tag{7.115}$$

$$U_2 = 2\rho I_1/(\pi \cdot \mathrm{sh}A_0) + \rho I_2 \coth A_0 \tag{7.116}$$

It is important to note that (7.115) implies a factor exp $(\mathrm{j}\omega t)$ and (7.116) a factor exp $\mathrm{j}(\Omega + \omega)t$. Apart from the existence of these two different factors, which take into account the time dependent nature of the circuit, both equations use the formalism of the phasors as it is applied to linear circuits. The factor $2/\pi$ of (7.115) results from the multiplication of sgn($\cos\Omega t$), the expansion of which is given in (7.8), by $\cos\omega t$ or $\cos(\omega + 2\Omega)t$.

The equations (7.115) and (7.116) are those of a symmetrical two-port, whose equivalent lattice has, according to (vol. IV, sec. 6.2.8), the impedances:

$$Z_1 = \rho\,(\coth A_0 + 2/(\pi \cdot \mathrm{sh}A_0)) \tag{7.117}$$

$$Z_2 = \rho\,(\coth A_0 - 2/(\pi \cdot \mathrm{sh}A_0)) \tag{7.118}$$

The image parameters are inferred by the formulas (vol. XIX, eqs. (3.17) and (3.20)):

$$W = \rho\sqrt{\coth^2 A_0 - 4/(\pi^2 \mathrm{sh}^2 A_0)} \qquad (7.119)$$

or, also,

$$\mathrm{th}\left(\frac{a}{2}\right) = \sqrt{\frac{\coth A_0 - 2/(\pi\,\mathrm{sh} A_0)}{\coth A_0 + 2/(\pi\,\mathrm{sh} A_0)}} \qquad (7.120)$$

$$\mathrm{th}^2\left(\frac{a}{2}\right) = \frac{\cosh A_0 - 2/\pi}{\cosh A_0 + 2/\pi} \qquad (7.121)$$

and

$$\cosh a = (\pi/2)\cosh A_0 \qquad (7.122)$$

In the case of ideal diodes, $\alpha = 0$ and $A_0 = 0$. As a result, $\cos A_0 = 1$ and, by (7.122), $a = \mathrm{arc}\cosh \pi/2 = 8.9$ dB. By (7.119) W is infinite since $\sinh A_0 = 0$. Consequently, this is a situation where the image attenuation a is finite, but it can only be effected between infinite image terminations. In practice, it is desirable to work between finite terminations to carry out the transfer of a nonzero power between the source and the charge. In that case, the actual attenuation becomes infinite. The explanation of this result is obvious. An ideal switch realizes the equation $i_2 = \pm i_1$. Now, these two currents are sinusoids of different angular frequencies; they cannot be equal, except for few isolated points, if $i_1 \equiv i_2 \equiv 0$.

Therefore, the ring modulator can only operate if we assume that the diodes are piecewise linear. In other words, the attenuator of lattices of impedances r_o and r_s is an essential part of the ring modulator between similar filters, and this does not give much hope for a very satisfactory operation.

For α and A_0 being small, we can expand (7.119)

$$W = \rho\,\frac{\sqrt{\cosh^2 A_0 - 4/\pi^2}}{\mathrm{sh} A_0} \qquad (7.123)$$

$$\cong (\rho/A_0)\sqrt{1 - 4/\pi^2} \cong (r_0/2)\sqrt{1 - 4/\pi^2} \qquad (7.124)$$

The expansion of (7.122) gives

$$A = \mathrm{arc}\cosh \pi/2 + \pi\alpha^2/\sqrt{\pi^2/4 - 1} \qquad (7.125)$$

$$= 8.9 + 22.6\alpha^2 \text{ dB} \qquad (7.126)$$

Thus, the attenuation is higher than that of the same modulator between resistive terminations, and this type of filtering should be prohibited.

7.5.8 Ring Modulator Between Inverse Filters

Let us assume, on the contrary, that the input has an ideal current filter and the output an ideal voltage filter. We thus have $i_1 = \sqrt{2}I_1\cos\omega t$ and $u_2 = \sqrt{2}U_2\cos(\Omega + \omega)t$. By solving (7.113) and (7.114) with respect to unfiltered quantities, we find

$$i_2 = u_2/(\rho \coth A_0) - ni_1/\cosh A_0 \tag{7.127}$$

$$u_1 = nu_2/(\cos A_0) + \rho i_1/(\coth A_0) \tag{7.128}$$

If we equal the amplitudes of the terms $\Omega + \omega$ and ω of the two equations, respectively, we find

$$I_2 = U_2/(\rho \coth A_0) - 2I_1/(\pi \cosh A_0) \tag{7.129}$$

$$U_1 = 2U_2/(\pi \cosh A_0) + \rho I_1/(\coth A_0) \tag{7.130}$$

By solving these two equations with respect to U_1 and U_2, we find the open circuit impedance matrix of the two-port:

$$Z = \rho \begin{pmatrix} \dfrac{\dfrac{4}{\pi^2} + \mathrm{sh}^2 A_0}{\mathrm{sh}A_0 \cosh A_0} & \dfrac{2}{\pi\,\mathrm{sh}A_0} \\[3mm] \dfrac{2}{\pi\,\mathrm{sh}A_0} & \coth A_0 \end{pmatrix} \tag{7.131}$$

from which we find a determinant equal to ρ^2. Then, $Y_{11} = Z_{22}/\rho^2$ and the image parameters are calculated by

$$\coth a = \sqrt{Z_{11}Y_{11}} = \sqrt{1 + 4/(\pi^2\,\mathrm{sh}^2 A_0)} \tag{7.132}$$

or, also,

$$\mathrm{sh}\,a = (\pi/2)\,\mathrm{sh}A_0 \tag{7.133}$$

which gives $\alpha = 0$ for $\alpha = A_0 = 0$. Therefore, modulator operation is possible with ideal diodes. For A_0 being small, we have $a = \pi A_0/2$, and the actual attenuation, which equals the image attenuation, is

$$A = 27.4\,\alpha\ \text{dB}$$

440

The image terminations are calculated by

$$W_1 = \sqrt{(Z_{11}/Y_{11})} = 2\rho/\pi \tag{7.134}$$

$$W_2 = \sqrt{(Z_{22}/Y_{22})} = \pi\rho/2 \tag{7.135}$$

Bibliography

[1] F. KRUMMENACHER, Micropower Switched Capacitor Biquadratic Cells, *Proc. ESSCIRC'81,* pp. 175–177, Freiburg, RFA, 1981.

[2] L.O. CHUA, Memristor—the Missing Circuit Element, *IEEE Trans. Circuit Theory,* vol. CT-18, pp. 507–519, 1971.

[3] L.W. NAGEL, SPICE 2: A Computer Program to Simulate Semiconductor Circuits, Memo UCB/ERL-M520, University of California, Berkeley, 1975.

[4] N. ROUCHE, J. MAWHIN, *Equations différentielles ordinaires,* Masson, Paris, 1973.

[5] H. GOLDSTEIN, *Classical Mechanics,* Addison-Wesley, Reading, Mass., 1950.

[6] G.J.O. JANESON, *Topology and Normed Spaces,* Chapman and Hall, London, 1974.

[7] T. MATSUMOTO, L.O. CHUA, H. KAWAKAMI, S. ICHIRAKU, Geometric Properties of Dynamic Nonlinear Networks: Transversality, Local Solvability and Eventual Passivity, *IEEE Trans. Circuits Syst.,* vol. CAS-28, pp. 406–428, May 1981.

[8] V. GUILLEMIN, A. POLLAK, *Differential Topology,* Prentice-Hall, Englewood Cliffs, NJ, 1974.

[9] L.O. CHUA, T. MATSUMOTO, S. ICHIRAKU, Geometric Properties of Resistive Nonlinear n-Ports: Transversality, Structural Stability, Reciprocity and Anti-reciprocity, *IEEE Trans. Circuits Syst.,* vol. CAS-27, pp. 577–603, July 1980.

[10] J.E. MARSDEN, *Elementary Classical Analysis,* Freeman, San Francisco, 1974.

[11] G.J. MINTY, Monotone Networks, *Proc. R. Soc. London, Ser. A,* vol. 257, pp. 194–212, September 1960.

[12] J. VANDEWALLE, L.O. CHUA, The Colored Branch Theorem and its Application in Circuit Theory, *IEEE Trans. Circuits Syst.,* vol. CAS-27, pp. 816–825, September 1980.

[13] A.N. WILLSON (ed.), *Nonlinear Networks: Theory and Analysis,* IEEE Press, New York, 1974.

442

[14] T. Nishi, L.O. Chua, Topological Criteria for Nonlinear Resistive Circuits Containing Controlled Sources to have a Unique Solution, *IEEE Trans. Circuits Syst.*, vol. CAS-31, pp. 722–741, August 1984.

[15] N.G. Lloyd, Degree Theory, *Cambridge Tracts in Mathematics 73*, Cambridge University Press, Cambridge, 1978.

[16] L.O. Chua, N.N. Wang, On the Application of Degree Theory to the Analysis of Resistive Nonlinear Networks, *Int. J. Circuit Theory Appl.*, vol. 5, pp. 35–68, January 1977.

[17] L.O. Chua, R.A. Rohrer, On the Dynamic Equations of a Class of Nonlinear RLC Networks, *IEEE Trans. Circuit Theory*, vol. CT-12, pp. 475–489, December 1965.

[18] S.W. Director, R.A. Rohrer, The Generalized Adjoint Network and Network Sensitivities, *IEEE Trans. Circuit Theory*, vol. CT-16, pp. 318–323, August 1969.

[19] L.A. Zadeh, C.A. Desoer, *Linear System Theory, the State Space Approach*, McGraw-Hill, New York, 1963.

[20] A. Sangiovanni-Vincentelli, Y.T. Wang, On Equivalent Dynamic Networks: Elimination of Capacitor Loops and Inductor Cutsets, *IEEE Trans. Circuits Syst.*, vol. CAS-25, pp. 174–177, March 1978.

[21] Ph. Verburgh, Unicité, multiplicité et absence de régimes périodiques en ferrorésonance, *Thesis No. 490*, EPFL, Lausanne, Switzerland, 1983.

[22] L.O. Chua, M. Hasler, J. Neirynck, Ph. Verburgh, Dynamics of a Piecewise-Linear Resonant Circuit, *IEEE Trans. Circuits Syst.*, vol. CAS-29, pp. 535–547, August 1982.

[23] A. Azzouz, R. Duhr, M. Hasler, Transitions to Chaos in a Simple Non linear Circuit Driven by a Sinusoidal Voltage Source, *IEEE Trans. Circuits Syst.*, vol. CAS-30, December 1983.

[24] N. Levinson, A Second Order Differential Equation with Singular Solutions, *Ann. Math. Stat.*, vol. 50, pp. 127–152, 1949.

[25] T.S. Parker, L.O. Chua, A Computer-Assisted Study of Forced Relaxation Oscillations, *IEEE Trans. Circuits Syst.*, vol. CAS-30, pp. 518–533, August 1983.

[26] D. Ruelle, Strange Attractors, *The Mathematical Intelligencer*, vol. 2, pp. 126–137, 1980.

[27] P. Collet, J.-P. Eckmann, *Iterated Maps on the Interval as Dynamical Systems*, Birkhauser, Basel, Amsterdam, Stuttgart, 1980.

[28] J. Guckenheimer, P. Holmes, Nonlinear Oscillations, Dynamical Systems, and Bifurcations of Vector Fields, *Applied Mathematical Sciences*, vol. 42, Springer-Verlag, New York, Berlin, Heidelberg, Tokyo, 1983.

[29] J.K. Hale, Ordinary Differential Equations, *Pure and Applied Mathematics*, vol. XXI, Wiley-Interscience, New York, 1969.

[30] N. ROUCHE, P. HABETS, M. LALOY, Stability Theory by Liapunov's Direct Method, *Applied Mathematical Sciences*, vol. 22, Springer-Verlag, New York, 1977.

[31] W. LEDERMANN (ed.), *Handbook of Applicable Mathematics, vol. 1: Algebra*, Wiley-Interscience, Chichester, 1980.

[32] NEMYTSKII, STEPANOV, *Qualitative Theory of Differential Equations*, Princeton University Press, Princeton, N.J., 1960.

[33] E. FREIRO, L.G. FRANQUELLO, J. ARACIL, Periodicity and Chaos in an Autonomous Electronic System, *IEEE Trans. Circuits Syst.*, vol. CAS-31, pp. 237–247, March 1984.

[34] L.O. CHUA, K.A. STROMSMOE, Lumped-Circuit Models for Nonlinear Inductors Exhibiting Hysteresis Loops, *IEEE Trans. Circuit Theory*, vol. CT-17, pp. 564–574, November 1970.

[35] L.O. CHUA, D.N. GREEN, Graph Theoretic Properties of Dynamic Nonlinear Networks, *IEEE Trans. Circuits Syst.*, vol. CAS-23, pp. 292–311, May 1976.

[36] L.O. CHUA, D.N. GREEN, A Qualitative Analysis of the Behavior of Dynamic Nonlinear Networks: Steady-State Solutions of Nonautonomous Networks. *IEEE Trans. Circuits Syst.*, vol. CAS-23, pp. 530–550, September 1976.

[37] M. HASLER, PH. VERBURGH, On the Uniqueness of the Steady-State for Nonlinear Circuits with Time-Dependent Sources, *IEEE Trans. Circuits Syst.*, vol. CAS-31, pp. 702–713, August 1984.

[38] A. GELB, W.E. VANDERVELDE, *Multiple-input describing functions and nonlinear system design*, McGraw-Hill, New York, N.Y., 1968.

[39] N. BOGOLYOUBOV, I. MITROPOLSKI, *Les méthodes asymptotiques en théorie des oscillations non linéaires*, Gauthier-Villars, Paris, 1962.

[40] L.O. CHUA, Y.-S. TANG, Nonlinear Oscillation via Volterra Series, *IEEE Trans. Circuits Syst.*, vol. CAS-29, pp. 150–168, March 1982.

[41] T. YOSHIZAWA, Stability Theory and the Existence of Periodic Solutions and Almost Periodic Solutions, *Applied Mathematical Sciences*, vol. 14, Springer-Verlag, New York, 1975.

[42] V. BELEVITCH, *Théorie des circuits non linéaires en régime alternatif*, Gauthier-Villars, Paris, 1959.

Select Bibliography

The Traité d'Électricité, listed below by volume number, is published by the Presses Polytechniques Romandes (Lausanne, Switzerland) in collaboration with the École Polytechnique Fédérale de Lausanne. The title of each volume is given with the year of publication in parenthesis. English translations by Artech House are denoted by an asterisk with the year of publication in parenthesis.

Vol.	Author	Title
I	Frédéric de Coulon & Marcel Jufer	Introduction à l'électrotechnique (1981).
II	Philippe Robert	Materiaux de l'électrotechnique (1979).
III	Fred Gardiol	Electromagnétisme (1979).
IV	René Boite & Jacques Neirynck	Theorie des reseaux de Kirchhoff (1983).
V	Daniel Mange	Analyse et synthèse des systèmes logiques (1979). *Analysis and Synthesis of Logic Systems (1986).
VI	Frédéric de Coulon	Theorie et traitement des signaux (1984). *Signal Theory and Processing (1986).
VII	Jean-Daniel Chatelain	Dispositifs à semiconducteur (1979).
VIII	Jean-Daniel Chatelain & Roger Dessoulavy	Electronique (1982).
IX	Marcel Jufer	Transducteurs électromécaniques (1979).
X	Jean Chatelain	Machines électriques (1983).
XI	Jacques Zahnd	Machines séquentielles (1980).
XII	Michel Aguet & Jean-Jacques Morf	Energie électrique (1981).
XIII	Fred Gardiol	Hyperfréquences (1981). *Introduction to Microwaves (1984).
XIV	Jean-Daniel Nicoud	Calculatrices (1983).
XV	Hansruedi Bühler	Electronique de puissance (1981).

446

Index